Springer Oceanography

The Springer Oceanography series seeks to publish a broad portfolio of scientific books, aiming at researchers, students, and everyone interested in marine sciences. The series includes peer-reviewed monographs, edited volumes, textbooks, and conference proceedings. It covers the entire area of oceanography including, but not limited to, Coastal Sciences, Biological/Chemical/Geological/Physical Oceanography, Paleoceanography, and related subjects.

More information about this series at http://www.springer.com/series/10175

Victor Egorov

Theory of Radioisotopic and Chemical Homeostasis of Marine Ecosystems

 Springer

Victor Egorov
A. O. Kovalevsky Institute of Biology
of the Southern Seas
Russian Academy of Sciences
Sevastopol, Russia

Translated by
Thomas Alexander Beavitt
Institute of Philosophy and Law
Ural Branch of the Russian Academy of Sciences
Yekaterinburg, Russia

Scientific Consultant to the Translation
Maria Streletskaya
Institute of Geology and Geochemistry
Ural Branch of the Russian Academy of Sciences
Yekaterinburg, Russia

ISSN 2365-7677   ISSN 2365-7685 (electronic)
Springer Oceanography
ISBN 978-3-030-80581-4   ISBN 978-3-030-80579-1 (eBook)
https://doi.org/10.1007/978-3-030-80579-1

This Springer imprint is published by the registered company Springer Nature Switzerland AG
The registered company address is: Gewerbestrasse 11, 6330 Cham, Switzerland

Preface

This monograph is devoted to the study and mathematical description of patterns of biogeochemical interaction between living and inert matter that comprise the radiochemical components of the marine environment. The semi-empirical theory of the radioisotope and mineral metabolism of aquatic organisms (hydrobionts), presented on the timescale of sorption and metabolic processes represented by trophic interactions, is parametrically compatible with contemporary methods for describing the balance of matter and energy in marine ecosystems. It is demonstrated that the regularities of radioisotopic and chemical homeostasis in the marine environment are realized in accordance with Le Chatelier's principle. Criteria for assessing the ecological carrying capacity, biogeocenosis assimilation capacity and radiocapacity of water masses, which form the basis of the radioisotope and mineral homeostasis theory applying to marine ecosystems, are substantiated. Methods for the sustainable development of critical and recreational zones of the Black Sea, which take into account marine pollution factors by regulating the balance between the consumption of water quality resources and their reproduction as a result of natural biogeochemical processes, are proposed.

The monograph is intended for biogeochemists, ecologists and specialists in marine nature management.

The work was carried out within the framework of the state assignment of the A. O. Kovalevsky Institute of Biology of the Southern Seas (IBSS) Federal Research Centre on the topic of "Molismological and Biogeochemical Fundamentals of Marine Ecosystem Homeostasis" (state registration number AAAA-A18-118020890090-2).

Sevastopol, Russia   Victor Egorov

Acknowledgements The author would like to express deep appreciation and gratitude to his teacher and colleague Academician G. G. Polikarpov and other colleagues at the Department of Radiation and Chemical Biology of the Federal Research Center of the Institute of Biology of the Southern Seas of the Russian Academy of Sciences (FRC IBSS) for their creative collaboration. He also thanks R. V. Gorbunov, Director of IBSS, Candidate of Geographical Sciences, and Yu. G. Korniychuk, Deputy Scientific Director, Candidate of Biological Sciences, for their valuable research support, as well as O. Yu. Kopytova and Yu. G. Marchenko for technical editing and proofreading of the monograph.

Contents

About the Author

Victor Egorov is Scientific Director of the RAS A. O. Kovalevsky Institute of Biology of the Southern Seas (IBSS) Federal Research Centre, Alumnus of the Department of Cybernetics and Computer Engineering of the Sevastopol National Technical University, Doctor of Biological Sciences, Professor, Honoured Scientist and Engineer of the Autonomous Republic of Crimea, Academician of the National Academy of Sciences of Ukraine and Academician of the Russian Academy of Sciences.

Abbreviations

BSRC	Black Sea rim current
CIL	Cold intermediate layer
DOM	Dissolved organic matter
FAO	Food and Agriculture Organization of the United Nations
GESAMP	Joint Group of Experts on the Scientific Aspects of Marine Environment Protection (IMO/FAO/UNESCO-IOC/WMO/WHO/IAEA/UN/UNEP)
IAEA	International Atomic Energy Agency
IMO	International Maritime Organization
IOC	Intergovernmental Oceanographic Commission
PAR	Photosynthetically active radiation
SOM	Suspended organic matter
UN	United Nations
UNEP	United Nations Environment Programme
UNESCO	United Nations Educational, Scientific and Cultural Organization
UQHL	Upper quasi-homogeneous layer
WHO	World Health Organization
WMO	World Meteorological Organization

Chapter 1
Introduction

The nuclear era began in 1945 with the testing of atomic bombs at the Alamogordo test site in the USA, followed by their first military use over Hiroshima and Nagasaki in Japan. Since then, the release of artificial radionuclides into seas and oceans in the form of atmospheric fallout and nuclear waste has become a highly significant environmental factor in marine pollution (Shvedov and Shirokov 1962). Following the first use of nuclear weapons and during the period of nuclear tests, new sciences were born: radioecology (Kuzin and Peredel'skii 1956), radiation biogeocenology (Timofeev-Resovskii 1957), marine radioecology (Polikarpov 1964), nuclear hydrophysics (Nelepo 1970). These new disciplines involved large-scale studies of the patterns of distribution and migration of radioactive substances in the marine environment, as well as the effect of ionising radiation on marine organisms (Bowen 1979; Marine Environmental Quality 1971; Shvedov and Patin 1968; Whicker and Schultz 1982). The concern of mankind with the problem of global radioactive contamination of the biosphere led to the conclusion of the 1963 Moscow Partial Test Ban Treaty (PTBT) banning nuclear explosions in space, in the atmosphere and under water. It was calculated that the release of fragment radionuclides ^{137}Cs, ^{90}Sr, ^{133}Xe, ^{131}I and transuranic elements (Summary report 1986) into the environment as a result of tests of atomic weapons in open environments, as well as due to nuclear accidents occurring for example at the Windscale (1957) and Chernobyl (1986) nuclear power plants, exceeded the contribution from the military use of atomic bombs by many orders of magnitude (Gudiksen et al. 1989). For this reason, the problem of studying the anthropogenic radioactive contamination of the marine environment has become one of the most urgent tasks of ecological research.

The second half of the twentieth century was characterised by the global industrial revolution, which is associated with the growing role of technological uses of chemicals and their compounds. On the basis of publications, it can be estimated that about 10 million chemical compounds are currently known to mankind. Meanwhile, their number continues to grow by about 1000 new compounds annually. When released into the seas and oceans, most of these chemicals can have a negative impact on the living components of ecosystems.

Following the end of the Second World War, the international community began to make practical efforts to protect the marine environment. The first UN Conference on the Law of the Sea (UNCLOS I) was held in Switzerland in 1958. From the 1960s onwards, a number of conventions aimed both at the prevention of marine pollution and the mitigation of its consequences were adopted: in 1974, the Helsinki Convention on the Protection of the Marine Environment of the Baltic Sea Area (revised in 1992); in 1976, the Barcelona Convention for the Protection of the Marine Environment and the Coastal Region of the Mediterranean; in 1978, the Kuwait Regional Convention for Co-operation on the Protection of the Marine Environment from Pollution; in 1992, the Bucharest Convention on the Protection of the Black Sea against Pollution, as well as a number of others. In December 1972, the United Nations Environment Programme (UNEP) was adopted in order to develop and consolidate environmental programmes in the United Nations system, based on UN General Assembly Resolution 2997 (XXVII) (Gureev et al. 2011). In 1977, according to the decision of the Joint Group of Experts on the Scientific Aspects of Marine Environmental Protection (GESAMP), bringing together the IMO, FAO, UNESCO, WMO, WHO, IAEA, UN and UNEP, the term "pollution" was adopted to refer to the direct or indirect introduction of substances or energy by humans into the marine environment resulting in harmful effects such as damage to living resources, risks to human health, as well as interference with marine activities such as fishing, deterioration of the quality of seawater consumed and aesthetic benefits (IMCO/FAO/UNESCO/WMO/WHO/IAEA/UN joint group of experts 1977). In 1982, the UN Convention on the Law of the Sea, which is of great importance for the protection of the marine environment, was adopted. This convention enshrined the international legal regulatory framework for the prevention of marine pollution by dumping of wastes and other materials, including the disposal of radioactive materials, as well as encoding the legal protection of natural renewable (living) marine resources.

According to contemporary estimates, the scale of the release of anthropogenic pollutants into the marine environment in a number of cases exceeded the intensity and scale of natural biogeochemical phenomena (Alekin and Lyakhin 1984; Vernadsky 1965). In this connection, the problem of determining the effect of pollution on the living matter of the marine environment, as well as assessing the degree of danger to mankind, led to the birth of new scientific directions: the radiation and chemical ecology of aquatic organisms (Polikarpov et al. 1972); the radiochemoecology of the Black Sea (Polikarpov and Risik 1977); marine dynamic radiochemoecology (Polikarpov and Egorov 1986).

The foundational approaches to solving the problem of the interaction of living and inert matter with radioactive and chemical components of the environment were set forth in the revolutionary theories of Vladimir Ivanovich Vernadsky concerning the living and inert matter of the biosphere. He wrote: "living matter participates in geochemical processes... with its mass, its chemical composition and its energy" (Vernadsky 1978), advancing the hypothesis that "life—living matter, as it were—creates the domain of life for itself" (Vernadsky 1965). It follows from Vernadsky's statements that the complexes of biogeochemical interactions in ecosystems possess the property of homeostasis, i.e., natural mechanisms for regulating their resistance to various perturbing factors.

Solving the problem of the effect of chemicals and their radioactive isotopes on marine biogeocenoses[1] requires a study of patterns of sorption and metabolic concentration by hydrobionts, the distribution of radionuclides and their isotopic carriers within cellular structures, as well as the effect of ionising radiation and toxic effects on the vital functions of hydrobionts and on the reorganisation of the structure and function of ecosystems. On the other hand, the impact of biogeochemical interaction mechanisms in ecosystems affects the formation of cycles of pollutants in the marine environment, as well as influencing the transformation flows of their physicochemical forms and their elimination into geological depots, i.e., bottom sediment strata. It is for these reasons that the problem of the interaction of living and inert matter with the radioactive and chemical components of the marine environment is considered with increasing urgency, reflected in the long-term international project entitled 'Interaction between Water and Living Matter' (Polikarpov et al. 1979).

The purpose of these studies, as formalised in international programmes, is to develop a methodology suitable for assessing the radioactive and chemical pollution of aquatic organisms (hydrobionts) and bodies of water, determining the conditioning capability in relation to the homeostasis of marine ecosystems according to the pollution factor, as well as predicting their evolution as a result of climate change and anthropogenic impact.

According to contemporary concepts of the structure and function of ecosystems, the most effective approach for their study involves modelling. The conceptual basis for constructing quantitative theories in ecology consists in the balance method, originally used in studies of the dynamics of changes in the biomass of biogeocenotic components (Lotka 1925; Volterra 1926). Subsequently, the method was supplemented with the principle of describing energy balance in ecosystems (Vinberg and Anisimov 1966; Zaika 1972; Petipa 1981). Studies have shown that the patterns of sorption and metabolic absorption, distribution and transformation of chemicals and their radionuclides depend both on their content in aquatic organisms and on their concentration in the aquatic environment (Polikarpov and Egorov 1986). It is for this reason that, in order to assess the radioactive and chemical pollution of aquatic organisms and waters, to determine the conditioning capability of ecosystems in relation to homeostasis by the pollution factor, as well as to predict their evolution as a result of climatic changes and anthropogenic impact, it becomes necessary to develop a theoretical framework that is parametrically compatible with contemporary theories explaining the material, energy and mineral balance in marine ecosystems.

An evidential requirement for such models consists in the need to link their parametric basis to the reflection of the natural laws that govern the interaction of living

---

[1] Translator's Note: Defined by an English translation of the Great Soviet Encyclopaedia (1979) as "an interrelated complex of living and inert components associated with each other by material and energy exchange", the concept of biogeocenosis is not readily translatable into a more common English term (such as "ecosystem") without destroying its particular meaning.

and inert matter with radioactive and chemical components of the marine environment; in other words, that their determination be based on the results both of empirical data and of theoretical concepts of sorption-, metabolic-, physiological-, trophic- and energetic processes occurring in ecosystems. Due to these requirements, the parametrisation of such models should be closely related to empirical data; therefore, we will refer to the theory that satisfies these requirements as semi-empirical. The foundations of the semi-empirical theory of radioisotope and mineral exchange of marine ecosystems, constructed according to the results of observations recorded in the course of experiments with a radioactive label of pollutants, were published in 1986 in the monograph by G. G. Polikarpov and V. N. Egorov entitled *Marine Dynamic Radiochemoecology*. Since then, new mechanisms for the formation of ecosystem homeostasis in compensation for factors relating to the pollution of the marine environment have been identified and mathematically represented, along with a substantiation of biogeochemical criteria for regulating anthropogenic impact, leading to the development of a theoretical basis for studying anthropogenic ecology and the biogeochemical cycles of marine ecosystems. This theoretical basis can be used to support the sustainable development of aquatic areas in terms of factors for maintaining the balance between consumption and reproduction of resources connected with the quality of the marine environment.

References

Alekin OA, Lyakhin YuI (1984) Khimiya okeana. Gidrometeoizdat, Leningrad, 343 p. (in Russian)

Bowen HJM (1979) Environmental chemistry of the elements. Academic Press, London, New York, p 333

Gudiksen PH, Harvey TF, Lange R (1989) Chernobyl source term, atmospheric dispersion and dose estimation. Health Phys 57(5):697–706. https://doi.org/10.1097/00004032-198911000-00001

Gureev SA, Zenkin IV, Ivanov GG (2011) Mezhdunarodnoe morskoe pravo. Yur. norma, INFRA-M Izdat. dom, Moscow, 432 p. (in Russian)

IMCO/FAO/UNESCO/WMO/WHO/IAEA/UN joint group of experts on the scientific aspects of marine pollution (GESAMP) (1977) United Nations, New York, 35 p

Kuzin AM, Peredel'skii AA (1956) Okhrana prirody i nekotorye voprosy radioaktivno-ekologicheskikh svyazei. Okhrana Prirody i Zapovednoe Delo v SSSR 1:65–78 (in Russian)

Lotka AJ (1925) Elements of physical biology. Williams & Wilkins Comp., Baltimore, 495 p

Marine Environmental Quality (1971) The report of a special study held under the auspices of the Ocean Sciences Committee of the NAS-NRC Ocean Affair Board, 9–13 Aug 1971. National Academy of Sciences, Washington D. C., 107 p

Nelepo BA (1970) Yadernaya gidrofizika. Atomizdat, Moscow, 224 p. (in Russian)

Petipa TS (1981) Trofodinamika kopepod v morskikh planktonnykh soobshchestvakh: zakonomer-nosti potrebleniya pishchi i prevrashcheniya energii u osobi. Naukova dumka, Kiev, 242 p. (in Russian)

Polikarpov GG (1964) Radioekologiya morskikh organizmov. Atomizdat, Moscow, 295 p. (in Russian)

Polikarpov GG, Risik NS (eds) (1977) Radiokhemoekologiya Chernogo morya. Naukova dumka, Kiev, 231 p. (in Russian)

Polikarpov GG, Egorov VN (1986) Morskaya dinamicheskaya radiokhemoekologiya. Energoat-omizdat, Moscow, 176 p. (in Russian)

Polikarpov GG, Zesenko AYa, Lyubimov AA (1972) Dinamika fiziko-khimicheskogo prevrashcheniya radionuklidov mnogovalentnykh elementov v srede i nakoplenie ikh gidrobiontami. In: Polikarpov GG (ed) Radiatsionnaya i khimicheskaya ekologiya gidrobiontov. Naukova dumka, Kiev, pp 5–42. (in Russian)

Polikarpov GG, Valyashko MG, Bowen VT et al (1979) Dolgosrochnaya mezhdunarodnaya programma issledovanii po probleme: "Vzaimodeistvie mezhdu vodoi i zhivym veshchestvom". In: Vzaimodeistvie mezhdu vodoi i zhivym veshchestvom: proceedings of the international symposium, Odessa, 6–10 Oct 1975, vol 2. Nauka, Moscow, pp 201–207. (in Russian)

Shvedov VP, Patin SA (1968) Radioaktivnost' morei i okeanov. Atomizdat, Moscow, 288 p. (in Russian)

Shvedov VP, Shirokov SI (eds) (1962) Radioaktivnye zagryazneniya vneshnei sredy. Atomizdat, Moscow, 233 p. (in Russian)

Summary report on the post-accident review meeting after the Chernobyl accident: a report by the International Nuclear Safety Advisory Group (1986) IAEA, Vienna, 537 p

Timofeev-Resovskii NV (1957) Primenenie izluchenii i izluchatelei v eksperimental'noi biogeotsenologii. Botanicheskii Zhurnal 42(2):161–194 (in Russian)

Vernadsky VI (1965) Khimicheskoe stroenie biosfery Zemli i ee okruzheniya. Nauka, Moscow, 374 p. (in Russian)

Vernadsky VI (1978) Zhivoe veshchestvo. Nauka, Moscow, 358 p. (in Russian)

Vinberg GG, Anisimov SI (1966) Matematicheskaya model' vodnoi ekosistemy. In: Fotosintez sistem vysokoi produktivnosti. Nauka, Moscow, pp 213–223. (in Russian)

Volterra V (1926) Variazioni e fluttuazioni del numero d'individui in specie animali conviventi. Memoria Della Reale Accademia Nazionale Dei Lincei 2:31–113

Whicker FW, Schultz V (1982) Radioecology: nuclear energy and the environment, vol I. CRC Press Inc., Boca Raton, p 212

Zaika VE (1972) Udel'naya produktsiya bespozvonochnykh. Naukova dumka, Kiev, 145 p. (in Russian)

Chapter 2
Biogeochemical Mechanisms of the Interaction of Living and Inert Matter with the Radioactive and Chemical Components of the Marine Environment

Introduction

The structure of biogeochemical mechanisms is considered in terms of the main abiotic and biotic factors of the interaction of living and inert matter with the radioactive and chemical components of the marine environment. The content of substances contained in the waters of the world's oceans[1] are presented in terms of their radioisotopic composition and physico-chemical forms. Taking abiotic interactions into account, it is shown that current rates of anthropogenic release of chemicals and their compounds constitute a significant factor in the depletion of the carrying capacity of the aquatic environment. On the example of the transport of fission radionuclides from the site of the accident at the Chernobyl nuclear power plant, it is demonstrated that the influence of hydrometeorological processes manifests itself at various spatial and temporal scales—from diurnal and synoptic in time to global in terms of space. The effect of hydrodynamic mechanisms depends on the state of systems considered at various temporal and spatial scales in terms of stationarity. For nonstationary conditions, the impact of waves, currents, convection and diffusion processes for chemicals or their compounds dissolved in (or not differing in specific density from) water is always directed towards a decrease in the distribution gradients of their concentration in the aquatic environment. Between 1986 and 2000, such trends were identified in the vertical distribution profiles of ^{90}Sr and ^{137}Cs in the waters of the western halistatic region of the Black Sea. For stationary conditions, the influence of interactive hydrodynamic mechanisms is determined by the capacity characteristics of the aquatic host medium for radionuclides and their isotopic and non-isotopic carriers.

---

[1] Translator's Note. The term "World Ocean" is used in the original Russian to suggest a unified oceanic system.

The role of the biotic factor in the formation of the radioisotopic and chemical composition of marine waters is considered. It is shown that the concentrating capacity of living and inert matter is most satisfactorily reflected by the concentration factor (CF), which is equal to the ratio of the concentrations of the radionuclide or its isotopic carrier in the hydrobiont and in the water. On the basis of a large amount of empirical material obtained from the results of field observations and aquarium experiments using radiolabelling, it is demonstrated that the value of CF depends on the pH, temperature and salinity of waters, as well as on the physicochemical state and concentration of isotopic and non-isotopic carriers of chemicals in the marine environment. However, observations carried out over a period of two decades led to the conclusion that the concentration factors of ^{90}Sr by the mass representative species of Black Sea algae, molluscs and fish do not depend on the concentration of these radionuclides in water. In general, it is demonstrated that the concentration functions of aquatic organisms are described by the equations of a straight line and Freundlich adsorption isotherm or Langmuir adsorption model, while the dependence of their uptake rate is described by the Michaelis–Menten equation. It is noted that the role of the biotic factor in the formation of the radioisotopic and chemical composition of marine waters largely depends on the dimensional spectra of living and inert matter, on the trophic and specific marine ecosystems, as well as on the intensity of production, reduction and sedimentation processes.

2.1   Basic Physical and Chemical Properties of Radionuclides and Their Isotopic and Non-isotopic Carriers

While the chemical properties of stable elements and their radionuclides are identical, their physical characteristics may differ. Each chemical substance has a different biological significance, toxicity and range of tolerance. At low concentrations in the marine environment, chemical substances can have a limiting effect on production processes in ecosystems, while at high concentrations, the effect can be toxic. Radioactive isotopes are always present in the aquatic environment in trace amounts relative to the mass concentrations of their isotopic and non-isotopic carriers. Only in some cases—for example, for ^{210}Po—is the impact of radionuclides on marine organisms limited by the degree of chemical toxicity. For the most part, this impact depends on the intensity of absorbed doses of ionising radiation, which is in turn determined by the type (alpha, beta or gamma) of radioactive decay, the radiation energy and penetrating ability, as well as geometric conditions and the residence time of aquatic organisms in the radioactive environment. In this connection, the prescience of the problem posed by Vladimir Vernadsky should be noted: "In the radiogeological and biogeochemical processes associated with living matter, we can clearly observe the ability of living organisms to exercise selectivity for certain isotopes, that is, to distinguish between atoms of different structures and weight in the same chemical element. Here we leave the world of chemical phenomena and enter the world of atoms—radiogeological and radiochemical phenomena" (Vernadsky 1965). This problem can be attributed to the "isotope effect", which manifests itself both at the physicochemical and biochemical levels. According to physical-organic chemistry, it is the quantum–mechanical factor that gives rise to differences in the rates of chemical reactions of isotopes of atoms of the same element. Conversely, from a biochemical point of view, the "isotope effect" is manifested in the differentiation of penetration fluxes through cell membranes of isotopes of the same element having different mass numbers.

The "sanitary-hygienic and biogeochemical behaviour" of radionuclides and their isotopic and non-isotopic carriers in the environment exhibits both common and heterogeneous regularities. For this reason, the patterns of interaction of living and inert matter with radioactive pollutants of the marine environment are often considered separately from chemical interactions, albeit within the frameworks of interrelated sciences including geochemistry, biogeochemistry, radiochemistry, radioecology, chemoecology, radiation and chemical biology, ecotoxicology.

2.2 Structure of Biogeochemical Mechanisms of the Interaction of Living and Inert Matter with the Radioactive and Chemical Components of the Marine Environment

Radioactive and chemical substances entering the marine environment interact with both abiotic and biotic factors (Polikarpov and Egorov 1986). A structure diagram of the biogeochemical mechanisms of the interaction of living and inert matter with the radioactive and chemical components of the marine environment is given in Fig. 2.1.

In the first instance, dissolved chemical substances and solid aggregates transported by currents are redistributed in the upper, well-mixed water layer under the influence of waves and turbulence (block 1 in Fig. 2.1). As a result of the action of the vertical current velocity component, including advection, diffusion and migration through the thermo- and halocline layers, the substances penetrate into deep waters. Physical transport processes (block 2) lead to vertical and horizontal distribution of pollutants. Simultaneously with the transport, radioactive decay of radionuclides (block 3) and abiotic transformation of physicochemical forms (block 4) of substances of various biological significance, including toxic pollutants of the marine environment, occur. Next, in the series of abiotic factors, are included sedimentation of terrigenous suspensions (block 5), chemogenic sedimentation (block 6), sorption

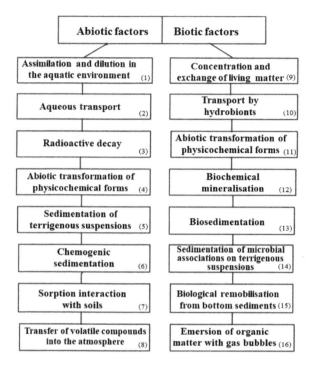

Fig. 2.1 Structure diagram of the biogeochemical mechanisms of the interaction of living and inert matter with the radioactive and chemical components of the marine environment

interaction of chemical compounds dissolved in water with soils (block 7) and the elimination of volatile compounds into the atmosphere (block 8).

The influence of biotic factors in the system of biogeochemical interactions is manifested in processes of sorption and metabolic exchange leading to the concentration of chemical substances and their isotopic and non-isotopic carriers by living and inert matter of the marine environment (block 9), in the transport of aquatic organisms (block 10), in the biotic transformation of their physicochemical forms (block 11), in biomineralisation (block 12), in gravitational biosedimentation (block 13) and in sedimentation on microbial associations on terrigenous suspensions (block 14) with further remobilisation from bottom sediments (block 15) and the emersion of organic matter with gas bubbles (block 16).

It should be noted that, while the impact of biogeochemical interactions of living and inert matter with the radiochemical components of the marine environment occurs within unified spatio-temporal scales, their influence on the fate of chemical substances—including pollutants—in the marine environment is variable. The mechanisms noted in boxes 1 and 2 in Fig. 2.1 lead only to the redistribution and dilution of chemical substances in marine waters. Blocks 5, 6, 12 and 13 cause deposition of contaminants in bottom sediments, block 8 reflects their release into the atmosphere, while blocks 3, 4, 10 and 11 are associated with mechanisms that ensure both entirely irreversible (blocks 3 and 11) and partially irreversible (blocks 4 and 10) removal of pollution from the marine environment. The purification of water is possible due to the sorption interaction of polluting water components with soils (block 7). Biological remobilisation (block 14) leads to re-pollution of waters, while the advection of organic matter to the surface (block 15) leads to contamination of the neuston biotope (air–water interface).

2.3 Radioisotope and Chemical Composition of the World's Oceanic Waters. Dilution and Uptake Capacity

The oceanic waters contain all known naturally occurring radionuclides along with their isotopic and non-isotopic carriers. The natural radioactivity of marine waters is determined by long-lived radionuclides having a half-life from 13.6×10^9 years (^{40}K) to 1.5×10^{17} years (^{124}Sn), the radioactive decay series of uranium, actinium and thorium, as well as cosmogenic radioisotopes having a half-life from 53 days (^7Be) to 2.5×10^6 years (^{10}Be) (Popov et al. 1979). The concentrations of naturally occurring radionuclides in the world's oceanic waters are presented in Table 2.1.

The concentrations of cosmogenic radionuclides are given in Table 2.2.

The totally natural radioactivity inventory of the world's oceanic waters is 1.474×10^{22}Bq. In the surface 10 m deep-water sediment layer, the corresponding figures is 2.726×10^{22}Bq (Baxter 1983).

Following the nuclear accident on April 26, 1986, a significant quantity of radionuclides from the Chernobyl nuclear reactor entered the natural environment. Summary data from the emergency that occurred at the fourth reactor is given in Table 2.3.

The characteristics of the accidental release of transuranic elements from this reactor are summarised in Table 2.4.

Comparative data on the total release of selected anthropogenic radionuclides into the environment are given in Table 2.5.

The concentrations of chemical elements in seawater are presented in Table 2.6.

Currently, about 10 million chemical compounds are known to science. Meanwhile, their number continues to grow by about 1000 new compounds annually. Under particular circumstances, most of these compounds can have a negative impact on the living components of ecosystems. The problems of chemical pollution of the marine environment were considered for the first time from a global perspective at the 1977 meeting of GESAMP experts (IMCO/FAO/UNESCO/WMO/WHO/IAEA/UN joint group of experts 1977). In influential monographs (Bowen 1979; Champ and Park 1982), it was shown that the anthropogenic chemical contamination of the marine environment has become a significant factor in the carrying capacity of the world's oceanic waters. Table 2.7 presents comparative data on the absorptive capacity of the oceanic waters and the annual transport of a number of chemical elements as a result of natural and anthropogenic processes.

Table 2.1 Concentrations of natural radionuclides in seawater (Popov et al. 1979)

Radionuclide	Half-life	Specific activity, decay dpm/L			Decay energy, J/s per Litre
		α	β	γ	
^{48}Ca	2×10^{16} years	–	7×10^{-4}	–	–
^{115}In	6×10^{14} years	–	3×10^{-4}	–	3.9×10^{-20}
^{40}K	1.36×10^{9} years	–	660	90	2.7×10^{-12}
^{138}La	7×10^{10} years	–	10^{-6}	2×10^{-5}	1×10^{-20}
^{87}Rb	6.6×10^{10} years	–	10	–	7.4×10^{-15}
^{124}Sn	1.5×10^{17} years	–	10^{-11}	–	
Uranium series					
^{238}U	4.5×10^{9} years	2	–	0.4	4.9×10^{-14}
^{234}Th	24.3 days	–	2	0.3	
^{234m}Pa	1.15 min	–	2	0.1	
^{234}Pa	6.7 h	–	0.06	0.006	
^{234}U	2.5×10^{5} years	2	–	0.5	
^{230}Th	8.1×10^{4} years	0.04	–	0.001	4.9×10^{-16}
^{226}Ra	1620 years	0.2	–	–	2×10^{-14}
^{222}Rn	3.825 days	0.2	–	–	
^{218}Po	3.05 min	0.2	0.006	–	
^{218}At	2 s	5×10^{-5}	–	–	
^{214}Pb	26.8 min	–	0.2	0.2	
^{214}Bi	19.7 min	7×10^{-5}	0.2	0.2	
^{210}Tl	1.4 min	–	7×10^{-14}	–	
^{214}Po	1.64×10^{-4} s	0.2	–	–	
^{210}Pb	22 years	–	0.2	0.2	
^{210}Bi	5 days	–	0.2	–	
^{210}Po	139 days	0.2	–	0.2	
Actinium series					
^{235}U	7.1×10^{8} years	0.1	–	0.1	1×10^{-15}
^{231}Th	25.5 h	–	0.1	0.1	
^{231}Pa	3.3×10^{4} years	0.005	–	0.005	4.9×10^{-16}
^{227}Ac	22 years	5×10^{-5}	0.005	–	
^{223}Fr	21 min	–	5×10^{-5}	–	
^{227}Th	18.7 days	0.005	–	0.005	
^{223}Ra	11.2 days	0.005	–	0.005	
^{219}Rn	3.92 s	0.005	–	0.002	
^{215}Po	1.83×10^{-3} s	0.005	–	–	
^{211}Pb	36 min	–	0.005	0.002	

(continued)

Table 2.1 (continued)

Radionuclide	Half-life	Specific activity, decay dpm/L			Decay energy, J/s per Litre
		α	β	γ	
^{211}Bi	2.16 min	0.005	–	2×10^{-5}	
^{211}Po	0.52 s	2×10^{-5}	–	–	
^{207}Tl	4.79 min	–	0.005	0.005	
Thorium series					
^{232}Th	1.4×10^{10} years	0.01	–	0.003	$> 1.5 \times 10^{-15}$
^{228}Ra	6.7 years	–	0.01	–	
^{228}Ac	6.13 years	–	0.01	0.01	
^{228}Th	1.9 years	0.01	–	0.04	
^{224}Ra	3.64 days	0.01	–	5×10^{-4}	
^{220}Rn	54.5 s	0.01	–	–	
^{216}Po	0.159 s	0.001	–	–	
^{212}Pb	10.6 h	–	0.001	0.01	
^{212}Bi	60.5 min	–	0.001	0.005	
^{208}Th	3.1 min	–	0.005	0.005	
^{212}Po	3.03×10^{-7} s	0.01	–	–	

Table 2.2 Concentrations of cosmogenic radionuclides in the surface layer of the oceans (Popov et al. 1979)

Radionuclide	Half-life	Specific activity	
		Decay dpm/L	Decay of the element dpm/g
^{3}H	12.5 years	0.036	3.3×10^{-4}
^{7}Be	53 days	0–0.75	–
^{10}Be	2.5×10^{6} years	10^{-6}	1.6×10^{-3}
^{14}C	5730 years	0.260	10
^{26}Al	7.4×10^{5} years	1.2×10^{-8}	1.2×10^{-3}
^{32}Si	500 years	2.4×10^{-5}	8.0×10^{-3}
^{36}Cl	3.1×10^{5} years	0.55×10^{-3}	3×10^{-5}
^{39}Ar	270 years	2.9×10^{-6}	5.0×10^{-3}

The data from these estimates indicate that the total annual transport of chemical elements as a result of natural and anthropogenic processes can range from $n \times 10^{-5}$ to $n \times 10^{1}\%$ of their content in the waters of the oceans.

Figure 2.2 shows the timescales of the turnover of chemical matter in the oceans as a result of the annual transport of elements. These data indicate that the turnover of such heavy metals as Cr and Cu takes place on a climatological timescale, while the turnover of Al, Fe, Mn and Pb occurs on an interannual timescale.

Table 2.3 Summary data on the content of some radionuclides in the fourth reactor of the Chernobyl NPP (specific activity figures are given as of 26/04/1986) and their release into the environment as a result of the accident (Il'in and Pavlovskii 1988; Belyayev et al. 1991; Gudiksen et al. 1991)

Radionuclide	Half-life	Reactor content		Contaminants released into the environment	
		kg	PBq	PBq	% of content in the reactor
^{137}Cs	30.2 years	8.1	210–260	37–100	14–48
^{134}Cs	2.1 years	3.2	140–150	48	32–34
^{90}Sr	29.1 years	4.3	160–220	1.3–8.1	0.6–5.0
^{106}Ru	368 days	6.9	860	6.3	0.7
^{144}Ce	284 days	3.3	3900	5.2	0.1
^{133}Xe	5.2 days	–	–	4400	90–100
^{131}I	8.04 days	–	6500	1300	60–90

Table 2.4 Characteristics of the main transuranic elements, forming an integral part of the radioactive release due to the accident at the fourth reactor of the Chernobyl nuclear power plant (activities are given as of 06/05/1986) (Summary Report 1986; Izrael 1990)

Radionuclide	Half-life	Reactor activity at the time of the accident, PBq	Emission activity, TBq
^{238}Pu	86.4 years	0.96	30.6
^{239}Pu	24,110 years	0.85	25.9
^{240}Pu	8553 years	1.22	37.0
^{241}Pu	14.7 years	170.2	5180.0
^{241}Am	433 years	0.14	4.1
^{242}Cm	162.8 days	25.9	777.0
^{243}Cm	28.5 years	0.037	1.1
^{244}Cm	18.1 years	0.096	3.4
Total	–	200	~ 6060

Table 2.5 Comparative data on the release of selected radionuclides into the environment (in PBq) (Gudiksen et al. 1989)

Radionuclide	Hiroshima	Windscale	Nuclear weapons testing	Chernobyl
^{137}Cs	0.1	4.4×10^{-2}	1300–1500	89
^{134}Cs	–	1.1×10^{-3}	–	48
^{90}Sr	8.5×10^{-2}	2.2×10^{-4}	650–1300	7.4
^{133}Xe	140	14	2.1×10^{6}	4400
^{131}I	52	0.6	7.8×10^{5}	1300

Table 2.6   Concentration of chemical elements in seawater, μg/L

Element	Average		Range	
	Bowen (1979)	Popov et al. (1979)	Bowen (1979)	Popov et al. (1979)
Ag	0.04	0.1	0.03–2.7	0.019–120
A1	2	5	1–8.4	1–350
Ar	450	–	–	–
As	3.7	2.3	0.5–3.7	0.46–80
Au	0.004	0.005	0.0005–0.027	0.0036–44
B	4440	4500	–	200–9300
Ba	13	30	2–63	6.2–60
Be	0.0056	0.0006	0.0006	0.0006–380
Bi	0.02	0.02	0.015–0.02	–
Br	67,300	68,000	–	12,100–66,300
C	28,000	28,000	–	–
Ca	412,000	422,000	–	–
Cd	0.11	–	< 0.01–9.4	< 0.01–0.5 to 4.7
Ce	0.0012	0.0012	–	0.0002–0.85
C1	19,350,000	19,870,000	–	–
Co	0.02	0.08	0.01–4.1	0.0018–4.1
Cr	0.3	0.6	0.2–50	0.005–2.5
Cs	0.3	0.5	0.15–0.42	0.28–10
Cu	0.25	3	0.05–12	0.8–200
Dy	0.00091	0.00091	–	0.0005–0.0014
Er	0.00087	0.0009	–	0.00061–0.00124
Eu	0.00013	0.00013	–	0.00009–0.00114
F	1300	1400	–	240–1500
Fe	2	3	0.03–70	4–200
Ga	0.03	0.03	–	0.0015–0.5
Gd	0.0007	0.0007	–	0.0005–0.00115
Ge	0.05	0.06	–	0.05–2
H	110,000,000	107,000,000	–	–
He	0.007	–	–	–
Hf	0.007	–	–	< 0.008
Hg	0.03	0.05	0.01–0.22	< 0.003–0.36
Ho	0.00022	0.03		0.00012–0.00059
I	60	60	50–70	4–75
In	0.00011	0.0001	–	0.00002–0.01
K	399,000	416,000	–	–

(continued)

Table 2.6 (continued)

Element	Average		Range	
	Bowen (1979)	Popov et al. (1979)	Bowen (1979)	Popov et al. (1979)
Kr	0.21	–	–	–
La	0.0034	0.0034	–	0.0025–2.5
Li	180	180	170–194	70–200
Lu	0.00015	0.00015	–	0.0001–0.00075
Mg	1,290,000	1,326,000	–	–
Mn	0.2	2	0.03–21	0.03–23
Mo	10	10	4–10	0.5–15
N	640	500	–	1–600 to 1300
Na	10,770,000	11,050,000	–	–
Nb	0.01	0.01	0.01–0.015	0.01–0.02
Nd	0.0028	0.0028	–	0.0013–0.0065
Ne	0.12	–	–	–
Ni	0.56	2	0.13–43	0.12–43
O	883,000,000	856,000,000	–	–
P	60	70	60–88	1–110 to 250
Pa	$\leq 5 \times 10^{-8}$	2×10^{-10}	5×10^{-8}–2×10^{-10}	2×10^{-10}–7×10^{-8}
Pb	0.03	0.03	0.03–13	0.02–5.2
Po	1.5–10^{-11}	2×10^{-11}	7×10^{-12}–2×10^{-11}	–
Pr	0.00064	0.0006	–	0.00041–0.002
Ra	8.9×10^{-8}	1×10^{-7}	$(3.2$–$9) \times 10^{-8}$	4.5×10^{-8}–9×10^{-7}
Rb	120	120	67–195	35–640
Re	0.004	–	0.004–0.0084	0.0027–0.0108 up to 0.021
Rn	6×10^{-13}	6×10^{-13}	–	–
Ru	0.0007	0.0007	–	–
S	905,000	928,000	–	–
Sb	0.24	0.2	0.18–5.6	0.12–1.76 to 6.7
Sc	0.0006	0.0015	0.0006–0.12	0.00001–0.04
Se	0.2	0.45	0.052–0.2	0.052–6
Si	2200	1000	2200–2900	84–5600 to 10,100
Sm	0.00045	0.00045	–	0.00026–0.001
Sn	0.004	0.01	0.002–0.81	0.009–1.22
Sr	7900	8500	7000–8500	7500–18,500
Ta	0.002	0.02	–	0.0025

(continued)

Table 2.6 (continued)

Element	Average		Range	
	Bowen (1979)	Popov et al. (1979)	Bowen (1979)	Popov et al. (1979)
Tb	0.00014	0.00014	–	0.00006–0.00036
Te	–	0.01?	–	–
Th	0.001	0.00004	0.0001–0.22	0.0000022–0.002
Ti	1	1	–	0.16–19
T1	0.019	0.01	–	0.0094–0.0166
Tm	0.00017	0.0002	–	0.00009–0.00037
U	3.2	3.3	0.04–6	0.15–4.7
V	2.5	1.5	0.9–2.5	0.3–5
W	0.1	0.12	0.001–0.7	0.11
Xe	0.05	–	–	–
Y	0.013	0.013	–	0.0112–0.3
Yb	0.00082	0.0008	–	0.00052–0.00172
Zn	4.9	5	0.2–48	0.7–3800
Zr	0.03	0.026	–	0.01–0.04

2.4 Meteorological Factor

The primary factors influencing the release of radioactive and chemical substances into marine ecosystems are comprised of meteorological and hydrometeorological processes. In case of technogenic—including nuclear—accidents, the impact is prevalent at small spatial intervals and synoptic timescales. In this regard, most coastal countries have detailed mobilisation plans in terms of measures aimed at mitigating such disasters. It has been established that atmospheric transport can, in a matter of days, contaminate sea areas located hundreds or even thousands of kilometres from the localisation sites of nuclear accidents, as happened following the nuclear disaster at the Chernobyl nuclear power plant (CNPP) on April 24, 1986 (Fig. 2.3).

In Fig. 2.3, peaks caused by the radiation of post-Chernobyl radionuclides ^{95}Zr–Nb and ^{140}Ba can be seen in the gamma spectra of green algae samples recorded in the Tunis Strait on 20th May 1986, less than one month after the accident at the CNPP. Already by 22nd May, a long-lived fission radionuclide ^{137}Cs was found in a seston ash sample in the Tyrrhenian Sea; on 28th May, the radionuclide content in the ash of an enteromorpha seaweed sample obtained in the Aegean Sea was supplemented by ^{140}La and the short-lived radioisotope of iodine ^{131}I.

Many studies have shown that the influence of meteorological processes can be traced over long distances. In Fig. 2.4a, the circles show the locations of *Mytilus galloprovincialis* mussels in 2004–2006 in various seas of the Mediterranean basin

Table 2.7 Characteristics of the absorptive capacity of the oceanic waters in relation to chemical substances (Bowen 1979)

Element	Absorptive capacity, t	Annual transport of elements, 10^3 t				Total annual transport		Turnover period in the world's oceans, years
		Natural processes		Anthropogenic processes		Tonnes/year	in % of the absorptive capacity	
		Weathering	Riverine efflux	Mining	Combustion			
Al	2.74×10^9	330,000	11,000	134,000	2800	4.78×10^8	1.75×10^1	5.7×10^0
B	6.03×10^{12}	40	550	2440	20	3.05×10^6	5.06×10^{-5}	2.0×10^6
Ba	1.78×10^{10}	2000	370	3000	56	5.43×10^6	3.00×10^{-2}	3.3×10^3
Ca	5.64×10^{14}	160,000	550,000	830,000	420	1.54×10^9	2.70×10^{-4}	3.7×10^5
Cd	0.15×10^9	0.4	3.7	7.7	0.065	1.19×10^4	7.00×10^{-3}	1.3×10^4
Co	0.03×10^9	80	7.4	22.7	1.1	1.11×10^5	4.00×10^{-1}	2.5×10^2
Cr	0.41×10^9	400	37	2500	2.8	2.94×10^6	7.10×10^{-1}	1.4×10^2
Cs	0.41×10^9	12	0.74	0	0.084	1.28×10^4	3.10×10^{-3}	3.2×10^4
Cu	0.34×10^9	200	110	6190	4.6	6.50×10^6	1.90×10^0	5.3×10^1
F	1.78×10^{12}	2800	3700	2000	220	8.72×10^6	4.90×10^{-4}	2.0×10^5
Fe	2.74×10^9	160,000	19,000	680,000	2200	8.61×10^8	3.14×10^1	3.2×10^6
Hg	0.04×10^9	0.2	3.7	8.4	8.4	2.07×10^4	5.00×10^{-3}	2.0×10^3
I	8.22×10^{10}	0.56	74	2	4.8	8.14×10^4	9.90×10^{-5}	1.0×10^6
K	5.46×10^{14}	84,000	81,000	18,500	840	1.84×10^8	3.37×10^{-5}	3.0×10^6
Li	2.47×10^{11}	8	74	4.4	2.8	8.92×10^4	3.62×10^{-5}	2.8×10^6
Mg	1.77×10^{15}	92,000	150,000	242	560	2.43×10^8	1.37×10^{-5}	7.3×10^6
Mn	0.27×10^9	3800	300	24,600	14	2.87×10^7	1.05×10^1	9.5×10^0
Mo	1.37×10^{10}	6	19	76	0.8	1.02×10^5	7.43×10^{-4}	1.3×10^5
Ni	0.77×10^9	320	19	560	5.6	9.05×10^5	1.18×10^{-1}	8.5×10^2

(continued)

Table 2.7 (continued)

Element	Absorptive capacity, t	Annual transport of elements, 10^3 t				Total annual transport		Turnover period in the world's oceans, years
		Natural processes		Anthropogenic processes		Tonnes/year	in % of the absorptive capacity	
		Weathering	Riverine efflux	Mining	Combustion			
P	8.22×10^{10}	4000	740	11,200	36	1.60×10^7	1.94×10^{-2}	5.1×10^3
Pb	0.04×10^9	56	110	3340	183	3.69×10^6	8.97×10^0	1.1×10^1
S	1.24×10^{15}	1000	140,000	51,700	84,000	2.77×10^8	2.23×10^{-5}	4.5×10^6
Sb	0.33×10^9	0.8	7.4	75	0.28	8.35×10^4	2.54×10^{-2}	3.9×10^4
Si	3.01×10^{12}	1,100,000	260,000	380,000	8400	1.75×10^9	5.80×10^{-2}	1.7×10^3
Sr	1.08×10^{13}	1500	2600	0	42	4.14×10^6	3.83×10^{-5}	2.6×10^6
U	4.38×10^9	10	15	23	0.28	4.83×10^4	1.10×10^{-3}	9.1×10^4
Zn	6.7×10^9	760	30	255	14	1.06×10^6	1.56×10^{-2}	6.3×10^3

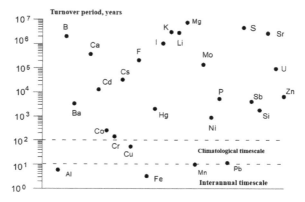

Fig. 2.2 Timescales of the turnover of chemical matter in the oceans as a result of the annual transport of elements

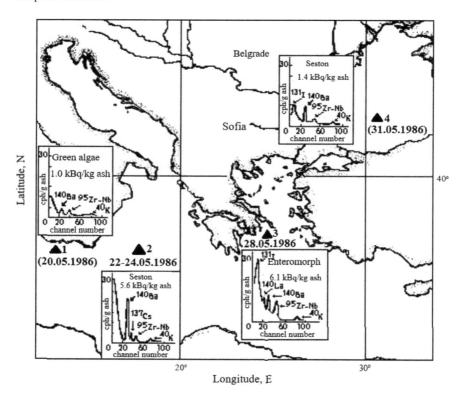

Fig. 2.3 Gamma spectra of samples of seston and aquatic organisms in the Mediterranean, Aegean and Black Seas in May 1986 (Egorov et al. 2008)

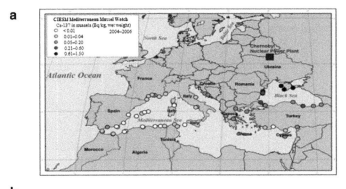

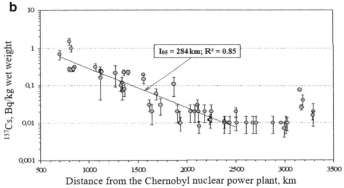

Fig. 2.4 **a** Location of sampling stations for mussels in the Mediterranean basin in 2004–2006; **b** concentration of ^{137}Cs in the soft tissues of mussels (*Mytilus galloprovincialis*) depending on distance from the CNPP (Thébault et al. 2008)

sampled under the international programme Mediterranean Mussel Watch. The location of the CNPP on the territory of Ukraine is shown in this figure by a shaded square.

Figure 2.4b shows the results of measurements of the concentration of ^{137}Cs in the soft tissues of mussels correlated with distance from the CNPP. An approximation of the observation results by an exponential function showed that the concentration of radiocaesium in mussels significantly decreased by half every 284 km from the CNPP. The materials presented in Fig. 2.4 show in general that the radioactive signal from the CNPP was traceable at a distance of 2500 km.

The studies of (Styro et al. 1991) showed that the influence of meteorological processes is significant on a global spatial scale. As a result of testing of nuclear weapons in open environments, it was established (Fig. 2.5) that the maximum radioactive fallout was confined to the middle latitudes of the northern and southern hemispheres of the earth.

Fig. 2.5 Latitudinal effect
of radioactive fallout on the
ocean surface (Styro et al.
1991)

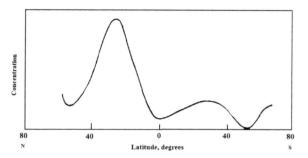

2.5 Hydrodynamic Factor

According to most of the influencing factors, the world's oceans comprise a non-stationary or quasi-stationary dynamical system (Lebedev et al. 1974). The main task of hydrophysical and bio-oceanographic marine sciences consists in the investigation of hydrodynamic processes, whose various aspects are considered in the specialised published literature. Theoretical approaches to their study are associated with the description of dynamic interactions in the marine environment at various space–time scales using Newton's second law of mechanics as implemented in the form of the Euler and Cauchy equations. The main hydrophysical mechanisms ensuring the formation of fields of distribution of chemicals and their isotopic carriers in spatial and depth dimensions consist in wave, current, convection and diffusion processes (Blatov and Ivanov 1992). The impact of these processes is one of the most significant factors of non-stationarity in the manifestation of the interaction mechanisms of living and inert matter with the radioactive and chemical components of the marine environment.

Figure 2.6 shows the results of measurements of the concentrations of ^{137}Cs and ^{90}Sr in the surface layer of the Black and Aegean Seas in 1987. One year after the Chernobyl nuclear disaster, maxima occurring as a result of the unevenness of atmospheric precipitation, as well as the influence of river runoff in the northwestern part of the sea and the discharge of waters from the North Crimean Canal, can be seen in the fields of the spatial distribution of radioactive contamination of the surface waters of the Black Sea.

An analogous survey of the distribution fields of (a) ^{137}Cs and (b) ^{90}Sr in the surface waters of the Black Sea in 1998–2000 is depicted in Fig. 2.7.

A comparison of the distribution field data presented in Figs. 2.6 and 2.7 shows that during the period from 1987 to 1998–2000 the influence of hydrodynamic processes manifested itself in a general decrease in the gradients of ^{137}Cs and ^{90}Sr in the surface waters of the Black Sea, as well as in the movement of maxima in these fields under the influence of sea currents.

The results of observations of the vertical distribution of ^{137}Cs and ^{90}Sr in the Black Sea in the post-Chernobyl period are shown in Fig. 2.8. It can be seen that the input of ^{90}Sr and ^{137}Cs to the surface layers of the Black Sea significantly exceeded

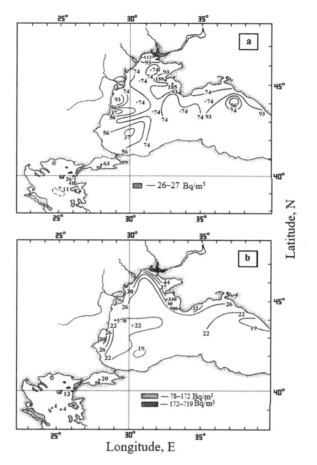

Fig. 2.6 Concentration distributions of **a** ^{137}Cs and **b** ^{90}Sr in the surface layer of the Black and Aegean Seas in 1987 (in Bq/m^3) (Stokozov 2003; Egorov et al. 2005)

the pre-Chernobyl levels of their concentration in water following the Chernobyl accident. In subsequent years, a decrease in the concentrations of ^{90}Sr and ^{137}Cs in surface waters was observed along with their transport to deep waters. In Fig. 2.8a, b, the solid lines in the vertical distribution profiles of radionuclide concentrations show the deepening trends in the gradient maxima. It can be seen that the slope angles of these trends are practically coincident irrespective of the biogeochemical characteristics of ^{90}Sr and ^{137}Cs. This indicates that the trends correspond to the average annual rate of vertical mixing of waters as a result of the combined impact of hydrodynamic processes, which was 10–12 m/year in the central part of the western cyclonic cycle of the Black Sea (Stokozov et al. 2008).

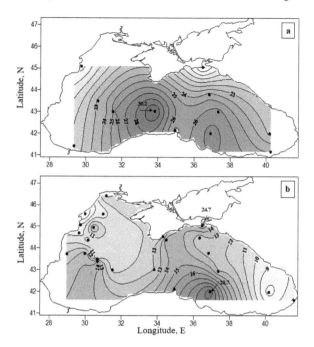

Fig. 2.7 Horizontal distributions of **a** ^{137}Cs and **b** ^{90}Sr in the surface layer of the Black Sea in the period 1998–2000 (in Bq/m^3). Dots indicate sampling stations (Stokozov 2003; Egorov et al. 2005)

2.6 Influence of Physicochemical Factors on the Radioisotopic and Chemical Composition of Living and Inert Matter

2.6.1 Concentrating Capacity of the Components of Marine Ecosystems

The majority of chemical elements and their radionuclides contained in water have been found in marine organisms as well as in inert matter (Vinogradov 1967; Lowman et al. 1971; Ancellin et al. 1979; Bowen 1979). The chemical and radioisotopic composition of the components of marine ecosystems depends on the influence of both abiotic and biotic factors. In inert matter, this composition is formed mainly as a result of geochemical processes, sorption interactions and waste products of living matter. For living matter, the basis of vital processes is formed by chemical elements and their compounds, which are used to form organs and tissues. Such chemical compounds determine the currently incalculable complex of biochemical reactions responsible for the formation of energy nutrition, somatic and generative growth, as well as the mineral metabolism of marine organisms. Nevertheless, in order to study the biogeochemical cycles of chemical substances and their isotopic

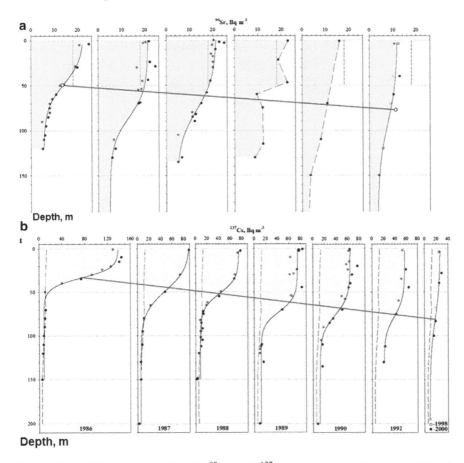

Fig. 2.8 Vertical distribution profiles of **a** ^{90}Sr and **b** ^{137}Cs in the waters in the western halistatic region of the Black Sea (dashed lines indicate pre-Chernobyl levels of radionuclide concentration) (Polikarpov and Egorov 2008)

and non-isotopic carriers of varying biological significance, it is sufficient to operate with such terms as the chemical and radioisotope composition of living and inert matter, the patterns of their concentration, exchange, circulation and mineralisation in marine ecosystems, as well as their elimination in geological depots.

Radioisotopic and chemical-analytical studies have revealed various formation mechanisms of the chemical composition of aquatic organisms and inert matter. Following the radioactive contamination of the Black Sea as a result of the nuclear accident at the Chernobyl nuclear power plant, it was found that the ^{137}Cs content of many species of algae differed by almost an order of magnitude (Fig. 2.9). Differences of several orders of magnitude were also recorded in the ^{239,240}Pu content of the Black Sea water, algae, molluscs, fish and sediments (Fig. 2.10).

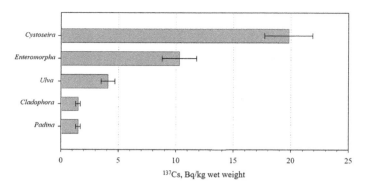

Fig. 2.9 Average concentrations of ^{137}Cs in algae of the Black Sea in 1986 According to: (Kulebakina 1996; Database of IBSS radiation and chemical biology department 2006)

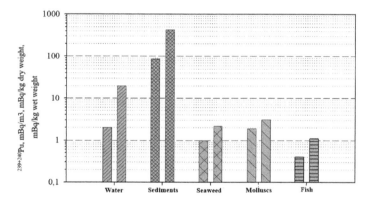

Fig. 2.10 Maximum and minimum values of ^{239,240}Pu concentrations in abiotic and biotic components of the ecosystem of the Sevastopol bays (Tereshchenko 2003, 2005, 2006)

Significant species-specific differences were found in the natural ^{210}Po radionuclide content of demersal (goby, round goby, black goby and toad goby), in benthopelagic (whiting, peacock wrasse and labrus wrasse), as well as in pelagic (atherina, horse mackerel, garfish, spicara, sprat and anchovy) species (Fig. 2.11).

It was determined (Fig. 2.12) that, with an increase in cell volume from 10^2 to 10^6 μm^3, the intracellular carbon content of unicellular diatoms decreased from 80–90% to 2.3% in terms of their dry weight (Taguchi 1976). It is shown (Fig. 2.13) that the concentration of copper in the dry mass of tissues in the 4–5 cm size group of mussels *Mytilis galloprovincialis* (a) rose with an increase in their specific mass in the body of bivalve mollusks but (b) decreased with an increase in the specific mass of the valves. Critical organs were additionally identified in which chemically hazardous pollutants of the aquatic environment were most concentrated. The maximum content of organochlorine compounds (PCB, HCH, DDE and DDT) was

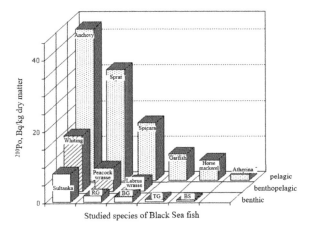

Fig. 2.11 ^{210}Po concentrations in fish from the coastal zone and bays of Sevastopol. RG—round goby; BG—black goby; TG—toad goby; BS—black scorpionfish (Lazorenko 2000, 2002)

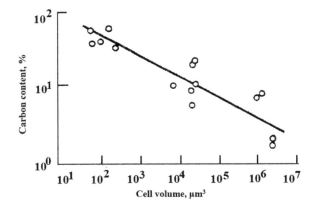

Fig. 2.12 Carbon content in diatom cells of different sizes (Taguchi 1976)

recorded in the subcutaneous fat of Black Sea bottlenose dolphins *Tursiops truncatus ponticus* (Fig. 2.14). The highest concentrations of mercury were found in the liver of females and in the white muscles of males of the Black Sea turbot *Psetta maxima maeotica* (Fig. 2.15).

The materials presented in Figs. 2.9, 2.10, 2.11, 2.12, 2.13, 2.14 and 2.15 generally indicate the existence of various biogeochemical mechanisms for the formation and concentration of the chemical and radioisotope composition of the components of marine ecosystems. The Academician Vladimir Vernadsky was the first to draw attention to the significance of the concentration of radioactive and chemical substances by aquatic organisms. He wrote: "By the concentration function of living matter, I

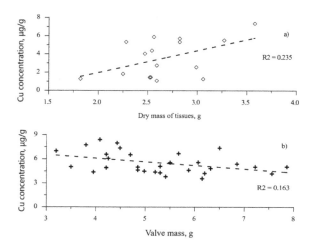

Fig. 2.13 Concentration of copper in dry mass of tissues (**a**) and valves (**b**) in the 4–5 cm size group of mollusks *Mytilis galloprovincialis* (Pospelova et al. 2018)

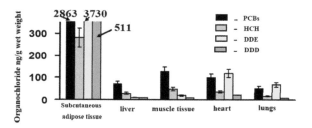

Fig. 2.14 Distribution of organochlorides in organs and tissues of the Black Sea bottlenose dolphin *Tursiops truncatus ponticus* (two-year-old female) (Malakhova 2006)

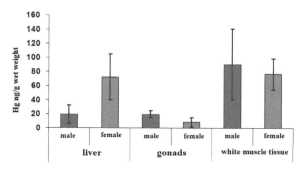

Fig. 2.15 Mercury content in the organs of the Black Sea turbot *Psetta maxima maeotica* in 2008–2009 obtained in the Sevastopol coastal area at a fishing depth of 60 m (based on materials by S. K. Svetasheva)

refer to those processes of a living organism, which, in essence, consist in the selection of certain chemical elements by the organism from its environment" (Vernadsky 1965). On the example of the content of radium in two species of duckweed in Kiev ponds, he introduced into science a measure of the bioaccumulation (in relation to a radioelement) capacity—or concentration factor (CF)—of aquatic organisms:

$$CF = C_h/C_w \qquad (2.1)$$

which was determined by the ratio of the concentrations of the radioelement in the hydrobiont (C_h) and the surrounding aquatic environment (C_w) (Vernadsky 1929). The relation (2.1), which subsequently found wide application in radiation hydrobiology, freshwater and marine radioecology, was referred to as the concentration factor of radionuclides, as well as that of their isotopic and non-isotopic carriers (Timofeev-Resovskii 1957; Timofeeva-Resovskaya et al. 1958; Trapeznikov et al. 2007; Polikarpov 1964, 1966). Information on the concentration factors of chemical compounds by components of marine ecosystems is presented in Table 2.8.

According to Polikarpov (1967), Bachurin (1968), the CF indicator characterises the level of (bio)accumulation of a radionuclide by a hydrobiont (aquatic organism), established as a result of dynamic equilibrium of the simultaneously occurring processes of the entry of radioisotopes into organisms and their elimination. For this reason, $CF = const$ only when the chemical and isotopic composition of the aquatic environment is stable. Under other conditions, the dependence of the change in CF over time reflects the kinetic patterns of the course of processes directed towards reaching a stationary or limiting level of CF.

2.6.2 Waterborne Radionuclide Concentration

It has been established that the concentration of radionuclides along with their isotopic and non-isotopic carriers by hydrobionts and inert matter depends on a number of abiotic and biotic factors. It is known from radiochemistry that the adsorption of trace amounts of elements onto the surface of a solid is proportional to their concentration in solution up to 10^{-4} mol/L while maintaining a constant value of the adsorption coefficient (Starik 1960). In research carried out between 1961 and 1964, Polikarpov discovered that the factors of concentration (CF) by hydrobionts of radioisotopes, whose introduction into the marine aquatic environment does not disturb the concentration of the corresponding chemical elements, do not depend on the concentration of these isotopes in terms of radioactivity (Polikarpov 1964). At very low concentrations of the isotopic carrier (or when it is entirely absent), this regularity can be preserved for isotope concentrations from 10^{-6} to 10^{-3} mol/L (Polikarpov 1966).

Research carried out from 1986 to 2010 within the framework of the programme of radioecological monitoring of the Black Sea following the Chernobyl accident, revealed the following findings. Immediately following the incident in 1986, the

Table 2.8 Concentration factors of chemical elements in the components of the marine ecosystem (Ancellin et al. 1979)

Component	Am	Sb	Ag	Ce	Cs	Cr	Co	Fe	I	Mn	Mo
Bottom sediments	25,000*	1000*	*	5000*	1000	*	3000*	1,000,000*	500*	1000*	250*
Seaweed	*	100*	5000*	5000	100	5000*	1000	20,000	10,000	5000	100
Crustaceans	1800*	100*	3000*	1500	50	1000*	1000	5000	100	5000	100
Molluscs	*	100*	40,000*	1500	50	1000*	1500	20,000	100	10,000	100
Fish	50*	100*	4000*	100	50	500*	200	1000	15	1000	20

Component	Nb	Pu	Ra	Ru	Sr	Te	Th	T	U	Zn	Zr
Bottom sediments	10,000*	10,000*	*	10,000	30	*	*	*	4000*	2000*	10,000*
Seaweed	100%	20,000*	150*	2000	100	1000*	20,000*	1*	30*	2000	2000
Crustaceans	19,	1000*	*	500	50	100*	*	1*	*	4000	500
Molluscs	1000	1000*	*	2000	20	100*	6000*	1*	30*	80,000	1000
Fish	50	100*	100	10	5	10*	1500*	1*	0.1*	5000	30

*Very large scatter of values (up to 100 times or more) or lack of published data

concentration of ^{90}Sr in the waters of Sevastopol Bay increased by an order of magnitude (Fig. 2.16a). Then, over the course of one year, the concentration decreased by almost five times thanks to the carrying capacity of the waters of the bay. Since 1987, there has been a trend towards a decrease in the concentration of ^{90}Sr in water to levels below the pre-Chernobyl levels, followed by secondary pollution of waters with radiostrontium to the pre-Chernobyl level (Fig. 2.16a). Figure 2.16b shows that the trend of changes in the concentration of ^{90}Sr in brown algae seaweed for the most part coincided with the long-term trend of changes in the concentration of radiostrontium in the water of the bay. At the same time, although it was only after 1987 that the dynamic regularities of the concentration of ^{90}Sr in mussels (Fig. 2.16c) coincided with the regularities of changes in the concentration of this radionuclide in water, no response was observed in terms of a change in the concentration of ^{90}Sr in whiting (Fig. 2.16d). The examination showed that the concentration factors (CF) of ^{90}Sr in *Cystoseira* brown algae during the post-Chernobyl period remained constant over all the years of observation (Fig. 2.17a) and did not depend on changes in the concentration of radiostrontium in water (Fig. 2.17b), which explains trends in the concentration characteristics of water (Fig. 2.16a) and *Cystoseira* brown algae (Fig. 2.16b). According to calculations, the concentration factors of ^{90}Sr in mussels also neither changed over the observation period (Fig. 2.18a) nor depended on changes in the concentration of this radionuclide in water (Fig. 2.18b). When considering the effect of non-coincident trends in the concentration of ^{90}Sr in the seawater of Sevastopol Bay (Fig. 2.16a) and in mussels (Fig. 2.16c), considered alongside the coinciding trends for brown algae, it should be taken into account that *Cystoseira* has a ratio of annual production (P) to biomass (B) $P/B \approx 1$, while the life cycle of mussels can be up to 10 years. This explains why the new-grown branches of Cystoseira seaweed can enter isotopic equilibrium with water on a smaller time scale than mussels.

The results of observations in the monitoring mode presented in Fig. 2.19a show that the concentrations in (C_h) and concentration factors (CF) of ^{90}Sr by whiting increased following the accident at the Chernobyl nuclear power plant during 1986–1998. Subsequently, the concentration of this radionuclide in fish began to decrease in proportion to the trend of its decreasing concentration in water, with concentration factors observed to reach stationary levels (Fig. 2.19b). Based on the results of the analysis of hydrobionts, it follows that the prediction of radioactive contamination of waters should take into account not only their concentration capacity, but also their kinetic mineral metabolism characteristics.

2.6.3   Influence of Temperature and Illuminance

It has been established that, for a number of radionuclides, the concentration factors do not depend on temperature. The temperature coefficient of brown algae in relation to ^{144}Ce was close to unity (Polikarpov 1967). The concentration factor of silver (^{110m}Ag) in marine isopods was somewhat dependent on temperature in the range of $+10$ to $+20\,°C$ (Sayhan et al. 1985). However, in a range from $+20$ to $+31\,°C$, water

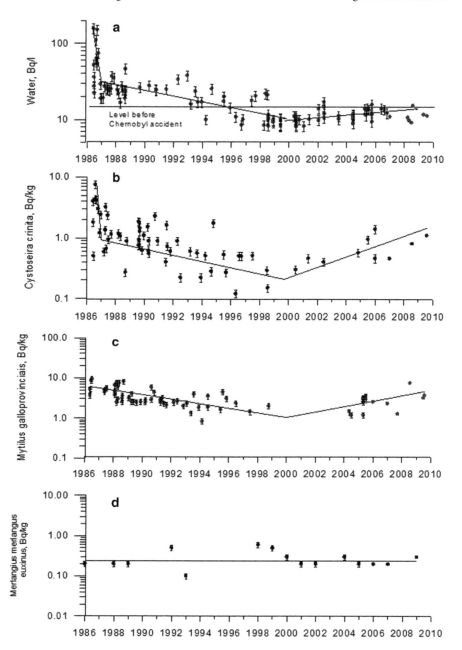

Fig. 2.16 Change in the post-Chernobyl concentration of ^{90}Sr in **a** water, **b** *Cystoseira* brown alga, **c** in mussels and **d** in whiting in the marine area of the Sevastopol Bay (Polikarpov and Egorov 2008)

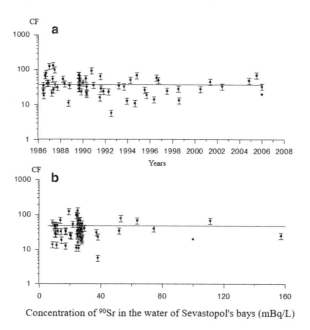

Concentration of ^{90}Sr in the water of Sevastopol's bays (mBq/L)

Fig. 2.17 Dynamics of change in **a** the concentration factors of ^{90}Sr by Cystoseira in the post-Chernobyl period (1986–2008), **b** depending on the concentration of radiostrontium in the water of the Sevastopol Bay (Database of IBSS radiation and chemical biology department 2006; Egorov et al. 1994; Kulebakina 1996; Mirzoeva et al. 2000, 2005, 2006)

temperature was not observed to have a significant effect on the sorption of radionuclides ^{54}Mn, ^{57}Co, ^{59}Fe, ^{65}Zn, ^{85}Sc and ^{137}Cs by fish larvae (Harvey 1971). The effect of temperature on the accumulation and excretion of ^{60}Co and ^{65}Zn radionuclides by shrimps has been found to be insignificant both with regard to parenteral and alimentary uptake pathways (Weers 1975). The literature also contains data indicating that the concentration factors of radionuclides by hydrobionts can significantly depend on temperature. In the experiment work carried out by I. Ivleva and her colleagues (Ivleva et al. 1984), temperature change was observed to significantly influence the concentration levels of ^{65}Zn in unicellular algae. An increase in temperature led to a rise in the concentration of ^{65}Zn from seawater by the biocenosis of oysters, crabs and mussels (Duke et al. 1969). In experiments with ^{204}Tl and green algae *Ulva rigida*, it was found that the ratio of the concentration factors of monovalent thallium at $+ 14$ and $+ 24$ °C was 1.6, while that of trivalent thallium, on average, did not differ from unity (Polikarpov et al. 1972). Despite changes in the dynamics of accumulation of radionuclides by hydrobionts at different temperatures, their limiting concentration factors were not observed to differ significantly (Kryshev and Sazykina 1986).

The accumulation of radionuclides by marine organisms can both be unaffected by light and closely depend on lighting conditions. It was shown (Polikarpov 1964) that the uptake kinetics of ^{32}P by *Ulva rigida* did not vary under heterogeneous

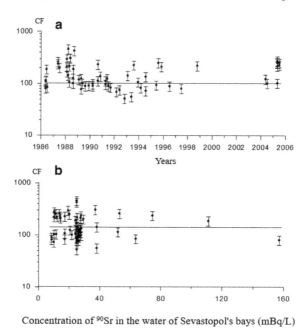

Concentration of ^{90}Sr in the water of Sevastopol's bays (mBq/L)

Fig. 2.18 a Dynamics of the concentration factors of ^{90}Sr in *Mytilus galloprovincialis* mussels and **b** the dependence of the concentration factor of ^{90}Sr in mussels on the concentration of this radionuclide in water in the period 1986–2006. (Database of IBSS radiation and chemical biology department 2006; Egorov et al.1994; Kulebakina 1996; Mirzoeva et al. 2000, 2005, 2006)

illumination conditions applied over a period of 32 days. A similar pattern was noted for ^{90}Sr and ^{144}Ce. Conversely, the factors of concentration of ^{60}Co and ^{137}Cs by this alga were observed to be 2–3 times higher under conditions of availability of light than in the dark. Higher concentration factors of ^{59}Fe were also determined for *Ulva pertusa* under conditions of increased illumination (Hiyama and Shimiza 1964). It was noted (Rice 1965) that an increase in the concentration factors may be due not to the increase in illumination itself, but rather to the formation process of new protoplasm, synthesised as a result of the uptake of photosynthetically active radiation by algae.

2.6.4 Influence of Water Salinity

The information available in the literature indicates that salinity has an ambiguous effect on the levels of radionuclide and isotopic carrier concentration by hydrobionts. For a number of closely related species of marine animals and plants, the concentration factors (*CF*) were not observed to significantly differ at water salinity equilibrium from 17 to 35‰ (Polikarpov 1967). Concentration factors of ^{90}Sr and ^{137}Cs were

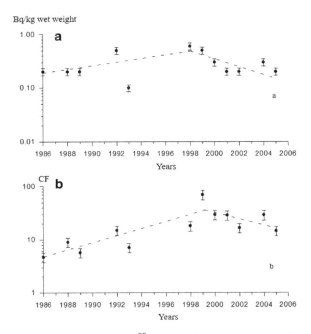

Fig. 2.19 a Dynamics of concentration of ^{90}Sr in whiting *Merlangius merlangus euxinus* and **b** concentration factors of ^{90}Sr in whiting in the period 1986–2005. (Database of IBSS radiation and chemical biology department 2006; Kulebakina 1996; Mirzoeva et al. 2000, 2005, 2006)

observed to decrease with increasing salinity in macrophytes, invertebrates and fish (Bryan and Ward 1962; Zlobin 1968; Luci and Jelisavciê 1970; Wright 1977). The effect of increased salinity on zinc, cadmium and copper concentration factors in mussels (Phillips 1977) and cadmium in crabs (Wright 1977) was also shown to be negative. The *CF* of ^{51}Cr, [106]Ru, [144]Ce, [125]Sb and ^{137}Cs was observed to be higher in freshwater fish in water having a salinity of 10‰ than at 0.1‰ (Nechaev et al. 1972). Kulebakina (1984) experimentally confirmed that the concentration factors of ^{90}Sr in the Black Sea and Mediterranean *Cystoseira*, as well as in Mediterranean *Sargassum*, decreased with increasing water salinity in accordance with a power-law dependence. Similar dependencies were obtained in experiments on the accumulation of [45]Ca, [54]Mg, [65]Zn, ^{137}Cs and [204]Tl by Black Sea algae (*Cystoseira*, *Ulva* and *Enteromorpha*). The same work provides data indicating that the concentration factors of [106]Ru and [144]Ce by *Cystoseira* and *Ulva* increased with increasing water salinity, while the concentration factors of [48]V and [207]Bi in Black Sea *Ulva* reached a maximum at a salinity of 18 ‰ compared with salinity values of 10 and 24 ‰. Salinity changes did not affect the kinetics of radionuclide excretion by macrophyte species. In marine isopods, the elimination rate of [110m]Ag was higher at a salinity of 21 ‰ than at 6.54‰ (Sayhan et al. 1985). The effect of salinity on the concentration factors of radionuclides by hydrobionts can be explained primarily in terms of the change in the content of isotopic and non-isotopic carriers in waters having

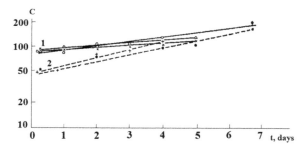

Fig. 2.20 Concentration factors of ^{65}Zn by the *Ulva rigida* alga from (1) Black Sea and (2) oceanic waters. Legends—data from different experiments (Ivanov and Rozhanskaya 1972)

varying salinity, the consequent effect on the osmotic mechanisms regulating the permeability of biological membranes and associated changes in the physiological functions of hydrobionts (Glasser 1962). Experiments have shown that, while the exchange rate of ^{65}Zn by *Ulva* green algae in the Black Sea (18‰) was higher than in ocean waters (35‰), stationary values of the factors of zinc concentration by algae did not vary (Fig. 2.20).

2.6.5 Physicochemical State of Isotopes and pH of the Medium

Summaries and reference books (Formy elementov i radionuklidov 1974; Bryan 1976; Bowen 1979; Popov et al. 1979) list many physicochemical forms of elements and their radionuclides that occur in the marine environment in the form of ions, in complex non-ionic forms, as well as in the form of pseudocolloids, colloids and suspensions. The physicochemical forms of chemical elements contained in the world's oceanic waters are presented below.

$Ag - AgCl_2^-$ [*1], $AgCl_3^{2-}$ [*2], $AgCl_4^{3-}$ [*3]

Al–colloid, $Al(OH)_4^-$ [*1, *3], $Al(OH_2)_2^+$? $Al(OH)_3$? [*2]

$Ar - Ar$[*2, *1]

$As - AsO_4H^{2-}$, organic [*1]$HAsO_4^{2-}$ [*2, *3], $H_2AsO_4^-$, H_3AsO_4, H_3AsO_3, organo complexes [*2]

$Au - AuCl_2^-$ [*2, *1]

$B - B(OH)_3$ [*2, *1], $B(OH)_4^-$, $B(OH)_3^{2-}$, organo complexes, complexes with Na$^+$, Mg^{2+}, Ca^{2+}[*2]

Ba–Ba^{2+} [*2, *1], BaSO$_4$ [*2]

Be–BeOH$^+$ [*1], hydroxo complexes? [*2]

Bi–BiO$^+$ [*1] BiO$^+$? BiOCl? BiCl$_4$? [*2]

Br–Br$^-$ [*2, *1]

$C - HCO_3^-$, CO_3^2, $CO_2 + HCO_3^- + CO_3^{-2}$ [*2]

Ca–Ca^{2+} *2 [*1]

Cd–colloid, CdCl$_2$ [*1, *3], CdCl$^+$ [*2, *3], Cd^{2+}, CdS0$_4$ [*2]

Ce–Ce^{3+} [*1]

C1–Cl$^-$ [*2, *1]

Co–Co^{2+} [*2, *1, *3], CoCO $_3$ *1, Co^{2+}, CoCl$^+$? CoOH$^+$? organo complexes [*2]

Cr–Cr(OH) $_3$, $Cr O_4^{2-}$ [*1], hydro complexes? [*2]

Cs–Cs$^+$ [*2, *1]

Cu–CuOH$^+$ [*2, *3], CuCO $_3$ [*1, *3], Cu^{2+} [*2, *3], organo complexes [*2]

Dy–DyOH^{2+} [*1], hydro complexes? [*2]

Er–ErOH^{2+} [*1]

Eu–EuOH^{2+} [*1], hydro complexes? [*2]

F–F$^-$, MgF$^+$ [*2, *1]

Fe–colloid, $Fe(OH)_2^{2+}$ [*1], Fe(OH)$_3$? [*3], FeOOH, organo complexes, $FeH_3SiO_4^{2+}$ [*2]

$Ga - Ga(OH)_4^+$ [*1], hydro complexes [*2]

Gd–GdOH^{2+} [*1], hydro complexes [*2]

$Ge - GeO_4H_3^-$ [*2, *1], Ge(OH)$_4$, $GeO_2(OH)_2^{2-}$ [*2]

H–H$_2$O [*2, *4], H$_2$, H$_2$O$_2$

He–He [*1]

Hf–Hf(0H)$_4$ [*1]

$Hg - HgCl_4^{2-}$ [*2, *1, *3], $HgCl_3^-$, $HgCl_3Br^{2-}$ [*3], organic [*1]

Ho–HoOH^{2+} [*1], hydro complexes [*2]

I–I$^-$, IO_3^-, CH$_3$I [*2, *1], I$_2$

In–hydro complexes [*2]

K–K$^+$ [*2, *1], KSO_4^-

Kr–Kr [*2, *1]

La–La^{3+} [*1], hydro complexes? [*2]

Li–Li$^+$ [*2, *1]

Lu–LuOH^{2+} [*1], hydro complexes? [*2]

Mg–Mg^{2+} [*2, *1], MgSO$_4$ [*2]

Mn–Mn^{2+}, MnCl$^+$ [*2, *4], Mn(OH)$_3$? Mn(OH)$_4$? [*2, *5], MnSO$_4$, MnOH$^+$, organo complexes [*2]

$Mo - Mo_4^{2-}$ [*2, *1, *3]

$N - NH_3 + NO_3^- + N_2$ [*2, *1], NH$_3$, N$_2$O, NH$_2$OH? [*2]

Na–Na$^+$ [*2, *1]

Nd–Nd^{3+} [*1], hydro complexes? [*2]

Ne–Ne [*2, *1]

Ni–Ni^{2+} [*2, *1, *3], NiCO3 [*1], NiOH$^+$? NiCl$^+$? [*2, *3]

O–OH$_2$ [*2, *1], O$_2$, H$_2$O$_2$

$P - HPO_4^{2-}$ [*2, *1], MgPO$_4$ [*1]

Pb–PbCO$_3$, colloid [*1], Pb^{2+} [*2], PbOH$^+$, PbCl$^+$ [*2, *3], PbCl$_2$ [*3]

Pr–Pr^{3+} [*1], hydro complexes? [*2]

Ra–Ra^{2+} [*2, *1]

Rb–Rb$^+$ [*2, *1]

$Re - ReO_4^-$ [*2, *1]

Rn–Rn [*2, *1]

$S - SO_4^{2-}$ [*2, *1], $NaSO_4^-$ [*1], $MgSO_4$, H_2S, S, $S_2O_3^{2-}$, SO_3^{2-} [*2]

$Sb - Sb(OH)_6^-$ [*3], $Sb(OH)_6^-$? Sb(V), Sb (III) [*2]

Sc–Sc(OH)$_3$ [*1], hydroxo complexes? [*2]

$Se - SeO_3^{2+}$ [*1], SeO_4^{2-} [*2, *5], Se (IV) [*2]

$Si - Si(OH)_4^-$ [*2, *4], colloid [*1]

Sm–SmOH^{2+} [*1], hydroxo complexes? [*2]

Sn–SnO$_4$H$_3$, organic [*1], hydroxo complexes? [*2]

Sr–Sr^{2+} [*2, *1]

Tb–TbOH^{2+} [*1], hydroxo complexes? [*2]

Te–TeO$_3$H$^-$ [*1]

Th–Th(OH)$_4$ [*1], hydroxo complexes? [*2]

Ti–Ti(OH)$_4$? [*2]

Tl–Tl$^+$ [*2]

Tm–TmOH^{2+} [*1], hydroxo complexes? [*2]

$U - UO_2(CO_3)_3^{4-}$ [*1], $UO_2(CO_3)_3^{4-}$? [*2]

V–H$_2$VO$_4$, HVO_4^{2-} [*1, *3], $VO_2(OH)_3^{2-}$ [*2], $(H_2V_4O_{13})^{4-}$, VO_3^- [*3]

$W - WO_4^{2-}$ [*2, *1]

Xe–Xe [*1]

Y–Y(OH)$_3$ [*1], hydroxo complexes? [*2]

Yb–YbOH^{2+} [*1], hydroxo complexes [*2]

Zn–Zn^{2+} [*2, *1, *3], ZnOH$^+$ [*2], ZnCl$^+$ [*2, *1, *3], ZnClOH?, hydroxo complexes [*2]

Zr–Zr(OH)$_4$ [*1], hydroxo complexes [*2]

Sources:

*1—Bowen (1979);

*2—Popov et al. (1979);

*3—Bryan (1976);

*4—Droop (1974);

*5—Fowler and Guary (1977).

It is noted that the spectrum of physicochemical forms of radionuclide intake into the marine environment depends on the type of sources. Radioactive waste from nuclear plants enters the ocean mainly in dissolved form, i.e., taking the form of simple or complex ions that can immediately participate in biochemical processes taking place in the marine environment (Gromov and Spitsyn 1975). In the case of pollution from nuclear explosions, radionuclides entering the surface of sea waters take the form of conglomerates fused with oxides and silicates of various elements (Aleksakhin 1963; Izrael 1968; Nelepo 1970). Studies of the adsorbability of [91]Y, [144]Ce, [144]Pr, [234]Th and [234]Pa on PTFE and glass showed that the equilibrium state of physicochemical forms of radionuclides depended on the "age" of solutions as well as the pH of the medium (Skulskii et al. 1974). The chemical "behaviour" of artificial radionuclides and natural isotopes of elements existing in the marine environment over long periods of time may not coincide or differ at all, since, due to radioactive decay, the period of circulation of artificial radionuclides in the environment is much shorter than the average persistence of stable nuclides in the ocean (Popov 1971). While the pH value in seawater does not change significantly (Popov et al. 1979), a higher pH increases the hydrolysability of elements, shifting the chemical equilibrium towards the formation of products having a higher molecular weight (Popov 1971). A change in the physicochemical form of an element can dramatically transform its toxicity, with the most toxic forms involving free metal ions (Florence 1983).

It has been established that the concentration of radionuclides by hydrobionts largely depends on the physicochemical form of their presence in seawater. A comparison of the factors of concentration of ^{90}Sr, ^{95}Zr–Nb, ^{106}Ru, ^{137}Cs and ^{144}Ce by marine organisms in the Irish Sea following contamination with wastes from the atomic plant at Windscale varied significantly from the concentration factors of radionuclides by marine organisms in waters polluted only by atmospheric precipitation (Mauchline 1963). The concentration factors of ^{91}Y in experiments with the Black Sea *Ulva* seaweed changed synchronously with the change in the sorption capacity of this radionuclide as its solution aged (Polikarpov et al. 1972). In the same work, it was shown that the concentration of trivalent ^{144}Ce in *Ulva rigida* was higher than its tetravalent form (Fig. 2.21), while ^{204}Tl sulphate accumulated to a higher level than monovalent and trivalent thallium hydroxide. A review carried out by Zesenko (1977) demonstrated that the concentration factors of hydrobionts are generally dependent on the physicochemical state of radionuclides.

It is reported that an increase in water alkalinity from 8 to 10 units can change the accumulation rates of radionuclides by hydrobionts almost twofold (Ivleva et al. 1984; Ring et al. 1985). The concentration factors of chemical elements by hydrobionts irrespective of their physicochemical forms are shown in Table 2.8.

Although deviations from this pattern were also noted, a decrease in pH was shown to generally result in the suppression of the toxicity of heavy metals (Linnik 1986). With decreasing pH, the chemical form of radionuclides changes in such a

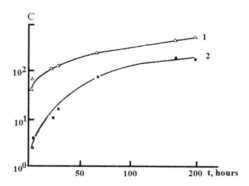

Fig. 2.21 Factors of concentration of trivalent (1) and tetravalent (2) cerium-144 by *Ulva rigida* alga (Polikarpov et al. 1972)

way that bioassimilation processes tend to increase, while sorption processes weaken (Kryshev and Sazykina 1986).

2.6.6 Concentration of Isotopic and Non-isotopic Carriers

A decrease in the concentrating function of hydrobionts was observed in the area of macroscopic concentrations of isotopic carriers (Stary and Zeman 1983). It was noted that the concentration factor (CF) of $HgCl_2$ by *Pernna veridis* molluscs having a mercury content of 0.05 mg/L was 304, while at 0.20 mg/L, it fell to 126 (Lakshmanan and Nabisan 1979). With a specific content of cadmium in water of $C_w = 20$ mg/L, its concentration in *Cerastoderma glaucum* bivalve molluscs was 12 mg/kg, while at $C_w = 40$ μg/L, the concentration was 18 mg/kg (Chabert 1984). It has been established that, across a wide range of changes in the concentration of an isotope carrier in water, the relationship between the stationary level of the concentration of a radioisotope or isotope carrier in a hydrobiont (C_h) and its concentration in an aqueous medium (C_w) can be described by two types of regularities: Henry's law (Titlyanova and Ivanov 1961; Brown et al. 1982; Stary and Zeman 1983), which coincides with the Freundlich equation, and the Langmuir equation (Nesmeyanov 1978).

It was determined that in the area of substrate microconcentrations in water (C_w), the rate of uptake of a radionuclide or isotopic carrier by an ecosystem component (v_u) directly from the aquatic environment is proportional to C_w (Corcoran and Kimball 1968; Bernhard 1971; Riziê 1972). The kinetic characteristics of the accumulation and excretion of radionuclides by living and inert components of the marine environment, which were shown to depend on the type of metabolic and sorption processes occurring over a wide range of changes in the concentration of chemical substrates in water, are mathematically described by the Michaelis–Menten equation (Patton 1968). For biogenic elements (P, N), the relationship between v_u and C_w for microorganisms and planktonic algae is described by the Michaelis–Menten equation (Dugdale 1967; Maclsaac and Dugdale 1969; Aizatullin and Leonov 1977; La Molta and Shich 1979; Egorov and Ivanov 1981; Hattori 1982; Rodach 1983; Levis et al. 1984).

At present, there are quite extensive data showing that the accumulation of some chemical elements by aquatic organisms is determined not only by the content of these elements in the environment, but also by the concentration in water of their macroelement analogues. This primarily refers to elements, such as Sr and Ca, Cs and K, and their radionuclides ^{90}Sr and ^{137}Cs, whose concentration by aquatic organisms depends on the content of calcium and potassium in the aqueous medium (Galli and Zattera 1979; Ramade 1980). Scott (1954) showed that, when replacing potassium with an equivalent amount of sodium, marine macrophytes compensate for the potassium deficiency by absorbing caesium. According to the observations of Pickering and Lucas (1962), when the calcium concentration in water is changed from 0.4 to 60 mg/L at a constant strontium concentration, the factors of calcium concentration in filamentous algae decrease from 18,600 to 97 units, while the factors of strontium concentration decrease from 10,700 to 100. The change in Sr concentration in water was observed to be inversely proportional to the level of Ca accumulation in *Cystoseira barbata* (Kulebakina and Parchevskaya 1973). Using experimental (Barinov 1965; Kulebakina and Parchevskaya 1973) and mathematical modeling methods (Bachurin 1968), it was shown that the kinetic regularities of the exchange of ^{90}Sr and ^{45}Ca by Cystoseira varied with a constant strontium and calcium content in the medium. With a changing concentration of an isotope carrier, it was proposed that the dependence of the rate of uptake of a radionuclide by a hydrobiont can be described by the Michaelis–Menten equation, in which the substrate affecting the metabolism of a hydrobiont is comprised of the total concentration of elements and their chemical analogues (Aizatullin et al. 1984). In the literature is noted not only the influence of non-isotopic carriers on the concentrating function of hydrobionts, but also that of elements that differ in terms of their chemical properties. It was found that the concentration factors of ^{203}Hg were significantly lower in those shrimps that were preliminarily kept under conditions of increased selenium content in seawater. The rate of elimination of ^{203}Hg by shrimp was not affected by selenium pretreatment (Skreblin et al. 1985). Increased phosphate concentrations in seawater were shown to enhance the excretion of Mn and Zn by molluscs (Miller et al. 1985).

2.6.7 Dimensional Characteristics of Hydrobionts

Studies carried out on mussels showed that, within the limits of observational accuracy, the total content of a number of metals (Mn, Cu, Zn, Ni, Fe and Pb) did not depend on the size and age of the molluscs (Beznosov and Plekhanov 1986). Conversely, the concentration of radionuclides of other metals in mussels decreased with an increase in the mass of the individual organisms. On the 13th day of an experiment carried out on mussels having an average weight of 2.1 g, the concentration factors of tetravalent selenium (^{75}Se^{+4}) were (46 ± 10); in molluscs weighing 21.8 g, the respective figures were (13 ± 2.5) units (Fowler and Benayen 1979). Decreases in the concentration factors of larger size group molluscs have also been reported in experiments involving mercury (Breittmayer and Zsurger 1983). An inverse dependence of the concentration of ^{40}K, ^{65}Zn, ^{95}Nb, ^{95}Zr, ^{103}Ru, ^{51}Cu and ^{144}Ce by euphausiids (krill) and Calanoida plankton on the mass of individual specimens has been noted (Osterberg et al. 1964). Similar dependencies were observed in experiments with ^{65}Zn, ^{106}Ru, ^{60}Co, ^{144}Ce and representatives of metapenaeid zooplankton species (Ivanov 1974); with ^{65}Zn and euphausiids (Small and Fowler 1973); with ^{57}Co, ^{45}Ca and *Idotea metallica* isopods (Ivanov 1974); with Co and *Macoma baltica* molluscs (McLeese and Ray 1984); as well as with Zn, Cd and shrimp species (Garsia and Fowler 1972). It has been shown that the elimination of Cu by oysters is proportional to their size, while accumulation is inversely proportional (Zaroogian 1979). However, the relative excretion of ^{75}Se was not observed to depend on the mass of mussel specimens. A prior experiment showed that mussels that accumulated ^{75}Se to various levels eliminated to the same degree in terms of percentage; after 97 days of observation, 16% of the initial amount of radionuclide remained in all size groups (Fowler and Benayen 1979). The noted kinetic patterns were explained by the fact that, within a specific taxon, a change in the size of marine organisms does not affect the metabolism of their individual cells (Bergner 1985), while the uptake of elements depends on surface-mass ratios (Osterberg et al. 1964; Ivleva et al. 1984).

2.6.8 Production Processes and the Specific Biomass of Hydrobionts

The experimental study of the uptake and exchange of radioactive and chemical matter in the processes of generative and somatic cell growth of aquatic organisms (hydrobionts) has specific features as a result of changes in the biotic characteristics of the system of the interaction of these organisms with substances in the marine environment. In and of themselves, the growth of individual animals or the division of unicellular algae or bacteria lead to nonstationarity of this system in terms of the concentration of a radioactive substance or its stable analogue in the medium and hydrobionts, as well as in the ratio of the mass of the hydrobiont to the volume of the medium.

In the literature, data are presented that indicate the presence of both negative and positive correlations between the concentration of elements and the growth rate of hydrobionts. It has been noted that mussels show a slight decrease in metal concentrations (except for Cd) in growing individuals (Amiard et al. 1986). Japanese researchers (Kumagai and Saeki 1983) found that the Cu and As content in starved rapa whelks was higher than in feeding individuals. Conversely, the concentration of Cd, Mn, Ni, and Co was shown to increase in feeding and growing molluscs. Rice (1965) found that, under equal illumination conditions, more rapid cell division in a unicellular marine algae culture led to the increased accumulation of ^{90}Sr; in a dividing culture, the intracellular concentration of this radioisotope was 64 times higher than the level of a non-dividing culture. Evidence concerning a decrease the concentration of ^{65}Zn in a dividing culture of marine unicellular algae is also presented in the literature (Davies 1973). The autoradiographic method was used to show that the uptake of nutrients and cell division are separated in time. The intracellular concentration of ^{14}C and ^{32}P was observed to decrease in the mitotic division phase of populations of unicellular marine algae (Tsytsugina and Lazorenko 1983).

In the marine environment, the specific biomass of hydrobionts can undergo significant changes. In oligotrophic regions of aquatic ecosystems, biological processes proceed at very low component value levels. As a rule, the biomass of phyto- and zooplankton in such regions does not exceed 50–100 mg/m^3 in terms of the wet weight of aquatic organisms. Nevertheless, at times of upwelling and during the spring algal bloom, plankton biomass can achieve values exceeding 2×10^5 mg/m^3

(Vedernikov and Starodubtsev 1971). The maximum biomass of zooplankton reaches 5–10 kg/m^3 (Vinogradov 1968).

A team led by Burlakova et al. (1979) experimentally established that unicellular sea algae accumulate mineral phosphorus to higher levels under conditions of low population density. At an algal density of 10^3 cells/L, intracellular phosphorus comprised 1.3×10^{-3} μgP/cell; with an increase in specific abundance to 10^5 cells/L, the intracellular phosphorus content fell to 3.9×10^{-5} μgP/cell. Similar relationships were found between organotrophy and population density in unicellular algae (Burlakova and Lavrent'ev 1975) and macrophytes (Khailov and Monina 1976). It has been shown that the relationship between the population density of peridinin dinoflagellates and the factors of their concentration of ^{91}Y and ^{65}Zn can be described by a power function (Parchevskii et al. 1977). It has been ascertained (Burlakova et al. 1979) that a decrease in the level of phosphate uptake with an increase in the population density of diatoms is also accompanied by a decrease in the rate of their photosynthesis upon light saturation. According to Khailov and Monina (1976), the inhibition of organotrophy with an increase in population density is the result of the accumulation of metabolic products in the environment or a lack of a limiting chemical substrate. It has been proposed that the relationship between population density and functional parameters of ecosystems should be regarded as a general biological phenomenon (Khailov and Popov 1983). At the same time, it is noted in the literature (Makarova 1984) that, since an increase in population density only inhibits the course of physiological processes when a certain limit is reached, which rarely occurs under natural conditions, there is no need to take this factor into account when using existing methods for assessing the production characteristics of populations.

2.6.9 Role of the Biotic Factor in the Formation of the Radioisotope and Chemical Composition of Waters

In works on the geochemistry of the oceans over the past decades, at least three main directions in the study of factors affecting the formation of the chemical and radioisotope composition of sea waters have been identified—physicochemical, sedimentation (mechanical and chemical differentiation) and biogeochemical (Vinogradov and Lisitsyn 1983). On a large-scale examination, it can be observed that the physical processes of mixing of waters lead to an equalisation of the concentrations of chemical substances in terms of aquatic zones and depth (Ozmidov 1968). If the intensity of such processes were inhibitory in relation to other processes, then the same ratio of the concentrations of dissolved inorganic substances would be observed in the sea. However, such an inhibition is not observed.

The contribution of the biotic factor to the formation of fields of radioactive and chemical substances in the ocean is investigated within the conceptual framework proposed by Vernadsky (1929). This biogeochemical paradigm is based on the study of the transformation of physicochemical forms leading to the redistribution and elimination of radionuclides from the aquatic environment as a result of the processes of their uptake, exchange and transfer by aquatic organisms, referred to as hydrobionts.

According to these contemporary concepts, the main redistribution process of radioactive and chemical substances in the ocean consists in the mobilisation of elements in the composition of biogenic particles and their release to the aquatic medium during the regeneration of biogenic material (Kuenzler 1965; Fowler and Small 1972; Bogdanov et al. 1983). Goldberg (1978) noted that: "The greatest spatial and temporal differences in the chemical composition of seawater are due to the primary production of organic matter in surface waters." The greater the biological significance of a chemical element or compound—i.e., the higher the levels of its concentration by aquatic organisms—the more significant the observed deficiency of this substance in the photic layer relative to the underlying waters. Lowman and co-authors (1971) determined the concentration factors of copper by phytoplankton and zooplankton to be 3.0×10^4 and 6.0×10^3 units, respectively, while the ratio of the content of all its forms in deep waters to the content in surface waters is 1.3. For zinc, this ratio is 1.8, while, for iron, it is 2.3. The CF of zinc by phytoplankton and zooplankton is 2.6×10^4 and 8.0×10^3 units, respectively, while, for iron, the corresponding figures are 4.5×10^4 and 2.5×10^4. The natural radionuclide ^{238}U, which is concentrated by aquatic organisms at low (10^1–10^2 units) concentration factors (Risik 1970), is distributed evenly throughout the waters of seas and oceans (Horne 1972). Conversely, ^{234}Th, a daughter product of the decay of ^{238}U, which is accumulated by aquatic organisms at a high CF (10^4–10^5 units) (Zesenko and Nazarov 1973), appears in the water of the photic layer in low concentrations relative to equilibrium (Bhat et al. 1968).

A biotic mechanism explaining the significantly uneven horizontal distribution of radioactive and chemical pollutants in the ocean is referred to as plankton "patchiness" (Mackas et al. 1985), due to the spatial intermittency of hydrophysical parameters of the environment (Ozmidov 1986).

Studies have shown that radioactive and chemical contamination entering the marine environment tends to be concentrated in the surface film (Timoshchuk et al. 1970; Pellenbarg and Church 1979; Kulebakina and Kozlova 1985). An additional biotic factor in the concentration of pollutants in the surface film consists in the phenomenon of "anti-rain" (Zaitsev 1967, 1970) consisting in a stream of organic particles rising with gas bubbles.

The physical removal of inorganic substances and their radionuclides dissolved in surface waters is due to the vertical component of the current velocity—advection, diffusion and the effect of internal waves. A limiting factor of vertical mixing consists in the formation of a density transition zone located at the depth of maximum temperature gradient (Ketchum and Bowen 1958; Belyaev et al. 1966; Ozmidov 1968).

The literature indicates several biotic mechanisms for the vertical transport of radioactive and chemical substances through the density transition zone of oceanic waters. It has been established that dinoflagellate plankton can migrate vertically with an amplitude of 5–10 m, up to a maximum of 20 m, rising during the day under conditions of increased illumination (Pavlova 1971; Kiselev 1980). Many zooplankton species migrate twice a day to vertical distances of up to 500 m in antiphase with phytoplankton, i.e., rising to the surface at night and descending during the day (Vinogradov 1968). During vertical migration, plankton can accumulate radioactive and chemical substances in areas of increased concentration and deposit them in layers with a low content, thereby acting as a "biological pump" (Polikarpov 1967).

Another important factor in the vertical transport of pollution consists in the biosedimentation resulting from the primary production of organic matter in the ecosystems of the photic layer. In the most productive oceanic zone located on the Peruvian shelf, primary production comprises 2.2 kg of organic carbon per square metre of the photosynthesising layer annually (Vedernikov and Starodubtsev 1971), corresponding to the synthesis of 61 g/m^2 of living biomass per day. In the oligotrophic regions of the oceans, the production of living matter can decrease by almost two orders of magnitude to 0.7 g/m^2 per day (Romankevich 1977). The proportion of organic matter mineralised in the photic zone varies across different regions of the ocean. Calculations indicate that, in the tropical waters of the Pacific Ocean, 12% of the detritus formed in 1 m^2 per day moves through the lower boundary of the surface 200 m layer towards the bottom; in the Kuril-Kamchatka region, the corresponding figure is 22%, while in the Sea of Japan it is 5% (Sazhina 1980). According to other estimates, the production of detritus by planktonic communities in the tropical Pacific Ocean reached 20% of primary production (Lebedeva 1986).

In rich coastal shelf waters, the mineralisation of primary production within the photic layer has been estimated at 50–65%; in eutrophic and mesotrophic waters, mineralisation reaches 82–89%, while in tropical oligotrophic waters, it can be up to 95% (Eppley and Peterson 1979). It has been estimated that biosedimentation flux from the photic oceanic layer can vary from 5 to 50% of the primary production rate (Skopintsev 1950; Lisitsyn and Vinogradov 1983; Down and Lorenzen 1985).

The biosedimentation flux primarily consists of faecal pellets from zooplankton (Fowler and Small 1972; Jefferson and Ferrante 1979; Paffenhöfer and Knowles 1979; Lisitsyn and Vinogradov 1983). In addition, it includes dead organisms, their molting skins and exuvia, as well as "sea snow"—colloidal flakes and aggregates of organic matter that arose during the transition from a dissolved form to a suspension with the participation of bacteria (Lisitsyn 1982; Bopaiah 1985; Youngbluth 1985). It has been suggested that the biotic factor is also essential in the transfer of terrigenous suspensions, which are rapidly overgrown with bacteria (Gorbenko and Kryshev 1985), as well as being absorbed by filter feeders, which form large pellet lumps from fine suspensions (Fowler and Small 1972).

Studies with mysid shrimps from the Barents Sea have shown that the rate of sinking of faecal pellets depends on both their size and composition (Arashkevich et al. 1986). At $+ 5$ °C, faecal lumps may remain intact for up to 35 days (Jefferson and Ferrante 1979). The rate of gravitational sinking of suspended matter can vary between within 10^1 and 10^3 m/day (Romankevich 1977; Rudyakov and Tseitlin 1980; Staresiniĉ et al. 1983). According to estimates by a number of authors (Menzel 1974;

Bogdanov et al. 1979; Jefferson and Ferrante 1979; Suess and Müller 1980; Tseitlin 1984), between 2 and 12% of the primary production synthesised in the photic layer reaches the bottom of the abyssal ocean. The suspended matter in surface waters has a higher proportion of organic material than in the bottom layer (Bogdanov et al. 1979; Lorenzen et al. 1983). The organic carbon content in bottom sediments correlates with the production of detritus in planktonic communities (Vinogradov 1986). Radionuclides of inorganic compounds are deposited on the seabed along with other sediment components (Baranova 1967; Gromov et al. 1979; Hornung et al. 1984; Kepkay 1986) where they may be subject to processes of remobilisation (Elder et al. 1979).

The influence of environmental factors on the fate of polluting materials varies significantly across different aquatic zones. In coastal and shallow waters where photosynthesis extends to the seabed, intensive processes of interaction of pollution with living matter occur throughout the depth of water as well as on the seabed itself. Waves generated by winds stir the waters and currents act to intensively transport polluting materials through the aquatic zone. In river estuaries, pollution can be deposited in soils along with river silt and contaminated salts that precipitate in the area of the hydrofront (Popovichev and Egorov 2008). This is especially true for such heavy metals as iron, manganese, zinc, copper, as well as for radionuclides (Demina 1982). In the open waters of seas and oceans, radioactive and chemical pollutants spread into aquatic zones and underlying layers under the influence of both abiotic and biotic factors. Within the photic layer, they are absorbed by the primary producers through sorption and food. Simultaneously with uptake, the mineralisation and transformation of their physicochemical forms take place as a result of desorption and the metabolism of hydrobionts. By being repeatedly passing through food webs, pollution is removed from the photic layer.

Many authors point to a lack of knowledge of the mechanisms that control the accumulation, retention and excretion of inorganic substances by hydrobionts and the migration of radionuclides and their isotopic carriers along the food chain (Polikarpov and Zaitsev 1969; Fontane 1972; Patin 1979; Morozov 1983). It is noted (Khailov 1974) that the determination of the main trophodynamic parameters in energy units reduces the biochemical specificity of food transformation, indicating the futility of studying the patterns of migration of pollutants along trophic chains only within the framework of the energy approach. In this regard, the role of individual biotic mechanisms in the formation of fields of radionuclides and isotopic carriers has not yet been sufficiently studied. According to a number of estimates, the contribution of daily migrations of zooplankton to the redistribution of radionuclides in the marine environment does not exceed 15% of biosedimentation transfer intensity (Lowman et al. 1971). The contribution of migratory fish species is considered negligible, since the share of fish in primary production in terms of energy equivalent does not exceed 7% (Skazkina and Danilevskii 1976; Petersen 1984).

Conclusion

The chemical and radioisotopic composition of the components of marine ecosystems depends on the complex influence of both abiotic and biotic factors. In inert matter, this composition is formed mainly as a result of geochemical processes, sorption interactions and the waste products of the vital functions of marine organisms. In terms of bioplasm, chemical elements and their various compounds form the basis for the mineral nutrition of the material components of ecosystems, as well as comprising elements of toxicological and ionising effects. Various chemical compounds are included in the composition of organs and tissues of marine organisms, which determine the course of a complex of biochemical reactions responsible for energy nutrition, somatic and generative growth, as well as the mineral metabolism of marine organisms. Biogeochemical interactions comprise the primary mechanism for the extraction of radionuclides dissolved in seawater, along with their isotopic and non-isotopic carriers, involving their binding with the living and inert matter of the marine environment. As a result, radionuclides acquire positive and / or negative buoyancy and are exposed to the hydrodynamic processes of water mixing; in addition, they can be carried to the coast, gravitationally transported either to the surface or into deep oceanic layers, as well as being deposited in sedimentary strata of bottom sediments. As a result of these mechanisms of interaction, the biogeochemical cycles of the turnover of radionuclides in the marine environment are significantly accelerated.

References

Aizatullin TA, Leonov AV (1977) Kinetika i mekhanizm transformatsii soedinenii fosfora i potrebleniya kisloroda v vodnoi ekologicheskoi sisteme. Vodnye Resursy 2:41–55 (in Russian)

Aizatullin TA, Lebedev VL, Khailov KM (1984) Okean. Fronty, dispersii, zhizn'. Gidrometeoizdat, Leningrad, 192 p. (in Russian)

Aleksakhin RM (1963) Radioaktivnye zagryazneniya pochvy i rastenii. Izdatel'stvo AN SSSR, Moscow, 167 p. (in Russian)

Amiard JC, Amiard-Triquet C, Berthet B, Métayer C (1986) Contribution to the ecotoxicological study of cadmium, lead, copper and zinc in the mussel *Mytilus edulis*. I.Field study. Mar Biol 90(3):425–431. https://doi.org/10.1007/BF00428566

Ancellin J, Guegueniat P, Germain P (1979) Radioécologie marine. Étude du devenir des radionucléides rejetés en milieu marin et applications à la radioprotection. Eyrolles, Paris, 256 p

Arashkevich EG, Vinogradov GM, Semenova TN (1986) Osedanie fekal'nykh pellet planktonnykh rakoobraznykh Barentseva morya. In: Ekologicheskaya i biologicheskaya produktivnost' Barentseva morya: abstracts of reports of the All-Union Conference, Murmansk. (in Russian)

Bachurin AA (1968) Matematicheskoe opisanie dinamiki protsessov radioaktivnogo zagryazneniya morskikh organizmov iz vodnoi sredy. Atomizdat, Moscow, 28 p. (in Russian)

Baranova DD (1967) Sravnitel'noe izuchenie sorbtsii i desorbtsii razlichnykh radionuklidov morskimi melkovodnymi gruntami. In: Voprosy biookeanografii. Naukova dumka, Kiev, pp 219–226. (in Russian)

Barinov GV (1965) Obmen ^{45}Ca, ^{137}Cs, ^{144}Ce mezhdu vodoroslyami i morskoi vodoi. Okeanologiya 5(1):111–116 (in Russian)

Baxter MS (1983) The disposal of high-activity nuclear wastes in the oceans. Mar Pollut Bull 14(4):126–132

Belyaev VI, Kolesnikov AG, Nelepo BA (1966) Opredelenie intensivnosti radioaktivnogo zarazheniya v okeane na osnove novykh dannykh o protsesse obmena. In: Proceedings of the III United Nations international conference on the peaceful uses of atomic energy, Geneva

Belyayev ST, Borovoy AA, Demin VF, Rimsky-Korsakov AA, Kheruvimov AN (1991) The Chernobyl source term. In: Proceedings of the seminar on comparative assessment of the environmental impact of radionuclides released during three major nuclear accidents: Kyshtym, Windscale, Chernobyl, vol 1. Luxembourg, 1–5 Oct 1990

Bergner PE (1985) On relation between bioaccumulation and weight of organisms. Ecol Model 27(3–4):207–220. https://doi.org/10.1016/0304-3800(85)90003-1

Bernhard M (1971) The utilization of simple models in radioecology. In: Marine radioecology symposium, Hamburg, F. R. Germany, 20–24 Sept 1971

Beznosov VN, Plekhanov SE (1986) Soderzhanie nekotorykh metallov v chernomorskikh midiyakh. Ekologiya 5:80–81 (in Russian)

Bhat SJ, Krishnaswamy S, Lal D, Rama MWS (1968) ^{234}Th/^{238}U ratios in the ocean. Earth Planet Sci Lett 5:483–491. https://doi.org/10.1016/S0012-821X(68)80083-4

Blatov AS, Ivanov VA (1992) Gidrologiya i gidrodinamika shel'fovoi zony Chernogo morya. Naukova dumka, Kiev, 244 p. (in Russian)

Bogdanov YuA, Gurvich EG, Lisitsyn AP (1979) Model' nakopleniya organicheskogo ugleroda v donnykh osadkakh Tikhogo okeana. Geokhimiya 6:918–927 (in Russian)

Bogdanov YuA, Gurvich EG, Lisitsyn AP (1983) Mekhanizm okeanskoi sedimentatsii i differentsiatsii khimicheskikh elementov v okeane. In: Biokhimiya okeana. Nauka, Moscow, pp. 165–200. (in Russian)

Bopaiah AB (1985) Microbial synthesis of macroparticulate matter. Mar Ecol Prog Ser 20:241–251. https://doi.org/10.3354/meps020241

Bowen HJM (1979) Environmental chemistry of the elements. Academic Press, London, New York, p 333

Breittmayer JP, Zsurger NV (1983) Accumulation du mercure dans les organes de la moule: Effets de la dose contaminante et de la taille des organismes. Revue Internationale D'océanographie Médicale 70–71:87–97

Brown MP, McLaughlin JA, O'Connor JM, Wyman K (1982) A mathematical model of PCB bioaccumulation in plankton. Ecol Model 15(1):29–47. https://doi.org/10.1016/0304-3800(82)900 66-7

Bryan GW (1976) Heavy metal contamination in the sea. In: Johnston R (ed) Marine pollution. Academic Press, London, New York, San Francisco, pp 185–302

Bryan GW, Ward E (1962) Potassium metabolism and the accumulation of ^{137}Caesium by decapod Crustacea. J Mar Biol Assoc UK 42(2):199–241. https://doi.org/10.1017/S0025315400001314

Burlakova ZP, Lavrent'ev NA (1975) K voprosu o skorosti oborota rastvorennykh organicheskikh veshchestv v morskoi vode kak funktsii plotnosti kletok v populyatsiyakh. In: VII All-Union conference on marine chemistry: abstracts. (in Russian)

Burlakova ZP, Krupatkina DK, Lanskaya LA, Yafarova DL (1979) Vliyanie plotnosti populyatsii morskikh odnokletochnykh vodoroslei na potreblenie fosfora i osnovnye fiziologicheskie pokazateli kletok. In: Vzaimodeistvie mezhdu vodoi i zhivym veshchestvom: proceedings of the international symposium, Odessa, vol 1, 6–10 Oct 1975. (in Russian)

Chabert D (1984) Bioaccumulation du cadmium chez un mollusque bivalve: *Cerastoderma glaucum*, Poiret 1789, après contamination. Vie Marine 6:57–61

Champ MA, Park PK (1982) Global marine pollution bibliography: ocean dumping of municipal and industrial wastes. IFI/Plenum Data Company, New York, p 399

Corcoran EF, Kimball JF (1968) Pogloshchenie, nakoplenie i obmen 90Sr u fitoplanktona otkrytogo morya. Voprosy radioekologii. Atomizdat, Moscow, pp 231–240

Database of IBSS radiation and chemical biology department (for the period 1964–2006): Water. Hydrobionts. Bottom sediments (2006) RCBD computer program, created in 1992 in Paradox, Sevastopol, 425 p. (in Russian)

Davies AG (1973) The kinetics of and a preliminary model for the uptake of radio-zinc by *Phaeo-dactylum tricornutum* in culture. In: Radioactive contamination of the marine environment: proceeding of a symposium, Seattle, 10–14 July 1972

Demina LL (1982) Formy migratsii tyazhelykh metallov v okeane (na rannikh stadiyakh okeanskogo osadkoobrazovaniya). Nauka, Moscow, 119 p. (in Russian)

Downs JN, Lorenzen CJ (1985) Carbon: pheopigment ratios of zooplankton fecal pellets as an index of herbivorous feeding. Limnol Oceanogr 30(5):1024–1036. https://doi.org/10.4319/lo.1985.30.5.1024

Droop MR (1974) The nutrient status of algal cells in continuous culture. J Mar Biol Assoc UK 54(4):825–855. https://doi.org/10.1017/S002531540005760X

Dugdale RC (1967) Nutrient limitation in the sea: dynamics, identification and significance. Limnol Oceanogr 12(4):685–695. https://doi.org/10.4319/lo.1967.12.4.0685

Duke T, Willis J, Price T, Fischler K (1969) Influence of environmental factors on the concentrations of ^{65}Zn by an experimental community. In: Nelson DJ, Evans FC (eds) Symposium on radioecology: proceedings of the 2nd national symposium, Ann Arbor, Michigan, 15–17 May 1967

Egorov VN, Ivanov VN (1981) Matematicheskoe opisanie kinetiki obmena tsinka-65 i margantsa-54 u morskikh rakoobraznykh pri nepishchevom puti postupleniya radionuklidov. Ekologiya Morya 6:37–43 (in Russian)

Egorov VN, Polikarpov GG, Stokozov NA, Kulebakina LG, Lazorenko GE, Mirzoeva NYu (1994) Some data on the fate of 90Sr in the aquatic system, including the region Chernobyl NPP accident, the Black Sea and the Aegean Sea. In: Proceedings of the seminar "European commission radiation protection 70": the radiological exposure of the population of the European community to radioactivity in the Mediterranean Sea. MARINA-MED project, Rome, 17–19 May 1994

Egorov VN, Polikarpov GG, Stokozov NA, Mirzoyeva NYu (2005) Estimation and prediction of ^{90}Sr and ^{137}Cs outflow from the Black Sea via the bosporus strait after The NPP Chernobyl accident. Morskoj Ekologicheskij Zhurnal 4(4):33–41 (in Russian)

Egorov VN, Polikarpov GG, Stokozov NA, Mirzoeva NYu (2008) Balans i dinamika polei kontsen-tratsii ^{137}Cs i ^{90}Sr v vodakh Chernogo morya. In: Polikarpov GG, Egorov VN (eds) Radioeko-logicheskii otklik Chernogo morya na chernobyl'skuyu avariyu. EKOSI-Gidrofizika, Sevastopol, pp 217–250. (in Russian)

Elder DL, Fowler SW, Polikarpov GG (1979) Remobilization of sediment-associated PCBs by the worm *Nereis diversicolor*. Bull Environ Contam Toxicol 21(1):448–452. https://doi.org/10.1007/bf01685451

Eppley RW, Peterson BJ (1979) Particulate organic matter flux and planktonic new production in the deep ocean. Nature 282(5740):677–680. https://doi.org/10.1038/282677a0

Florence TM (1983) Trace element speciation and aquatic toxicology. TrAC, Trends Anal Chem 2(7):162–166. https://doi.org/10.1016/0165-9936(83)87023-X

Fontane MA (1972) A new science-marine molysmology. In: Proceedings of the 2nd international ocean development conference, Tokyo, 05–07 Oct 1972

Formy elementov i radionuklidov v morskoi vode (1974) Nauka, Moscow, 173 p. (in Russian)

Fowler SW, Benayen J (1979) The influence of factors of surroundings on the flow of selenium through the marine organisms. In: Interaction between water and living matter: proceedings of the international symposium, Odessa, vol 1, 6–10 Oct 1975

Fowler SW, Guary JC (1977) High uptake efficiency for ingested plutonium in crabs. Nature 266(5605):827–828. https://doi.org/10.1038/266827a0

Fowler SW, Small LF (1972) Sinking rates of euphausiid fecal pellets. Limnol Oceanogr 17(2):293–296. https://doi.org/10.4319/lo.1972.17.2.0293

Galli C, Zaretta A (1979) Accumulation of cesium by some marine phytoplankters. Rapports Et Procès-Verbaux Des Réunions Commission Internationale Pour L'exploration Scientifique De La Mer Méditerranée 25–26(5):57–61

Garsia AP, Fowler SW (1972) Analysis de microelementos en invertebrados marinos del Golfo de California. In: Mexico: national oceanographic congress, Mexico, pp 140–169

Glasser R (1962) Beitrag zur Frage der Tragerabhangigkeit bei der Anreicherung Radioaktiver Isotope durch Wasserorganismen aus Wasseriger Losung. Biologisches Zentralblatt 81(5):17–27

Goldberg ED (1978) Modelirovanie khimicheskikh protsessov. In: Modelirovanie morskikh sistem. Gidrometeoizdat, Leningrad, pp 163–182. (in Russian)

Gorbenko YuA, Kryshev II (1985) Statisticheskii analiz dinamiki morskoi ekosistemy mikroorganizmov. Naukova dumka, Kiev, 143 p. (in Russian)

Gromov VV, Spitsyn VI (1975) Iskusstvennye radionuklidy v morskoi srede. Atomizdat, Moscow, 223 p. (in Russian)

Gromov VV, Shakhova NF, Emel'yanov EM (1979) Adsorbtsionnyi zakhvat tsinka donnymi otlozheniyami Atlanticheskogo okeana. Okeanologiya 19(5):835–839. (in Russian)

Gudiksen PH, Harvey TF, Lange R (1989) Chernobyl source term, atmospheric dispersion and dose estimation. Health Phys 57(5):697–706. https://doi.org/10.1097/00004032-198911000-00001

Gudiksen PH, Harvey TF, Lange R (1991) Chernobyl source term estimation. In: Proceedings of the seminar on comparative assessment of the environmental impact of radionuclides released during three major nuclear accidents: Kyshtym, Windscale, Chernobyl, vol 1. Luxembourg, 1–5 Oct 1990

Harvey RS (1971) Temperature effects on the maturation of midges (Tendipedidae) and their sorption of radionuclides. Health Phys 20(6):613–616. https://doi.org/10.1097/00004032-197106000-00008

Hattori A (1982) The nitrogen cycle in the sea with special reference to biogeochemical processes. J Oceanogr Soc Jpn 38(4):245–265. https://doi.org/10.1007/BF02111107

Hiyama Y, Shimizu M (1964) On the concentration factors of radioactive J Co, Fe and Ru in marine organisms. Rec Oceanogr Works Japan 7(2):43–77

Horne R (1972) Morskaya khimiya. Mir, Moscow, 400 p. (in Russian)

Hornung H, Krumgalz BS, Cohen Y (1984) Mercury pollution in sediments, benthic organisms and inshore fishes of Haifa Bay, Israel. Mar Environ Res 12(3):191–208. https://doi.org/10.1016/0141-1136(84)90003-5

Il'in LA, Pavlovskii OA (1988) Radioekologicheskie posledstviya avarii na Chernobyl'skoi AES i mery, predprinyatye s tsel'yu ikh smyagcheniya. Atomnaya energiya 65(2):119–129. (in Russian)

Ivanov VN (1974) Nakoplenie ^{54}Mn, ^{59}Fe, ^{60}Co, ^{65}Zn, ^{106}Ru, ^{144}Ce okeanicheskim zooplanktonom. In: Polikarpov GG (ed) Khemoradioekologiya pelagiali i bentali (metally i ikh radionuklidy v gidrobiontakh i srede). Naukova dumka, Kiev, pp 211–246. (in Russian)

IMCO/FAO/UNESCO/WMO/WHO/IAEA/UN joint group of experts on the scientific aspects of marine pollution (GESAMP) (1977) United Nations, New York, 35 p

Ivanov VN, Rozhanskaya LI (1972) Povedenie Zn-65 v morskoi vode i nakoplenie ego gidrobiontami. In: Polikarpov GG (ed) Radiatsionnaya i khimicheskaya ekologiya gidrobiontov. Naukova dumka, Kiev, pp 42–51. (in Russian)

Ivleva EV, Parchevskii VP, Lanskaya LA (1984) Vliyanie temperatury, solenosti i pH na nakoplenie ^{65}Zn morskimi odnokletochnymi vodoroslyami. In: Polikarpov GG (ed) Morskaya radiokhemoekologiya i problema zagryaznenii. Naukova dumka, Kiev, pp 78–82. (in Russian)

Izrael YuA (ed) (1968) Radioaktivnye vypadeniya ot yadernykh vzryvov (translation from English) Mir, Moscow, 344 p. (in Russian)

Izrael YuA (ed) (1990) Chernobyl': radioaktivnoe zagryaznenie prirodnykh sred. Gidrometeoizdat, Leningrad, 296 p. (in Russian)

Jefferson T, Ferrante JG (1979) Zooplankton fecal pellets in aquatic ecosystems. Bioscience 29(11):670–677

Kepkay PE (1986) Microbial binding of trace metals and radionuclides in sediments: Results from an in situ dialysis technique. J Environ Radioact 3(2):85–102. https://doi.org/10.1016/0265-931 X(86)90030-5

Ketchum BU, Bowen VT (1958) Biological factors determining the distribution of radioisotopes in the sea. In: Proceedings of the 2nd United Nations international conference of the peaceful uses of atomic energy, Geneva, 1–3 Sept 1958, vol 33. United Nations Publ., pp 429–433

Khailov KM (1974) Trofodinamika i biokhimiya v evolyutsii morskoi ekologii. In: Biokhimicheskaya trofodinamika v morskikh pribrezhnykh ekosistemakh. Naukova dumka, Kiev, pp 3–13. (in Russian)

Khailov KM, Monina TL (1976) Organotrofiya u morskikh makrofitov kak funktsiya plotnosti ikh populyatsii v usloviyakh eksperimenta. Biologiya Morya 6:29–34 (in Russian)

Khailov KM, Popov AE (1983) Kontsentratsiya zhivoi massy kak regulyator funktsionirovaniya vodnykh organizmov. Ekologiya Morya 15:3–16 (in Russian)

Kiselev IA (1980) Plankton morei i kontinental'nykh vodoemov, vol 2. Raspredelenie, sezonnaya dinamika, pitanie i znachenie. Nauka, Leningrad, 440 p. (in Russian)

Kryshev I, Sazykina T (1986) Matematicheskoe modelirovanie migratsii radionuklidov v vodnykh ekosistemakh. Energoatomizdat, Moscow, 255 p. (in Russian)

Kuenzler EJ (1965) Zooplankton distribution and isotope turnover during operation swordfish. US AEC Document NYO-3145–1 (New York Operations Office). New York, 12 p

Kulebakina LG (1984) Vliyanie solenosti na nakoplenie radionuklidov morskimi makrofitami. In: Polikarpov GG (ed) Morskaya radiokhmoekologiya i problema zagryaznenii. Naukova dumka, Kiev, pp 40–65. (in Russian)

Kulebakina LG (1996) Izuchenie migratsii 90Sr i 137Cs v ekosistemakh shel'fa Chernogo morya i nizhnego Dnepra posle Chernobyl'skoi avarii. In: Radioekologiya: uspekhi i perspektivy: materials of the international scientific seminar, Sevastopol, 3–7 Oct 1994. (in Russian)

Kulebakina LG, Kozlova SI (1985) Raspredelenie rastvorennoi i vzveshennoi form rtuti v Atlanticheskom okeane i v Sredizemnom more v sloe 0–100 m. Okeanologiya 25(2):248–253 (in Russian)

Kulebakina LG, Parchevskaya DS (1973) Vliyanie solenosti i kontsentratsii Sr i Ca v vode na nakoplenie ^{90}Sr tsistoziroi. In: Radioekologiya vodnykh organizmov, vol 2. Zinatne, Riga, pp 305–307. (in Russian)

Kumagai H, Saeki K (1983) The variations with growth in heavy metal contents of rock shell. Nippon Suisan Gakkaishi 49(12):1917–1920. https://doi.org/10.2331/suisan.49.1917

La Molta EJ, Shich WK (1979) Diffusion and reaction in biological nitrification. J Environ Eng Div 105(4):655–673

Lakshmanan PT, Nambisan PNK (1979) Accumulation of mercury by the mussel *Perna viridis* Linnaeus. Current Sci (india) 48(5):672–674

Lazorenko GE (2000) Molismologicheskoe issledovanie vodnoi ekosistemy Severo-Krymskogo kanala. In: Chteniya pamyati N. V. Timofeeva–Resovskogo: 100-letiyu so dnya rozhdeniya N. V. Timofeeva-Resovskogo posvyashchaetsya. EKOSI-Gidrofizika, Sevastopol, pp 100–107

Lazorenko GE (2002) Accumulation of ^{210}Po by the Black Sea fishes. In: High levels of natural radiation and radon areas: radiation dose and health effects: proceedings of the 5th international conference, Munich (Germany), 4–7 Sept 2000

Lebedev VL, Aizatullin TA, Khailov KM (1974) Okean kak dinamicheskaya sistema. Gidrometeoizdat, Leningrad, 203 p. (in Russian)

Lebedeva LP (1986) Produktsiya detrita v planktonnykh soobshchestvakh tropicheskikh raionov Tikhogo okeana. In: Biodifferentsiatsiya osadochnogo veshchestva v moryakh i okeanakh. Rostov-on-Don, pp 117–122. (in Russian)

Levis DI, Harvey W, Hodson RE (1984) Application of single- and multiphasic Michaelis–Menten kinetics to predictive modeling for aquatic ecosystems. Environ Toxicol Chem 3(4):563–574. https://doi.org/10.1002/etc.5620030406

Linnik PN (1986) Formy migratsii tyazhelykh metallov i ikh deistvie na gidrobiontov. Eksperimental'naya Vodnaya Toksikologiya 11:114–154 (in Russian)

Lisitsyn AP (1982) Lavinnaya sedimentatsiya. In: Lavinnaya sedimentatsiya v okeane. Izdatel'stvo Rostovskogo universiteta, Rostov-on-Don, pp 3–58. (in Russian)

Lisitsyn AP, Vinogradov ME (1983) Global'nye zakonomernosti raspredeleniya zhizni v okeane i biogeokhimiya vzvesi i donnykh osadkov. In: Biogeokhimiya okeana. Nauka, Moscow, pp 112–127. (in Russian)

Lorenzen CJ, Welscheyer NA, Copping AE, Vernet M (1983) Sinking rates of organic particles. Limnol Oceanogr 28(4):766–769. https://doi.org/10.4319/lo.1983.28.4.0766

Lowman FG, Rice TR, Richards FA (1971) Accumulation and redistribution of radionuclides by marine organisms. In: Radioactivity in the marine environment. The National Academies Press, Washington, DC, pp 162–199. https://doi.org/10.17226/18745

Luci C, Jelisavciê O (1970) Uptake of ^{137}Cs in some marine animals in relation to the temperature, salinity, weight and moulting. Internationale Revue Der Gesamten Hydrobiologie Und Hydrographie 50(5):783–796. https://doi.org/10.1002/iroh.19700550506

Mackas DL, Denman KL, Abbott MA (1985) Plankton patchiness: biology in the physical vernacular. Bull Mar Sci 37(2):652–674

Maclsaac JJ, Dugdale RC (1969) The kinetics of nitrate and ammonium uptake by natural populations of marine phytoplankton. Deep-Sea Res Oceanogr Abstr 16(1):45–57. https://doi.org/10.1016/0011-7471(69)90049-7

Makarova NP (1984) O metodakh izucheniya plotnostnoi regulyatsii funktsii u gidrobiontov i interpretatsii dannykh. Ekologiya Morya 18:88–93 (in Russian)

Malakhova LV (2006) Soderzhanie i raspredelenie khlororganicheskikh ksenobiotikov v komponentakh ekosistem Chernogo morya. Dissertation abstract, Sevastopol, 24 p. (in Russian)

Mauchline J (1963) The biological and geographical distribution in the Irish Sea of radioactive effluent from Windscale Works 1959 to 1960: Technical report, Rep. no. AHSB (RP) R 27, 37 p

McLeese DW, Ray S (1984) Uptake and excretion of cadmium, CdEDTA, and zinc by *Macoma baltica*. Bull Environ Contam Toxicol 32(1):85–92. https://doi.org/10.1007/bf01607469

Menzel DW (1974) Primary productivity, dissolved and particulate organic matter, and the sites of oxidation of organic matter. In: Goldberg E (ed) The sea, vol 5. Wiley Interscience Publishers. New York, London, pp 659–678

Miller WL, Blake NJ, Byrne RH (1985) Uptake of Zn^{65} and Mn^{54} into body tissues and renal concretions by the Southern Quahog, *Mercenaria campechiensis* (Gmelin): effects of elevated phosphate and metal concentrations. Mar Environ Res 17(2–4):167–171. https://doi.org/10.1016/0141-1136(85)90072-8

Mirzoeva NYu, Polikarpov GG, Egorov VN, Arkhipova SI, Korkishko NF (2000) Radioekologicheskii monitoring vodnykh ekosistem sevastopol'skikh bukht posle avarii na ChAES. In: Chteniya pamyati N. V. Timofeeva–Resovskogo: 100-letiyu so dnya rozhdeniya N. V. Timofeeva-Resovskogo posvyashchaetsya. EKOSI-Gidrofizika, Sevastopol, pp 131–138. (in Russian)

Mirzoeva NYu, Egorov VN, Polikarpov GG (2005) Soderzhanie ^{90}Sr v donnykh otlozheniyakh Chernogo morya posle avarii na Chernobyl'skoi AES i ego ispol'zovanie v kachestve radiotrassera dlya otsenki skorosti osadkonakopleniya. In: Sistemy kontrolya okruzhayushchei sredy: sredstva i monitoring. EKOSI-Gidrofizika, Sevastopol, pp 276–282. (in Russian)

Mirzoeva NYu, Arkhipova SI, Korkishko NF, Migal' LV (2006) Sravnitel'naya otsenka raspredeleniya 90Sr v gidrobiontakh vodoemov blizhnikh i znachitel'no udalennykh ot avariinoi ChAES. In: Problemy biologicheskoi okeanografii XXI veka: abstracts of international conference dedicated to the 135th anniversary of IBSS, Sevastopol, 19–21 Sept 2006. (in Russian)

Morozov NP (1983) Khimicheskie elementy v gidrobiontakh i pishchevykh tsepyakh. In: Biogeokhimiya okeana. Nauka, Moscow, pp 128–164. (in Russian)

Nechaev LN, Lyapin EN, Gusev DI et al. (1972) Kinetika obmena nekotorykh radioaktivnykh izotopov v tkanyakh ryb pri razlichnom temperaturno-solevom rezhime. In: Biologicheskoe deistvie vneshnikh i vnutrennikh istochnikov radiatsii. Meditsina, Moscow, pp 348–352. (in Russian)

Nelepo BA (1970) Yadernaya gidrofizika. Atomizdat, Moscow, 224 p. (in Russian)

Nesmeyanov AN (1978) Radiokhimiya. Khimiya, Moscow, 560 p. (in Russian)

Osterberg CL, Patullo J, Pearcy W (1964) Zinc-65 in euphausiids as related to Columbia river water off the oregon coast. Limnol Oceanogr 9(2):249–257. https://doi.org/10.4319/lo.1964.9.2.0249

Ozmidov RV (1968) Gorizontal'naya turbulentnost' i turbulentnyi obmen v okeane. Nauka, Moscow, 199 p. (in Russian)

Ozmidov RV (1986) Peremezhaemost' gidrofizicheskikh parametrov i ee rol' v formirovanii osobennostei gidrokhimicheskikh i gidrobiologicheskikh polei okeana. In: Antropogennaya evtrofikatsiya i izmenchivost' ekosistem Chernogo morya: materials of the international symposium, Moscow, 16–19 Oct 1984. (in Russian)

Paffenhöfer GA, Knowles SC (1979) Ecological implications of fecal pellet size, production and consumption by copepods. J Mar Res 37(1):35–49

Parchevskii VP, Burlakova ZP, Khailov KM, Lanskaya LA (1977) Vliyanie plotnosti populyatsii morskikh odnokletochnykh vodoroslei na nakoplenie radionuklidov. Ekologiya 3:96–98 (in Russian)

Patin SA (1979) Vliyanie zagryazneniya na biologicheskie resursy i produktivnost' Mirovogo okeana. Pishchevaya promyshlennost', Moscow, 304 p. (in Russian)

Patton A (1968) Energetika i kinetika biokhimicheskikh protsessov. Mir, Moscow, 159 p. (in Russian)

Pavlova EV (1971) Dvizhenie peridinievykh vodoroslei. In: Ekologicheskaya fiziologiya morskikh planktonnykh vodoroslei. Naukova dumka, Kiev, pp 143–167. (in Russian)

Pellenbarg RE, Church TM (1979) The estuarine surface microlayer and trace metal cycling in a salt marsh. Science 203(4384):1010–1012. https://doi.org/10.1126/science.203.4384.1010

Petersen G (1984) Energy flow in comparable aquatic ecosystems from different climatic zones. Rapports Et Procès-Verbaux Des Réunions Commission Internationale Pour L'exploration Scientifique De La Mer Méditerranée 183:119–125

Phillips DJH (1977) Effects of salinity on the net uptake of zinc by the common mussel *Mytilus edulis*. Mar Biol 41(1):79–88. https://doi.org/10.1007/BF00390584

Pickering DC, Lucas JW (1962) Uptake of radiostrontium by an alga and the influence of calcium ion in the water. Nature 193(4820):1046–1047

Polikarpov GG (1964) Radioekologiya morskikh organizmov. Atomizdat, Moscow, 295 p. (in Russian)

Polikarpov GG (1966) Radioecology of aquatic organisms. Reinhold Publ. Co., New York, p 314

Polikarpov GG (1967) Problemy radiatsionnoi i khimicheskoi ekologii morskikh organizmov. Okeanologiya 7(4):561–570 (in Russian)

Polikarpov GG, Egorov VN (1986) Morskaya dinamicheskaya radiokhemoekologiya. Energoatomizdat, Moscow, 176 p. (in Russian)

Polikarpov GG, Egorov VN (eds) (2008) Radioekologicheskii otklik Chernogo morya na chernobyl'skuyu avariyu. EKOSI-Gidrofizika, Sevastopol, 667 p. (in Russian)

Polikarpov GG, Zaitsev YuP (1969) Gorizonty i strategiya poiska v morskoi biologii: report at the Presidium of the Academy of Sciences of the Ukrainian SSR, 16 May 1968. Naukova dumka, Kiev, 31 p. (in Russian)

Polikarpov GG, Zesenko AYa, Lyubimov AA (1972) Dinamika fiziko-khimicheskogo prevrashcheniya radionuklidov mnogovalentnykh elementov v srede i nakoplenie ikh gidrobiontami. In: Polikarpov GG (ed) Radiatsionnaya i khimicheskaya ekologiya gidrobiontov. Naukova dumka, Kiev, pp 5–42. (in Russian)

Popov NI (1971) Khimicheskie aspekty morskoi radioekologii. In: Radioekologiya. Atomizdat, Moscow, pp 385–395 (in Russian)

Popov NI, Fedorov KN, Orlov VM (1979) Morskaya voda: spravochnoe rukovodstvo. Nauka, Moscow, 327 p (in Russian)

Popovichev VN, Egorov VN (2008) Obmen mineral'nogo fosfora vzveshennym veshchestvom v foticheskoi zone Chernogo morya. In: Polikarpov GG, Egorov VN (eds) Radioekologicheskii otklik Chernogo morya na Chernobyl'skuyu avariyu. EKOSI-Gidrofizika, Sevastopol, pp 548–574. (in Russian)

Pospelova NV, Egorov VN, Chelyadina NS, Nekhoroshev MV (2018) Soderzhanie medi v organakh i tkanyakh *Mytilus galloprovincialis* Lamarck, 1819 i potok ee sedimentatsionnogo deponirovaniya v donnye osadki v khozyaistvakh chernomorskoi akvakul'tury. Morskoi Biologicheskii Zhurnal 3(4):64–75. https://doi.org/10.21072/mbj.2018.03.4.07

Ramade P (1980) Radioecologie des milieux aquatiques. Masson, Paris, New York, Barcelone, Milan, p 191

Rice TR (1965) The role of plants and animals in the cycling of radionuclides in the marine environment. Health Phys 11(9):953–964

Ring M, Heerkloss R, Schnese W (1985) Einfluss von temperatur, pH-wert und nahrungsqualität unter laborbedingungen auf *Eurytemora affinis*. Wissenschaftliche Zeitschrift der Wilhelm-Pieck-Universität Rostock, Mathematisch-naturwissenschaftliche Reihe 34(6):22–25

Risik NS (1970) Mikroraspredelenie i nakoplenie urana v morskikh organizmakh. Dissertation abstract, Sevastopol, 32 p. (in Russian)

Riziê I (1972) Two-compartment model of radionuclides accumulation into marine organisms. I. Accumulation from a medium of constant activity. Mar Biol 15(2):105–113. Doi https://doi.org/10.1007/BF00353638

Rodach G (1983) Simulations of phytoplankton dynamics and their interactions with other system components during FLEX'76. In: Sündermann J, Lenz W (eds) North sea dynamics. Springer, Berlin, Heidelberg, pp 584–610. https://doi.org/10.1007/978-3-642-68838-6_40

Romankevich EA (1977) Geokhimiya organicheskogo veshchestva v okeane. Nauka, Moscow, 256 p. (in Russian)

Rudyakov YuA, Tseitlin VB (1980) Skorost' passivnogo pogruzheniya planktonnykh organizmov. Okeanologiya 20(5):732–738 (in Russian)

Sayhan T, Erdener B, Ünlü Y (1985) Biokinetics of silver (^{110m}Ag) in marine isopod. Rapports Et Procès-Verbaux Des Réunions Commission Internationale Pour L'exploration Scientifique De La Mer Méditerranée 29(7):235–337

Sazhina LI (1980) Plodovitost', skorost' rosta nekotorykh kopepod Atlanticheskogo okeana. Biologiya Morya 3:56–61 (in Russian)

Scott R (1954) A study of cesium accumulation by marine algae. In: Proceedings of the 2nd radioisotope conference, vol 1. Medical and Physiological Applications, Oxford, 19–23 July 1954

Skazkina EP, Danilevskii NN (1976) Ob ispol'zovanii khamsoi kormovoi bazy Chernogo morya. Trudy VNIRO 116(2):36–42 (in Russian)

Skopintsev BA (1950) Organicheskoe veshchestvo v prirodnykh vodakh (vodnyi gumus). Gidrometeoizdat, Leningrad, 290 p (Trudy Gosudarstvennogo okeanograficheskogo instituta 17(29)). (in Russian)

Skreblin M, Stegnar P, Prosenc A (1985) Effect of selenium on the uptake of mercury from seawater by the shrimp *Palaemon elegans*. In: 7 Journées d'études sur les pollutions marines en Mediterranée, Lucerne, 11–13 Oct 1984, pp 827–830

Skulskii IA, Lyubimov AA, Glazunov VV (1974) Protsessy gidroliza i adsorbtsii. In: Khemoradioekologiya pelagiali i bentali. Naukova dumka, Kiev, pp 192–201. (in Russian)

Small LF, Fowler SW (1973) Turnover and vertical transport of zinc by the euphausiid *Meganyctiphanes norvegica* in the Ligurian Sea. Mar Biol 18(4):284–290

Staresiniê N, Farrington J, Gagosian BR, Clifford CN, Hulburt EM (1983) Downward transport of particulate matter in the Peru coastal upwelling: role of theanchoveta, *Engraulis ringens*. In: Suess E, Theide J (eds) Coastal upwelling and its sediment record. Plenum, New York, London, pt A, pp 225–240

Starik IE (1960) Osnovy radiokhimii. Izdatel'stvo AN SSSR, Moscow, Leningrad, 459 p. (in Russian)

Stary J, Zeman A (1983) Radionuclides in the investigation of the cumulation of toxic elements on algae and fish. Isotopenpraxis Isotopes Environ Health Stud 19(7):243–244. https://doi.org/10.1080/10256018308544900

Stokozov NA (2003) Dolgozhivushchie radionuklidy ^{137}Cs i ^{90}Sr v Chernom more posle avarii na Chernobyl'skoi AES i ikh ispol'zovanie v kachestve trasserov protsessov vodoobmena. Dissertation abstract, Sevastopol, 21 p. (in Russian)

Stokozov NA, Egorov VN, Mirzoeva NYu (2008) Otsenki krupnomasshtabnogo vertikal'nogo vodoobmena v Chernom more s ispol'zovaniem 134Cs, 137Cs i 90Sr v kachestve trasserov. In: Polikarpov GG, Egorov VN (eds) Radioekologicheskii otklik Chernogo morya na chernobyl'skuyu avariyu. EKOSI-Gidrofizika, Sevastopol, pp 448–464. (in Russian)

Styro DB, Bumyalene ZhV, Kadzhene GI, Kleiza IV, Lukinskine MV, Pogrebnyak EV (1991) Ob"emnaya aktivnost' radionuklidov iskusstvennogo proiskhozhdeniya v poverkhnostnykh vodakh Baltiiskogo morya v avguste – dekabre 1988 g. Atomnaya Energiya 70(6):405–408 (in Russian)

Suess E, Müller PJ (1980) Productivity sedimentation rate and sedimentary organic matter in the oceans. II. Elemental fractionation. In: Colloques Internationaux du CNRS, Paris, France, pp 17–26

Taguchi S (1976) Relationship between photosynthesis and cell size of marine diatoms. J Phycol 12(2):185–189. https://doi.org/10.1111/j.1529-8817.1976.tb00499.x

Tereshchenko NN (2003) Izuchenie soderzhaniya radionuklidov plutoniya v donnykh otlozheniyakh Streletskoi bukhty. In: Radiatsionnaya bezopasnost' territorii. Radioekologiya goroda: abstracts of international conference, Moscow, 24–26 Nov 2003

Tereshchenko NN (2005) Radionuklidy plutoniya v komponentakh pribrezhnykh chernomorskikh ekosistem v akvatorii Sevastopolya. Naukovi zapiski Ternopil's'koho natsional'noho pedahohichnoho universitetu. Seriya: Biologiya. Spetsialnyi vypusk: Hidroekolohiya 4(27):243–247

Tereshchenko NN (2006) Akkumulirovanie izotopov plutoniya gidrobiontami Chernogo morya. In: Abstracts of the V congress on radiation research (radiobiology, radioecology, radiation safety), Moscow, 10–14 Apr 2006

Thébault H, Rodriguez y Baena AM, Andral B, Barisic D, Benedicto Albaladejo J, Bologa A, Boudjenoun R, Delfanti R, Egorov V, El Khoukhi T, Florou H, Kniewald G, Noureddine A, Patrascu V, Pham MK, Scarpato A, Stokozov N, Topcuoglu S, Warnau M (2008) ^{137}Cs baseline levels in the Mediterranean and Black Sea: a cross-basin survey of the CIESM mediterranean mussel watch programme. Mar Pollut Bulle 57(6–12):801–806. https://doi.org/10.1016/j.marpolbul.2007.11.010

Timofeev-Resovskii NV (1957) Primenenie izluchenii i izluchatelei v eksperimental'noi biogeotsenologii. Botanicheskii Zhurnal 42(2):161–194 (in Russian)

Timofeeva-Resovskaya EA, Popova EI, Polikarpov GG (1958) O nakoplenii presnovodnymi organizmami khimicheskikh elementov iz vodnykh rastvorov. I. Kontsentratsiya radioaktivnykh izotopov fosfora, tsinka, strontsiya, ruteniya, tseziya i tseriya raznymi vidami presnovodnykh mollyuskov. Byulleten' MOIP. Otdelenie biologii 63(3):65–78. (in Russian)

Timoshchuk VI, Kulebakina LG, Filippov IA (1970) Raspredelenie strontsiya-90 v poverkhnostnom sloe Sredizemnogo morya. In: Radioekologicheskie issledovaniya Sredizemnogo morya. Naukova dumka, Kiev, pp 150–156. (in Russian)

Titlyanova AA, Ivanov VI (1961) Pogloshchenie tseziya tremya vidami presnovodnykh rastenii iz rastvorov razlichnoi kontsentratsii. Doklady AN SSSR 136(3):721–722 (in Russian)

Trapeznikov AV, Molchanova IV, Karavaeva EN, Trapeznikova VN (2007) Migratsiya radionuklidov v presnovodnykh i nazemnykh ekosistemakh, vol 1. Izdatel'stvo Ural'skogo universiteta, Ekaterinburg, 480 p. (in Russian)

Tseitlin VB (1984) Raschet velichiny potoka organicheskogo veshchestva, postupayushchego na dno okeana. In: Biogeokhimiya prikontinental'nykh raionov okeana: abstracts of reports of the All-Union meeting, Nal'chik. (in Russian)

Tsytsugina VG, Lazorenko GE (1983) Rol' mitoticheskogo deleniya v pogloshchenii biogennykh elementov prirodnymi populyatsiyami fitoplanktona. Ekologiya Morya 12:30–34 (in Russian)

Vedernikov VI, Starodubtsev ER (1971) Pervichnaya produktsiya i khlorofill v yugo-vostochnoi chasti Tikhogo okeana. Trudy Instituta Okeanologii AN SSSR 89:33–42 (in Russian)

Vernadsky VI (1929) O kontsentratsii radiya zhivymi organizmami. Doklady Akademii Nauk SSSR. Seriya A 2:33–34 (in Russian)

Vernadsky VI (1965) Khimicheskoe stroenie biosfery Zemli i ee okruzheniya. Nauka, Moscow, 374 p. (in Russian)

Vinogradov AP (1967) Vvedenie v geokhimiyu okeana. Nauka, Moscow, 213 p. (in Russian)

Vinogradov ME (1968) Vertikal'noe raspredelenie okeanicheskogo zooplanktona. Nauka, Moscow, 319 p. (in Russian)

Vinogradov ME (1986) Zavisimost' protsessov detritoobrazovaniya ot prostranstvenno-vremennykh izmenenii planktonnykh soobshchestv. In: Biodifferentsiatsiya osadochnogo veshchestva v moryakh i okeanakh. Rostov-on-Don, pp 66–76. (in Russian)

Vinogradov ME, Lisitsyn AP (eds) (1983) Biogeokhimiya okeana. Nauka, Moscow, 367 p. (in Russian)

Weers AW (1975) The effects of temperature on the uptake and retantion of ^{60}Co and ^{65}Zn by the common shrimp Crandon crandon (L.). In: Combined effects of radioactive, chemical and thermal releases to the environment, Stockholm, 2–5 June 1975

Wright DA (1977) The effect of salinity on cadmium uptake by the tissues of the shore crab Carcinus maenas. J Exp Biol 67:137–146

Youngbluth MJ (1985) Investigations of soft-bodied zooplankton and marine show using manned research submersibles. Bull Mar Sci 27(2):782–783

Zaitsev YuP (1967) Giponeiston i ego radioekologicheskoe znachenie. In: Voprosy biookeanografii. Naukova dumka, Kiev, pp 180–184. (in Russian)

Zaitsev YuP (1970) Morskaya neistonologiya. Naukova dumka, Kiev, 264 p. (in Russian)

Zaroogian GE (1979) Studies on the depuration of cadmium and copper by the American oyster Crassostrea virginica. Bull Environ Contam Toxicol 23(1):117–122. https://doi.org/10.1007/bf0 1769928

Zesenko AYa (1977) Izuchenie povedeniya i nakopleniya gidrobiontami razlichnykh fiziko-khimicheskikh form radioaktivnykh i stabil'nykh radionuklidov pervoi – vos'moi grupp Periodicheskoi sistemy elementov. In: Polikarpov GG, Risik NS (eds) Radiokhemoekologiya Chernogo morya. Naukova dumka, Kiev, pp 21–52. (in Russian)

Zesenko AYa, Nazarov AB (1973) O nakoplenii toriya-234 morskimi organizmami. In: Radiokhemoekologiya vodnykh organizmov, vol 2. Zinatne, Riga, pp 274–280. (in Russian)

Zlobin VS (1968) Dinamika nakopleniya radiostrontsiya nekotorymi burymi vodoroslyami i vliyanie solenosti morskoi vody na koeffitsienty nakopleniya. Okeanologiya 8(1):78–85 (in Russian)

Chapter 3
Semi-empirical Theory of Mineral and Radioisotopic Exchange of Living and Inert Matter in the Marine Environment

Introduction

This chapter presents a description of the semi-empirical theory of the mineral and radioisotopic exchange of living and inert matter in the marine environment.

In the first section, promising theoretical developments based on the empirical parametrisation of models are considered, a review of differential models of the interaction of aquatic organisms (referred to hereinafter as hydrobionts) with radioactive and chemical components of the marine environment is carried out, and methods for interpreting the results of natural observations and experiments using a radioactive tracer (radiolabelling) for the parameterisation of models are substantiated. Due to the structural complexity of the interactions of living and inert matter with radioactive chemical components of the marine environment belonging to the class of complex geosystems, mathematical models are appropriate approaches for their study and management. While the upper levels of the hierarchical structure of semi-empirical models should be comprised of theoretical blocks, at the lower levels, the main structural elements should be empirical. The polymorphism of empirical modeling is demonstrated on the basis that each finite series of observations can be described by a variety of models to a given accuracy based on various hypotheses about interaction mechanisms. For this reason, the main task of semi-empirical modelling consists in the unambiguous parametrisation of the compatibility of all of its component blocks at theoretically- and empirically reflected scales of interaction.

The second section of the chapter is devoted to empirical compartment models of the parenteral sorption (direct absorption of chemicals and their radionuclides from the aquatic environment) and metabolic exchange of living and inert components of marine ecosystems. It describes methods of parametrisation and verification of differential models of closed systems based on the

V. Egorov, *Theory of Radioisotopic and Chemical Homeostasis of Marine Ecosystems*, Springer Oceanography, https://doi.org/10.1007/978-3-030-80579-1_3

results of field observations and aquarium experiments carried out using radio-labelling. The scope of applicability of compartment-based balance models is substantiated on the basis of assumptions that the sorption of chemical substrates proceeds in accordance with metabolic reactions of the first or zero orders and that their desorption or intravital excretion by hydrobionts is proportional to their content in living or inert matter. Empirical models have been developed that take into account radioactive decay, the size spectra of allochthonous particles and hydrobionts, the generative and somatic growth of individual marine organism specimens, production processes and the specific mass of living and inert matter in the marine environment, the concentration of radionuclides with their isotopic and non-isotopic carriers, physical and chemical sorption processes, the biotic transformation of physicochemical forms of pollutants, as well as the limitation of primary production processes by biogenic elements.

The third section of the chapter is devoted to the description of compartment-based conceptions of the alimentary pathway of the mineral nutrition of aquatic organisms. It presents features of empirical parametrisation and verification of differential models of closed systems when changing the concentration of radionuclides and their isotopic carriers in food, taking into account the somatic and generative growth of hydrobionts.

The fourth section considers the application of generalised characteristics of differential semi-empirical models to describe the kinetics of the conjoint parenteral and alimentary pathways of the mineral nutrition of aquatic organisms. The balance equalities of the generalised model are obtained taking into account mass transfer in open systems. It is shown that the parametric basis of the model is compatible with contemporary methods for describing the material, energetic and mineral balance in marine ecosystems. In determining the scope of applicability of the generalised model, it is shown that this approach can form the basis for a semi-empirical theory of mineral and radioisotope exchange of living and inert matter in the marine environment.

3.1 Prospects for Theoretical Development Based on the Empirical Parametrisation of Models

It follows from the previous section that the study of the interaction of living and inert matter with the radioactive and chemical components of the marine environment requires a complex solution of problems associated with the absorption, physico-chemical transformation, migration and elimination into geological depots of chemicals of various biological and ecotoxicological significance. The physical, chemical and biological interactions taking place in marine ecosystems are studied within the various subdisciplines of oceanography. For example, the transport of pollutants by currents, water advection, turbulent and molecular diffusion is the proper subject of hydrophysical research (Nelepo 1970). Biogeochemistry, on the other hand, studies the processes of transfer and deposition of chemical compounds carried in sediments (Romankevich 1977). Factors affecting the energetic and chemical limitation of production processes in the ocean, as well as trophodynamic connections in ecosystems, fall within the competence of hydrobiology (Petipa 1981; Sushchenya 1972). The concentration characteristics of radioactive and chemical contaminants affecting hydrobionts and production processes in ecosystems are studied by specialists working in the field of aquatic radioecology (Polikarpov 1966) and toxicologists (Patin 1979). It is evident, therefore, that the solution of the indicated tasks requires the use of a systematic approach, resulting in their classification as complex geosystems (Belyaev 1978).

In accordance with the relevant underlying theoretical concepts (Belyaev 1978), geosystems belong to the class of complex systems, whose management is based on the use of mathematical models. In terms of their type, mathematical models of systems can comprise both empirical and theoretical elements. Empirical models consist of mathematical expressions that take various approaches to approximate experimental data on the dependences of the system state parameters on the parameters of the factors influencing them. In order to construct empirical models, it is not necessary to rely on ideas concerning the structure or internal mechanism of connections in the system. At the same time, the problem of determining the mathematical expression of the empirical model for a given array of observations within the selected accuracy of phenomenological description is ambiguous. At the same time, the problem of determining the mathematical expression of the empirical model for a given array of observations within the selected accuracy of phenomenological description is ambiguous, since there are an infinite number of mathematical expressions that approximate, within a given accuracy, the same experimental data on the dependence of parameters.

In radiation- and chemical biology, empirical models can be constructed on the basis of observations of nature carried out within the framework of monitoring, data on the concentration of chemicals distributed in the environment and marine organism microcosms, as well as the results of laboratory experiments.

Empirical models of natural objects constructed from observational data reflect changes in biotic and abiotic environmental parameters under the influence of the currently observed rate of intake of radionuclides and their isotopic and non-isotopic carriers. In order to be able to predict the level of pollution affecting hydrobionts and the environment, models can be constructed based on natural observation data. However, the predictive estimates obtained using such models can be objective only insofar as they reflect the interaction mechanisms responsible for the concentration and exchange of chemical or radioactive substances by hydrobionts at anticipated rates and levels of pollution of the marine environment.

Another approach to studying and predicting the state of ecosystems is associated with microcosms. Microcosms are understood as bodies of water in which special controls of biotic and abiotic factors are carried out alongside an assessment of the rate of pollution of water areas in order to construct models for forecasting purposes. The discharge of pollutants into microcosms tends to exceed the natural rate of their release to the marine environment. It is often possible to assign or regulate the rate of release of pollutants into microcosms. For the purposes of studying ecological microcosms, locations of sewage effluent, as well as water areas adjacent to areas of pollution discharge by chemical or nuclear industries, can be used. Following the testing of nuclear weapons, it became possible to consider the vast waters of the world's oceans as microcosms. The study of microcosms permits an examination of the mechanisms of biotic and abiotic interactions of pollutants with living matter, as well as informing the construction of functional models of ecosystem response to anthropogenic impacts. This approach was used to study the radioecological response of the Black Sea to the Chernobyl nuclear accident (Polikarpov and Egorov 2008). However, models constructed according to the results of observations carried out with regard to microcosms are only applicable for predicting the states of geosystems to the extent that they are identical, which is not always the case (Belyaev 1978). In addition, even when the biotic characteristics of microcosms and natural geosystems are identical, differences in the physical scales of studied phenomena may result in the inaccuracy of ecological forecasting based on this observational data.

Different mechanisms of interactions between geosystem parameters are distinguished in the course of laboratory experiments. In this regard, consistent patterns are established to govern changes occurring in parameters of interest influenced by specified factors applying under stationary environmental conditions. Empirical models built on the basis of laboratory observations reflect the interrelationships between the studied factors across the entire range of influences considered in the experimental setup, in the general case, with strictly fixed values of other biotic and abiotic parameters. However, the results obtained from considering such models cannot be fully transferred to natural geosystems due to the deliberately non-identical experimental conditions.

In general, empirical models can only reflect the behaviour of observed geosystems. At the same time, it is important to be able to predict the state of those geosystems that either do not yet exist or cannot be assumed. Such problems can arise, for example, when enumerating options for the location of nuclear power plants and related environmental forecasting, when considering various situations of emergency discharges of pollution, or during the construction of hydraulic structures that fundamentally change the hydrological or biological structure of geosystems.

Thus, predicting the behaviour of geosystems that are not identical to those available for observation is possible only on the basis of theoretical models (Belyaev 1978). Theoretical models of geosystems are constructed by synthesising generalised ideas about the individual processes and phenomena that make them up based on fundamental laws that describe the behaviour of matter and energy. Since a theoretical model reflects the functioning of an abstract geosystem, no observational data on the parameters of a particular geosystem is required for the initial derivation of its ratios. Rather, such a model is based on generalised a priori concepts about the structure of the geosystem, as well as the mechanism of connections between its constituent elements. For this reason, the suitability of using a theoretical model for solving forecasting problems is determined by the degree to which the a priori concepts reflect the actual mechanisms of interactions in the geosystem. Theoretical models describe not one object, but a particular class of phenomena, among which there can be both observable and unobservable phenomena. The extent to which the theoretical model reflects phenomena observed to occur in nature, as well as in microcosms, can serve as a measure of its adequacy (Belyaev 1978).

In order to construct theoretical models that represent dynamic problems of the interaction of living and inert matter with the radioactive and chemical components of the marine environment, it is necessary to study the mechanisms of parenteral and alimentary pathways of their absorption and exchange by aquatic organismsperforming the ecological role of hydrobiont, which mechanisms depend on the biotic and abiotic parameters of geosystems, as well as on the of distributive concentration characteristics and toxic effects of pollutants in the environment.

Theoretical models that contain empirical blocks at lower levels can be described as semi-empirical. For such models, mathematical expressions can be theoretically derived to an accuracy limited by empirically determined constants. As a rule, the models use empirical data in combination with theoretical expressions (Belyaev 1978). Semi-empirical models are applicable for describing both observable and unobservable phenomena. However, it is important to keep in mind the following: like most complex systems, geosystems have a hierarchical structure, in which the highest level is made up of the largest elements into which it can be subdivided. This is what determines the relativity of the model types. If a semi-empirical model has empirical blocks at some level, then in relation to the mechanisms of interactions that appear at higher levels, it has the properties of a theoretical model, while, at lower levels, it has the properties of an empirical model.

3.1.1 Dynamic Problems

In accordance with contemporary understandings, the model of a complex geosystem comprises a set of dependencies L_j, connecting the parameters g_j (Belyaev 1978):

$$L_j(g_1, g_2, \ldots g_j, \ldots g_n) = 0, \quad j = 1, n, \tag{3.1}$$

where n is the number of independent parameters.

The value of n depends on the degree of smoothing of the parameter of practical interest g_j by frequency within the selected scale and specified accuracy. With an increase in forecasting accuracy, implying the need to take smaller fluctuations of the parameter g_j into account along with an increase in the prediction period, the value of n also rises. This model consists in a formalisation with an unformalisable remainder. The model will only be able to reflect all the properties of an object when it becomes identical to it. For this reason, the model can only reflect individual properties of an object. The target versatility of the model's use may require such an expansion of the number of considered parameters n, for which at this research level it is impossible to obtain a sufficient number n of known or plausibly selected operators L_j, reflecting the mechanisms of interaction of the components of a complex geosystem. An unjustified increase in the number of operators accepted for consideration L_j complicates the analysis of the model. In addition, it can reduce predictive accuracy due to an increase in the proportion of operators accepted a priori inadequate to the geosystem. In the model, not all parameters have equal status. The predicted parameters are commonly referred to as forming the centre of the model. It is considered that versatility meets the requirements of adequacy if it allows the prediction problem to be solved with respect to the parameters forming the centre of the model, smoothed over the required frequency within the selected space–time scale at a given accuracy (Belyaev 1978). An optimal model is one that adequately reflects the objective complexity of the analysed interactions taking the smallest number of parameters and operators into consideration. The choice of time scales of research in solving dynamic problems of radio-chemoecology is of particular importance.

The particular importance of the selection of the temporal scale of research when solving dynamic problems of radiochemoecology can be illustrated with an example. When present in food, penetrating cell membranes or being adsorbed on biological surfaces, radicals or whole molecules of chemicals—including toxicants—are in contact with aquatic organisms (hydrobionts) for some time, after which they are returned to the aquatic environment, often in the form of other compounds (as a result of metabolism or desorption). During direct contact with a hydrobiont, atoms or molecules of the same substance can enter into various chemical compounds, participate in biochemical reactions, be incorporated into tissues and be transported through the organs of animals. In order to consider these processes on a microscale of time, it is necessary to be able to trace the chain of transformations that each atom or molecule of a substance undergoes. In order to describe the relationship between the concentration of a substance in a medium and its exchange by a hydrobiont on a microscale, it is necessary to use models with lagging arguments that take the characteristics of all transformation chains into account differentially. When considering the regularities of exchange at a mesoscale, the time intervals for the passage of atoms or compounds through the transformation chains are levelled to allow the entire process to be considered as a single flux or a small number of fluxes flowing at different rates. The description of mesoscale metabolic processes boils down to establishing a relationship between the concentration of a substance in a medium or food and the intensity of the exchanged fluxes, depending on the physiological state, population, as well as trophic and ecological characteristics of hydrobionts. According to the macroscale consideration, the substance exchanged by the hydrobiont is analysed in terms of the dependence of a single flux—or fluxes—on the integral characteristics of the accumulation of this substance in the medium and on the biotic parameters of the hydrobiont. It is self-evident that the molecular, physiological, population or ecological level of research is determined by the relationship between space–time scales, smoothness of frequency parameters and accuracy requirements. Semi-empirical models are considered for the mathematical description of interactions between living matter and radioactive or chemical substrates, as well as for solving problems connected with predicting the pollution and self-purification of the marine environment. The empirical blocks of these models were obtained on the basis of laboratory—usually radio-tracer—experiments on the study of alimentary and parenteral pathways for the absorption of substances by aquatic organisms. Serving as the lowest levels of the semi-empirical models, these blocks reflect bio- and abiotic interactions in the ecosystem.

In the present work, in order to construct mathematical models, the balance method for expressing their operators L_j was used. Balance equations that close the ecosystem in terms of matter and energy, limiting the biogenic substrate along with any environmental pollutants, are considered. In our view, the balance method used to close an ecosystem for chemical or radioactive substances is historically conditioned. Initially, models were developed on the basis of a material balance. These models primarily consisted of representations of trophic relationships of the predator–prey type (Volterra 1972; Lotka 1925). Subsequently, it transpired that biological tissues have varying energy equivalents. This resulted in the requirement to close the ecosystems in terms of their energetic and material balance (Vinberg and Anisimov 1966). Studies of biochemical life forms have distinguished between the different biogeochemical cycles of the various chemicals participating in ecosystems (Lowman et al. 1971). This is determined not only by differences in the effect of abiotic factors, but also by variations in the transfer of matter along food chains, which required, along with the balance in matter and energy, the development of methods for closing ecosystems in terms of chemical compounds. The task of creating the above empirical models followed from the requirements for developing a basis for the derivation of balance equalities that close ecosystems for chemical or radioactive substances.

Within the framework of this work, the structure and parameters of empirical models were determined from the results of experiments using a radioactive tracer (radiolabeling). This was motivated by the possibility of studying the kinetic regularities of chemical effects without violating the stationarity of the biotic and abiotic characteristics of experimental geosystems. The results of such experiments allowed the determination of parameters characterising the exchange processes of stable chemical analogues using the observational data on the behaviour of radionuclides in hydrobionts and the aquatic environment. The main task of the experimental research consisted in the development of empirical models applicable for the optimal solution by the balance method of problems of the mathematical description of the kinetics of the interaction of living and inert matter with radioactive or chemical components of the marine environment. The specified task involved two main aspects of the interaction between hydrobionts and pollutants of the marine environment. The first of these was associated with the study of the kinetic characteristics of the absorption, concentration and metabolism of pollutants by hydrobionts, as well as the effect of living matter on the accumulation and migration of chemical or radioactive substances in the environment. The second aspect was due to the mutagenic and toxic effects of pollutants on the biota. Within the framework of the present thesis, the problem of determining the allowable anthropogenic pressure on aquatic organisms was solved for those levels at which such toxic and mutagenic effects did not lead to undesirable or irreversible changes in the structural and production characteristics of ecosystems.

It is for this very reason that only the kinetic characteristics of interactions between living matter and pollutants of the marine environment are considered within the scope of this work.

3.1.2 Mathematical Models of the Interaction Kinetics of Hydrobionts with Radioactive and Chemical Components of the Marine Environment

Studies of the absorption of radionuclides by marine organisms in experiments with unchanging biotic and abiotic conditions have led to the conclusion that the kinetics of these processes in the non-stationary phase can be described by a power function (Barinov 1965). For practical purposes, it is often necessary to determine the rate of absorption of a radioactive tracer—or radiolabel—by a hydrobiont at an initial ($t = 0$) moment in time. Since the power function at $t = 0$ is not differentiable, it was necessary to select an approximation function in order to describe the complete series of experimental observations. It has been shown that the entire process at the accumulation stage (Bachurin 1968; Bloom and Raines 1971; Weers 1973; Gromov et al. 1979; Beasley et al. 1982), and deductions (Zlobin 1968; Cranmore and Harrison 1975; Guary and Fowler 1979) is adequately described by the sum of exponential terms:

$$C_h(t) = \sum_{i=1}^{n} B_i e^{-p_i t}, \tag{3.2}$$

where

$C_h(t)$ is the concentration of the radionuclide or its carrier in the hydrobiont at the moment of time t;

n is the number of exponential terms;

B_i and p_i are parameters.

The value $B_0 = 0$ corresponds to experiments on the elimination of radionuclides by aquatic organisms. The parameter p_i is calculated by the formula:

$$p_i = \frac{\ln 2}{\tau_i}, \tag{3.3}$$

where τ_i are the half-decay periods of exponentials.

With respect to the initial conditions of the experiment on the distribution of radionuclides by hydrobionts, at $C_h(t = 0)$ and $C_w(0) = C_{w0}$, where C_{w0} is the concentration of the radionuclide or isotopic carrier in water C_w at the initial moment in time $t = 0$, Eq. (3.2) has the form:

$$CF(t) = CF_S - \sum_{i=1}^{n} B_i e^{-p_i t}, \qquad (3.4)$$

where $CF(t)$ is the concentration factor (Vernadsky 1929; Timofeev-Resovskii 1957; Polikarpov 1964) of the radionuclide by the hydrobiont at the present moment in time (t), equivalent to the ratio of the radionuclide's concentration in the hydrobiont (C_h) to its concentration in the water (C_w);

B_i and p_i are parameters.

CF_S is the factor of concentration in the stationary phase at $CF(t \to \infty)$, referred to as statistical or limiting.

In relation (3.4):

$$CF_S = \sum_{i=1}^{n} B_i. \qquad (3.5)$$

Functions of the form (3.2)–(3.5) have found application for describing experimental data on parenteral and sorption intake of chemicals and radionuclides by living and inert matter, as well as their alimentary absorption by hydrobionts (Kowal 1971; Pentreath 1973; Hakonsonet al. 1975).

From the point of view of experimental theory (Nalimov 1971; Krug et al. 1977), the integral model of the form (3.2) is polynomial. The coefficients of the polynomials can be interpreted as members of the Taylor series, i.e., as the values of the partial derivatives at the point around which the expansion of the function that specifies the solution of unknown differential equations is performed. Nalimov (1971) notes: "Knowing the numerical coefficients of the segment of the Taylor series, it is impossible to restore the original function, whose analytical expression remains unknown to the researcher; even more so, it is impossible to restore the original differential equations that describe the mechanism of the process… This circumstance explains the variety of possible models applicable for the description of a single object."

The movement from integral (3.2)–(3.5) to differential models are based on a number of current assumptions. Having assumed that the dependence (3.4) is the reaction of the radionuclide in the marine environment/hydrobiont system to a single jump (threshold change in radioactivity from $C_w = 0$ to $C_w = 1$), Alexander Bachurin ascertained the transfer function of the system; at $n = 2$, according to the "black box" method, he obtained the differential equation of the dynamics of isotopic exchange between marine organisms and the environment (Bachurin 1968):

$$T_1 T_2 \frac{d^2 C_h}{dt^2} + (T_1 T_2) \frac{d C_h}{dt} + C_h = (B_1 T_1 + B_2 T_2) \frac{d C_w}{dt} + CF_S C_w, \qquad (3.6)$$

Where $T_i = p_i^{-1}$ are exponential time constants.

An alternative interpretation of relations (3.2) and (3.4) led V.I. Belyaev (1972) to obtain the isotope exchange equation by methods for the description of dispersed systems using Lagrangian variables (Belyaev 1964). Belyaev's model was built on the assumption that a hydrobiont contains several phases characterised by the concentrations of the isotope C_{hi} and its stable analogue C'_{hi}, as well as rates of exchange g_i, which determine the flux of the element per unit time between the phases of the organism, by referring to the unit mass of the phase. Under these assumptions, the exchange equation for each phase is:

$$\frac{dC_{hi}}{dt} + \frac{g_i}{C'_{hi}}C_{hi} = \frac{g_i}{C'_w}C_w, \tag{3.7}$$

where C'_w is the concentration of a stable structural analogue of an isotope in water.

The total isotope content in a hydrobiont ($C = \sum C_{hi}$) under initial conditions $C_{Hi} = C_{hi}(0)$ is expressed by the ratio:

$$C_h = \frac{C_w C'_h}{C'_w} - \sum_{i=1}^{n}\left[\left(\frac{C_w C'_h}{C_w} - C_{hi}(0)\right)e^{-\frac{g_i}{C'_{hi}}t}\right]. \tag{3.8}$$

The solution to Eq. (3.8) is identical to Eq. (3.4). At $n = 2$, the parameters of Eqs. (3.7)–(3.8) take the form:'

$$B_1 = \frac{C'_{h1}}{C'_w} - \frac{C_{h1}(0)}{C_h}; \quad B_2 = \frac{C'_{h2}}{C_w} - \frac{C_{h2}(0)}{C_h}; \tag{3.9}$$

$$p_1 = \frac{g_1}{C'_{h1}}; \quad p_2 = \frac{g_2}{C'_{h2}}; \quad CF'_S = \frac{C'_h}{C'_w} \tag{3.10}$$

where CF'_S are the concentrationfunctions of the stable analogue of the radionuclide by the hydrobiont.

Compartment models are widely used for describing the kinetics of the processes of exchange of radioactive and chemical substances by hydrobionts (Sheppard1948; Sheppard and Householder 1951; Solomon 1960; Berman 1965; Brownell et al. 1968; Fried 1968; Atkins 1969; Bernhard 1971; Bloom and Raines 1971; Conover and Francis 1973; Bernhard et al. 1975; Sukal'skaya and Likhtarev 1976). The basis for the application of compartment models consisted in the concept of hydrobionts having exchangeable (and non-exchangeable) resources—or pools—of chemical elements (radionuclides), which interact with the environment or with each other via metabolic reactions of various orders. In terms of compartments, whole marine organisms were considered, as well as their individual organs and tissues, and the various physicochemical states of radionuclides they contain. At the same time, discrete organs or tissues can be represented by more than one compartment.

Studies have shown (Riziê 1972; Botov 1975) that when constructing a compartment model, it is necessary to match its structural complexity to the accuracy of the exponential parameters determined from the experiment, which can be used to parameterise the models. Those compartment models that took this circumstance into account were classified as empirical differential models; otherwise, they were considered as theoretical models.

At $n = 2$, the kinetics of the interaction of exchangeable resources of a hydrobiont with a radionuclide in the medium is generally described by a compartment model having a structure as shown in Fig. 3.1. The compartments of this model are consistent with the radioactive or chemical component in the water (1 in Fig. 3.1) and in the hydrobiont (2 and 3 in Fig. 3.1). The specified block diagram is implemented by differential equations (Riziĉ 1972):

$$\frac{dA_1}{dt} = -A_1(r_{12} + r_{13}) + r_{21}A_2 + r_{31}A_3;$$

$$\frac{dA_2}{dt} = r_{12}A_1 - (r_{21} + r_{23})A_2 + r_{32}A_3;$$

$$\frac{dA_3}{dt} = r_{13}A_1 + r_{23}A_2 - (r_{31} + r_{32})A_3, \tag{3.11}$$

where A_1, A_2 and A_3 are the content of the radionuclide or isotopic carrier in the compartments corresponding to the water and the hydrobiont;

r with indices are the rates of exchange of a chemical or radioactive substrate between compartments.

With reference to a closed system (Fig. 3.1) having initial conditions $A_i(0) = A_{i0}$, where i is the compartment number, the general solution of Eqs. (3.11) takes the form:

$$A_i(t) = \sum_{i=0}^{2} a_i e^{-p_i t}, \tag{3.12}$$

where a_i and p_i are parameters.

Fig. 3.1 Structure of the model of radionuclide exchange in the water-hydrobiont system

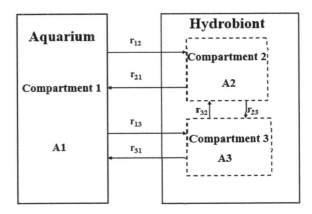

Since in Eq. (3.2) $p_0 = 0$, expressions (3.2) and (3.12) are identical when the biomass of an hydrobiont in a medium remains constant. In this regard, the determination of the parameters of Eq. (3.12)—A_1, A_2, A_3 and r with indices—an empirical compartment model boils down to the establishment of rates of exchange from experimental observations using the coefficients of Eqs. (3.2) and (3.4).

Studies have shown (Fried1968; Riziĉ 1972) that, for a three-compartment model possessing a full set of transport communications, this problem is unsolvable without the use of additional data, since, with $n = 2$, the theory of compartment analysis (Sheppard 1948) gives independent equations for establishing only four of the six indicators of metabolic rates determined by the structure of the model (Fig. 3.1). In this case, the compartment reflecting the kinetic laws of the behaviour of a radioactive or chemical substance in water must be connected by at least two transport communications with the compartments of the hydrobiont. This is due to the fact that one of these communications should reflect the processes of influx of the element into the hydrobiont, while the other reflects its elimination. The empirical dependence (3.2) does not carry information about transport communications, but only describes the kinetics of the total exchange between the environment and the hydrobiont. Therefore, in order to substantiate the structure of communications between the cameras, it is necessary to rely on information about exchange mechanisms or advance additional hypotheses. When the research task is invariant with respect to various sets of transport communications, the so-called mammillary model (Riziĉ 1972) has the simplest structure, reflecting the independent interaction of each compartment of the hydrobiont with the environment (at$r_{23} = 0$ and $r_{32} = 0$). For a mammillary model having different ratios of the substrate in the hydrobiont and in water, the rates of exchange of the compartments are calculated using the formulas (Egorov 1975):

$$r_{12} = m_{sp}B_1 p_1 p_2 / r_{13}; \quad r_{21} = p_1 p_2 (1 - m_{sp}CF_S)/r_{31}; \quad r_{13} = m_{sp}B_2 p_1 p_2 / r_{21};$$
$$r_{31} = (p_1 p_2 - S)/2(1 - m_{sp}B_1)/(1 - m_{sp}CF_S);$$
$$S = \sqrt{(p_1 + p_2)^2 - 4p_1 p_2 (1 - m_{sp}B_1)(1 - m_{sp}B_2)/(1 - m_{sp}CF_S)}. \tag{3.13}$$

Equations (3.13) follow from the main condition for the applicability of the model (3.11): $m_{sp} \cdot CF_S \ll 1$. This condition indicates that the model can be used to reflect the states of systems in which the metabolic pool of a chemical or radioactive substance in an hydrobiont is significantly lower than its pool in the aqueous medium. If this pool can be disregarded, then the rate of exchange indicators, normalised to the unit of biomass of the hydrobiont in the environment, are equal to:

$$r_{12} = B_1 p_1; \quad r_{21} = p_1; \quad r_{13} = B_2 p_2; \quad r_{31} = p_2. \tag{3.14}$$

In this case, the balance equalities of the model (3.11) in terms of C_h and C_w will look like:

$$\frac{dC_1}{dt} = p_1(C_w B_1 - C_1);$$

$$\frac{dC_2}{dt} = p_2(C_w B_2 - C_2);$$

$$\frac{dC_w}{dt} = m_{sp}[p_1 C_1 + p_2 C_2 - C_w(B_1 p_1 + B_2 p_2)]. \tag{3.15}$$

This explains why the problem of parametrising the model boils down to determining the parameters of Eq. (3.15) from the results of approximating experimental observations according to functions (3.2) or (3.4).

The closed system of a radioactive or chemical substrate in the marine environment and in a hydrobiont corresponds to the initial conditions $C_1(0) = C_{10}$, $C_2(0) = C_{20}$ and $C_w(0) = C_{w0}$. If the condition $m_{sp} \cdot CF_S \ll 1$ is preserved, then the solution of Eq. (3.15) with respect to C_h has the following form:

$$C_h(t) = CF_S C_{w0} + (C_{10} - C_{w0}B_1)e^{-p_1 t} + (C_{20} - C_{w0}B_2)e^{-p_2 t}. \tag{3.16}$$

This expression shows that, for any changes in the state of the system, the direction of the kinetic processes after the cessation of exposure is such that the ratio of C_h to C_w in the aquatic environment always tends to the value of the concentration factor in the stationary phase CF_S.

3.1.3 Methods for the Interpretation of the Results of Physical Observations and Experiments Using a Radioactive Tracer to Parametrise Semi-Empirical Models

In the previous sections, we substantiated the thesis that the parametrisation of semi-empirical models requires the development and adaptation of methods for interpreting natural and experimental observations associated with determining the dependences of concentration and exchange of living and inert matter by the chemical and radioactive components of the marine environment. In Sect. 2.6 it is noted that the relationship between the stationary levels of concentrations of chemical and radioactive substances in living and inert matter and their content in the marine environment can be described by the Freundlich and Langmuir equations.

The Freundlich equation has the form:

$$C_h = A \cdot C_w^{-b}, \tag{3.17}$$

where A and b are coefficients;

This equation comprises a hyperbolic function, which corresponds to the equation of a straight line on a logarithmic scale along the ordinate axes. The parameters of this straight line can be set by approximating the observation results using the least squares method. For this reason, the hypothesis of the correspondence of empirical observations to Freundlich's adsorption isotherm is usually tested by statistical criteria for linearity and scatter of observation results, plotted in logarithmic scales along the coordinate axes. It is taken on the basis of an estimate of the value of the coefficient of determination R^2. In this case, the parameters of Eq. (3.17) are determined from biased estimates of the coefficients of the equation of the straight line:

$$Log C_h = Log(A) - b \cdot Log C_w, \tag{3.18}$$

due to the fact that they were established from the condition of minimising not the sum of the squared deviations themselves, but rather the sum of the logarithms of the squared deviations.

The Langmuir equation (Nesmeyanov 1978) has the following form:

$$C_h = \frac{C_w C_{max}}{1/k_L + C_w}, \tag{3.19}$$

where k_L is the parameter of the Langmuir equation;

C_{max} is the equilibrium saturation concentration of a chemical element in a living or inert substance.

It should be noted that the C_{max} parameter of the Langmuir equation characterises the sorption or metabolic saturation of the components of living or inert substance by the substrate. It is for this reason that the theoretical interpretation of the data using the Langmuir equation is of higher practical importance in solving the problems of sanitary and hygienic regulation of the maximum permissible pollution of marine aquatic areas.

Studies of the regularities of mineral metabolism have shown that the dependence of the change in the rate of absorption of mineral elements by hydrobionts (V) corresponds to the kinetics of enzymatic reactions described by the Michaelis–Menten equation (Patton 1968):

$$V = \frac{V_{max} C_w}{K_m + C_w}, \tag{3.20}$$

where V_{max} is the maximum absorption rate of the substrate involved in the enzymatic reaction;

C_w is the concentration of the substrate in the aquatic environment;

K_m is a parameter having the dimension of units of the concentration of the substrate in water (C_w), referred to as the constant of the Michaelis–Menten equation.

The K_m value is numerically equal to the C_w value, at which the substrate absorption rate is half of the maximum. In the region of microconcentrations, it can be seen from relation (3.20) that, when $C_w \ll K_m$, the value of V is proportional to C_w with a constant proportionality coefficient value V_{max}/K_m, that is, the first order of metabolic reactions is observed. In the range of large values of C_w, when $C_w \ll K_m$, the value V_{max}/K_m—and hence the rate of substrate absorption—also does not depend on its quantity in the aqueous medium, which corresponds to zero-order metabolic reactions (Patton 1968).

The widespread use of the biological interpretation of the Michaelis–Menten equation was determined by its convenience, leading to the development of auxiliary linearised expressions that allow for a graphical definition of its applicability and parameters. It was proposed (Patton 1968) that these parameters be estimated according to the degree of linearity of empirical dependences constructed in the coordinates of the linearised Lineweaver–Burk equations:

$$\frac{1}{V_a} = \frac{K_m}{V_{max}} \frac{1}{C_w} + \frac{1}{V_{max}} \qquad (3.21)$$

or the modified Lineweaver–Burk equation:

$$\frac{C_w}{V_a} = \frac{1}{V_{max}} C_w + \frac{K_m}{V_{max}}. \qquad (3.22)$$

It should be noted that, while the Langmuir (3.19) and Michaelis–Menten Equations (3.20) differ in meaning, at $1/k_L = K_m$, they are parametrically identical. For this reason, the Lineweaver–Burk equations may be used to verify both laws.

In the experimental study of the uptake of radionuclides by living or inert matter directly from the aquatic environment due to sorption or metabolic processes, a radioactive tracer or label is introduced into the water of an aquarium in a physicochemical form identical to that present in the natural marine environment, as well as in a trace amount in relation to the concentration of an isotopic or a non-isotopic carrier in water.

The methodology for studying the kinetic characteristics and concentrating ability of hydrobionts in experiments involving a radiolabel is based on the introduction of radioactive analogues of chemical substances into the water of experimental aquariums and on the determination of the rate and stationary levels of absorption of labeled atoms as a result of establishing equal specific radioactivity in hydrobionts and in water. The point in time $t = 0$ in aquariums is characterised by the following initial conditions: $C_{wst} = C_{wst0}$; $C_{hst} = C_{hst0}$; $C_{wr} = C_{wr0}$ and $C_{hr} = 0$, where C_{wst0} and C_{hst0} are the concentrations of the chemical element and its radioactive label C_{wr0} and C_{hr0} in water and hydrobionts, respectively at $t = 0$. In this case, the stationary value of the concentration factor of a stable element in a hydrobiont $CF_{Sst} = C_{hst0}/C_{wst0}$, while that of a radionuclide is $CF_S = C_{hr0}/C_{wr0}$. Subsequently, as a result of sorption and/or mineral exchange under experimental conditions, equalisation of the specific radioactivity of the tag in water and in hydrobionts continues up to the moment when $C_{hr}/C_{hst} = C_{wr}/C_{wst}$. When stationarity of the mineral exchange system is attained, the ratio of the concentration of the label in the hydrobiont $C_{hr}(t_\infty \to \infty)$ to its concentration in water $C_{wr}(t_\infty \to \infty)$ is equalised both for the radionuclide and for its isotopic carrier $C_{hs}(t_\infty \to \infty)$ and $C_{hr}(t_\infty \to \infty)$, but $CF_S = CF_{Sst}$.

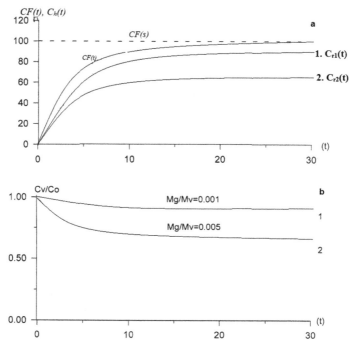

Fig. 3.2 Kinetic dependences of changes in the concentration of $C_{h1}(t)$ and $C_{h2}(t)$ and the concentration factor $CF(t)$ of the radionuclide in the hydrobiont or in inert matter (**a**) and its specific concentration C_v/C_o in water (**b**) at different specific masses of matter ($M_g/M_v = 0.001$ and $M_g/M_v = 0.005$, respectively) in an experimental aquarium

A typical pattern of the transition process, in which the concentrations of the radionuclide in the hydrobiont change in the water of the experimental aquarium, is shown in Fig. 3.2. Figures 3.2a, b show that the dependences of the change in time t of the concentration of the radioactive label in the hydrobiont $C_{hr}(t)$ and in water $C_{wr}(t)$ are largely determined by the specific mass of the hydrobiont in the aquarium. In aquariums where a hydrobiont has a relatively low specific mass (m_{sp1}), the trends in changes in the concentration of the radionuclide in the hydrobiont $C_{hr}(m_{sp1})$ and in water $C_{wr}(m_{sp1})$ will be higher than the corresponding trends in changes in the concentration of the radionuclide in the hydrobiont and in the water of the aquarium having a higher specific mass of the hydrobiont ($m_{sp2} > m_{sp1}$). This is due to the fact that, in the process of mineral exchange, the higher the specific mass of the hydrobiont in the aquarium the greater, the extent to which it extracts the radiolabel from the water.

In Sect. 3.1.2 of this chapter, it is shown that verification of models of the type (3.11) and (3.15) is generally performed according to the results of experimental observations, approximated by exponential dependences of the form (3.2) or (3.4). Meanwhile, the experimentally obtained estimates of B_i and p_i in these equations depend on the value of m_{sp}, which imposes restrictions on the design of radiolabel experiments based on the use of a radioactive tracer (Polikarpov and Bachurin 1970). These limitations are due to the fact that the trend of $CF(t)$ shown in Fig. 3.2 coincides with Eqs. (3.2) and (3.4) at different values of m_{sp1} and m_{sp2} only when $m_{sp} \cdot CF_S \ll 1$. Thus, the observance of the condition $m_{sp} \cdot CF_S \ll 1$ is required both when setting up experiments with a radioactive label and when parametrising models of the form described in (3.11) and (3.15).

The presented analysis indicates that, although the problem of separating n exponential functions (3.2) or (3.4) from experimental data has an exact solution, exponential functions do not possess the property of orthogonality. Since the matrix of normal equations for the approximation of observations by the least squares method is characterised by a large skew, it is necessary to consider numerical aspects of the problem in addition to its purely mathematical solution. Citing a mathematical method for extracting exponential functions, Cornelius Lánczos (Lánczos 1961) pointed out that initial data must be specified with exceptional accuracy in order to determine their coefficients. For example, to isolate four or five exponential terms, an observation accuracy of six to eight significant figures is required (Lánczos 1961). In their study into the accuracy of determining the parameters of the double-exponential function (3.4) depending on the error in the initial data, Glass and Garetta (1967) obtained the results shown in Table 3.1. It can be seen that the accuracy of determining the coefficients of the exponential function depends on the error of observations and their number, as well as on the ratio of their parameters.

In studies using the radioactive tracer method, the variability of the data is both due to the accuracy of sample mass and radiometry measurements, as well as to the biological variability. The data scatter is characterised by the coefficient of variation V_{var}, typically 10–30% (Ivanov et al. 1978). The number of replicates of measurements n_{meas} corresponding to each fixed point in time, depending on the accuracy of determining the analysed value, is estimated by the formula (Urbakh 1964):

Table 3.1 Accuracy of determining the parameters of the double-exponential function depending on the observation error (Glass and Garetta 1967)

Number of observation points	Average relative observation error, %	Accuracy of determining the parameters depending on the ratio p_1/p_2, %		
		4:1	3:1	2:1
11	2	6.34	13.42	36.06
	3	9.68	23.28	22.06
	5	21.66	32.72	52.34
	10	23.18	61.40	–
12	2	–	–	6.34
	10	–	–	36.10

The table includes the mean error values (double dispersion, 2σ) of four parameters B_1, B_2, and p_1, p_2 functions (3.16) depending on the average error of the function values at observation points

$$n_{meas} = \frac{u_\alpha^2}{\epsilon^2} V_{var}^2, \tag{3.23}$$

where ϵ is the limiting relative error;

u_α is a parameter selected from tables based on the significance level α.

From Eq. (3.23), it follows that, even with 10% variability for observations with a 1% error at a significance level of $\alpha = 0.05$ ($u_\alpha = 1.96$),it is necessary to carry out at least 386 replications when measuring radioactive samples of hydrobionts at each fixed point in time. However, it is notusually possible to take so many samples for analysis within a narrow time interval. Consequently, the problem of identifying more than two exponential functions from the data of radiolabel experiments is ill-defined. For this reason, observational data are usually approximated by one or two exponential terms. If the results of observations of the elimination of a radionuclide by a hydrobiont in time fall on the graph on a straight line in a logarithmic scale along the ordinate axes, then one exponential term is sufficient to interpret the data (at $n = 1$). Otherwise, it becomes necessary to determine the parameters of the two exponential terms. In this regard, it is of practical interest to estimate the parameters p_1,p_2 and B_1,B_2, the values of CF_S and the relative speed (V_{ex}) of the intake or exchange of a radionuclide by a hydrobiont, which is calculated by the formula obtained by differentiating expression (3.4):

$$V_{ex} = B_1 \cdot p_1 + B_2 \cdot p_2. \tag{3.24}$$

At present, the graphical method is that most widely used to determine the parameters of exponential functions (Troshin 1957). This method is based on the property of exponential functions having the form $B_{ie}^{-p_i t}$ and tending to zero at $t \to \infty$. This occurs most slowly for the term with the smallest modulo value of the parameter p_i. Starting from a particular moment in time, an approximate equality is valid (Bachurin 1968):

$$CF_S - CF(t) \approx B_{ie}^{-p_i t}. \tag{3.25}$$

The logarithm of expression (3.25) produces a dependence, which at $t \to \infty$ tends to a straight line:

$$\ln(B_{ie}^{-p_i t}) \approx \ln B_i - p_i t. \tag{3.26}$$

By setting this dependence aside on a logarithmic scale along the ordinate axis and drawing an asymptote to it, we can determine the values of B_1 and p_1 from the position of the asymptote to it, since the point of intersection of the asymptote with the ordinate axis has coordinates $(0, \ln B_1)$, while the tangent of its inclination to the abscissa axis is p_1. Eliminating the found term $B_{ie}^{-p_i t}$ from expression (3.25) allows a new function to be obtained comprised of a straight line at $n = 2$, along which the remaining values B_2 and p_2 are determined by similar operations on the asymptote.

The values p_i can also be determined by Eq. (3.25) through periods of half-decay of exponents $\tau_i = 0.693/p_i$, equal to the time interval at which the ordinate of any segment on the asymptote is halved. From expression (3.4) it can be seen that the parameters B_1, B_2 and p_1, p_2 can only be determined by the graphical method if the estimate of the value CF_S is known. For this reason, the duration of experiments intended for mathematical interpretation should be such that the process of concentration can be traced to a level close to a stationary process.

Estimates of V_{ex} are usually established by numerical differentiation of the dependence of the accumulation of the radionuclide by the hydrobiont according to the points of experimental observations. The accuracy of numerical differentiation is determined by the truncation error caused by replacing dependence (3.4) with an interpolation polynomial, as well as by the rounding error due to the difference between the experimentally determined sample $CF(t)$ and their general values (Kopchenova and Maron 1972). These errors are dependent on the time intervals Δt between $CF(t)$ measurements. With their reduction, the truncation error, as a rule, decreases, while the rounding error increases.

3.2 Parenteral Sorption and Mineral Exchange of Living and Inert Components of Ecosystems

In the context of the present work, parenteral absorption is understood as the entire set of sorption and metabolic processes of non-food intake of radioactive and chemical substances into the biotic and abiotic components of ecosystems. This absorption path is associated with the direct acquisition from the aquatic environment through the sorbing surfaces of inert substance or shells or through the integument of marine organisms, as well as with ingested water. The aim of the research was the parametrisation and verification of compartment models in relation to their use to reflect the concentrating and exchange functions of living and inert matter under conditions of changes in biotic and abiotic environmental factors.

3.2.1 Empirical Parametrisation and Verification of Compartment Models

In works devoted to the use of compartment models to describe the mineral metabolism kinetics of aquatic organisms (hydrobionts), the models were verified according to experimental observations, which observations also used to determine the parameters of this verification (Bernhard 1971; Riziĉ 1972). In this regard, it became necessary to develop methodological approaches to determine the unambiguity of their values with respect to changes in the initial conditions of open and closed

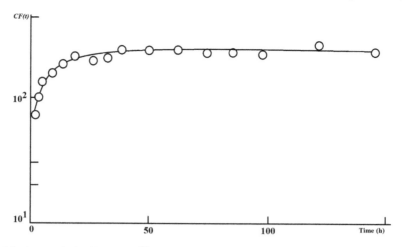

Fig. 3.3 Accumulation kinetics of ^{90}Sr *by Cystoseira barbata*; CF(t) is the concentration factor at the moment of time t (solid lines show the calculated data) (Egorov and Kulebakina 1974)

systems comprised of living or inert matter and a chemical or radioactive component of the marine environment. In subsequent works (Polikarpov and Egorov 1986), the problem of assessing the adequacy of a model (3.15) was solved by comparing empirical data approximated by the relation (3.4) with the results of solutions for implementing a model of differential equations with initial conditions corresponding to those obtained experimentally. In this case, the results of one series of experiments were used to determine the parameters of the model, while an additional series of experiments was carried out in order to assess its adequacy.

In experiments with ^{90}Sr and the Black Sea brown alga *Cystoseira barbata* (Egorov and Kulebakina 1973), an empirical dependence was obtained (Fig. 3.3), whose approximation made it possible to determine the parameters of the model (3.4):

$$B_1 = 93.0; \quad B_2 = 195; \quad p_1 = 1.7340 \ (\text{day}^{-1}); \quad p_2 = 0.0771 (\text{day}^{-1}). \quad (3.27)$$

In another series of experiments, algae were sequentially placed at regular intervals in aquariums containing ^{90}Sr, as well as in water where no radioactivity was present. The kinetics of absorption and excretion of ^{90}Sr by cystose in these experiments is described by the Eq. (3.16). When recalculating the data expressed through the concentrating ability of hydrobionts, the $CF(t)$ values were calculated by the formula:

$$CF(t) = ((C_{10} - C_{w0} \cdot B_1) + (C_{20} - C_{w0} \cdot B_2))/C_{w0}, \quad (3.28)$$

where C_{w0} is the concentration of ^{90}Sr in the water of the aquarium, in which the parameters of Eq. (3.16) were determined according to the experimental data.

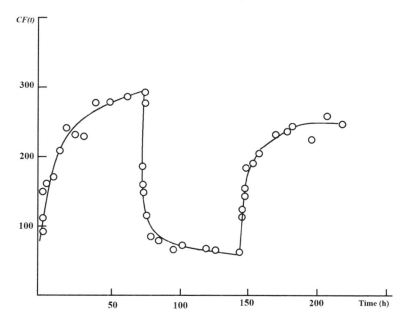

Fig. 3.4 Experimental results (white circle) and calculated data (solid lines) of the model of exchange ^{90}Sr between *Cystoseira barbata* and an aqueous medium; $CF(t)$ is the concentration factor calculated relative to the concentration of the radionuclide in the water of the first aquarium (Egorov and Kulebakina 1974)

The results of observations and theoretical description of radioisotope exchange kinetics of ^{90}Sr between *Cystoseira barbata* and the aquatic environment are illustrated in Fig. 3.4,

In analogous experiments (Fig. 3.5) carried out with Black Sea *Ulva* and ^{65}Zn (Ivanov et al. 1978), the dependence of the form (3.4) was obtainedhaving parameters $B_1 = 740$; $B_2 = 190$ ($CFs = 930$); $p_1 = 0.119$/day; $p_2 = 2.330$/day. The results of the description of the kinetics of absorption and excretion of ^{65}Zn by algae in an experiment carried out under different initial conditions are shown in Fig. 3.5. The comparison of the experimental results and their theoretical description confirmed the hypothesis that the mathematical model (3.15) adequately reflects the kinetic laws of the exchange of radionuclides by macrophytes.

The simulation showed good agreement with the experimental results (Fig. 3.6).

An assessment of the validity of the model (3.15) was also carried out in relation to microorganisms and zooplankton. In experiments with heterotrophic bacteria *Bacterium album* and ^{86}Rb (Parkhomenko and Egorov 1979), it was established that the data on the elimination of the radionuclide fell in time on a straight line on a logarithmic scale along the ordinate (y-)axis (Fig. 3.7b).

The results of experiments with *Bacterium album* attest to the possibility of interpreting the entire set of processes responsible for the exchange of rubidium by bacteria as the accumulation of a radionuclide in one exchangeable resource of a

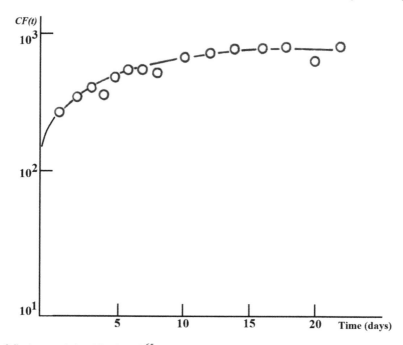

Fig. 3.5 Accumulation kinetics of ^{65}Zn by *Ulva rigida*; $CF(t)$ is the concentration factor at the moment of time t (solid lines show the calculated data) (Ivanov et al. 1978)

hydrobiont; that is, in order to describe the kinetic regularities, it was possible to use a model of the form (3.4) at p_2 and $B_2 = 0$. According to the empirical data on the elimination of ^{86}Rb by bacteria (Fig. 3.7b), after its preliminary accumulation for 60 min, $p_2 = 0.0151$/min was obtained. The value of $B_1 = CF_S = 7.74$ was calculated at $t_1 = 60$ min. and CF(t_1) = 4.8 The adequacy of the model description of the kinetics of uptake of ^{86}Rb by bacteria in the experiment is illustrated in Fig. 3.7a.

Consideration of the distribution information and elimination of ^{65}Zn by a nondividing culture of marine unicellular algae *Stephanopixis palmeriana* in experiments carried out by Z.P. Burlakova and her colleagues (Burlakova et al. 1980) used the formula (3.4) to define the model parameters $B_1 = 300$; $B_2 = 230$ ($CF_S = 530$); $p_1 = 0.5770$/h; $p_2 = 0.0152$/h. The adequacy of the description to the model (3.15) of experimental observations is illustrated in Fig. 3.8,

The adequacy of a model having a double-compartment structure in connection with zooplankton was investigated in experiments with ^{54}Mn and crustaceans *Hyperoche medusarum* and *Megalopa* sp., with ^{65}Zn and *H. medusarum* и *Clausocalanus mastigophorus* (Egorov and Ivanov 1981), as well as with ^{65}Zn and *Euchirella bella* (Piontkovsky et al. 1983). The results of these studies are illustrated by the example of experiments aimed at studying the absorption and excretion of ^{54}Mn by megalopa crab larvae. The description of kinetic regularities of exchange of ^{54}Mn by megalopaaccording to a model having parameters $B_1 = 3.3$; $B_2 = 90.3$ ($CF_S = 93.6$); p_1

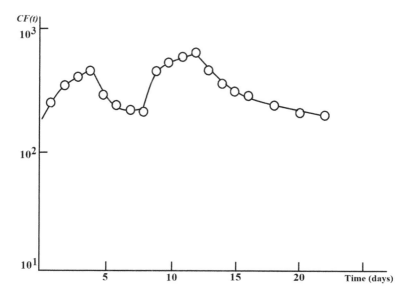

Fig. 3.6 Experimental results (white circle) and calculated data (solid lines) of the model of exchange ^{65}Zn between *Ulva rigida* and an aqueous medium; $CF(t)$ is the concentration factor calculated relative to the concentration of the radionuclide in the water of the first aquarium (Ivanov et al. 1978)

$= 0.1390/h$; $p_2 = 0.0021/h$, which isshown in Fig. 3.9, supported the hypothesis of the adequacy of the model in relation to zooplankton.

The experiments considered above were set up according to the technique described in the previous chapter, in which independent experiments are carried out in order to determine the parameters of the model and establish its adequacy. As such, they all adhered to the following conditions:

(a)   the physiological state of hydrobionts was maintained unchanged;

(b)   during the observation period, no significant generative or somatic growth of the studied organisms took place;

(c)   the temperature, salinity, pH and illumination of the environment were maintained in intervals corresponding to the natural conditions of the hydrobionts;

(d)   radionuclides were introduced into the experimental solutions in quantities that did not distort the salt composition of the waters, with their physicochemical forms corresponding to the forms of their stable analogues in water.

The correspondence of the differential compartment model (3.15) to the kinetic regularities of the exchange of minerals by hydrobionts under conditions in which biotic and abiotic components remain constant indicates that, during parenteral absorption of radioactive substances, the closed system comprised of marine environment/hydrobionts is self-identical at any microconcentrations of radionuclides and water. These data allow us to conclude that, regardless of the biological significance of mineral elements, the kinetics of the exchange of their radionuclides by

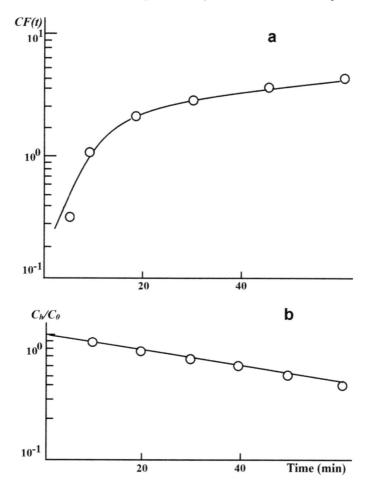

Fig. 3.7 Kinetics of accumulation (**a**) and excretion (**b**) ^{86}Rb *Bacterium album*; $CF(t)$ is the concentration factor at the moment of time t; C_h/C_0 is the ratio of radioactivity in the hydrobiont at the current and initial moments of time (solid lines show the calculated data) (Parkhomenko and Egorov 1979)

hydrobionts of different trophic levels can be interpreted in terms of a process of concentration of elements in one or two hydrobionts exchanged with rates of first-order metabolic reactions. Within the variation interval of concentrations of radionuclides in the aquatic environment, the relative volumes of reserves (B_i) and the rates of their exchange (p_i) remain unchanged.

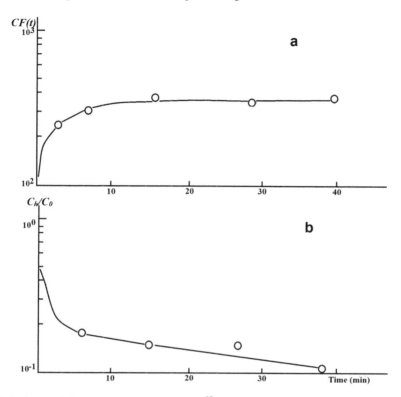

Fig. 3.8 Accumulation (**a**) and excretion (**b**) of ^{65}Zn by *Stephanopixis palmeriana*. Solid lines correspond to the calculated model data; $CF(t)$ is the concentration factor at time t; C_h/C_0 is the ratio of radioactivity in the hydrobiont at the current and initial moments in time (Burlakova et al. 1980)

3.2.2 Radioactive Decay

The mathematical models of the closed system comprised of radionuclides in a marine environment and hydrobionts, considered above along with methods for determining their parameters, tend to assume that the influence of radioactive decay on the time scale of the processes under study is negligible. In actuality, some of the radioactive isotopes used in the described experiments have a short half-life comparable to the duration of experimental observations. For example, the half-lives of isotopes of such biologically significant chemical substances as ^{32}P and ^{42}K are 14.3 days and 12.36 h, respectively. The half-life of iodine isotopes, comprising environmentally hazardous short-lived products of nuclear explosions, ranges from 54 min for ^{134}I to 8.06 days for ^{131}I (Artsimovich 1963). The decay of the radioactive isotopes used in experiments can manifest itself both independently and against the background of the decay of daughter atoms in parent-daughter radionuclide chains in the medium. In both cases, the radionuclide–hydrobiont system in the marine environment becomes

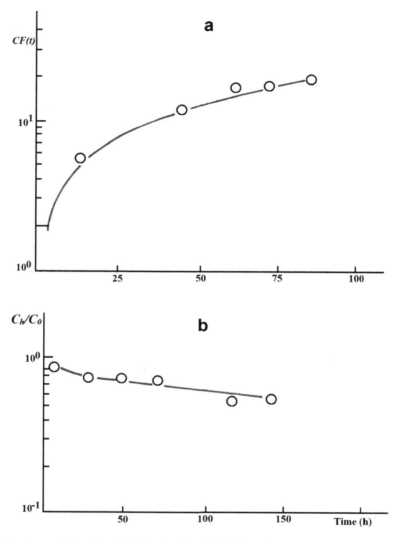

Fig. 3.9 Accumulation (**a**) and elimination (**b**) of ^{54}Mn by crab megalopa (individuals weighing 5.5 mg). Solid lines correspond to the calculated model data; $CF(t)$ is the concentration factor at time t; C_h/C_0 is the ratio of radioactivity in the hydrobiont at the current and initial moments in time (Egorov and Ivanov 1981)

open in terms of elements. In the first case, radioactive decay leads to a decrease in the amount of radionuclide in the hydrobiont and aqueous medium. In the second case, simultaneously with the decay of the isotope, "hot" daughter atoms are produced both in the aqueous medium and in the hydrobiont, along with their subsequent decay.

The problem of studying the effect of radioactive decay on the kinetic characteristics of the exchange of radionuclides by hydrobionts was solved theoretically using compartment models (Egorov and Zesenko 1977) under the assumption that radioactive decay does not affect the biotic characteristics of the system, while daughter atoms, which result from the decay of parent isotopes, are involved in the metabolism of hydrobionts as well as their stable analogues. The kinetic equations for the exchange of the parent and daughter radionuclides by the hydrobiont are presented on the model with one exchange resource pool for each radionuclide, taking both radioactive decay and the "birth" of daughter atoms into account:

$$\frac{dC_{hP}}{dt} = p_P(C_{wP}CF_{SP} - C_{hP}) - C_{hP}\beta_P;$$

$$\frac{dC_{wP}}{dt} = p_P m_{sp}(C_{hP} - C_{wP}CF_{SP}) - C_{wP}\beta_P;$$

$$\frac{dC_{hD}}{dt} = p_D(C_{wD}CF_{SD} - C_{hD}) + \beta_D(C_{hP} - C_{hD});$$

$$\frac{dC_{wD}}{dt} = p_D m_{sp}(C_{hD} - C_{wD}CF_{SD}) + \beta_D(C_{wP} - C_{wD}), \qquad (3.29)$$

where p_P, p_D and CF_{SP}, CF_{SD} are the exchange rate constants and static concentration factors of the parent and daughter radionuclides respectively;

β_P and β_D are the constant rates of decay of the parent and daughter isotopes;

m_{sp} is the specific biomass of hydrobionts in the environment;

$C_{wP}, C_{wD}, C_{hP}, C_{hD}$ are the concentrations of parent and daughter radionuclides in the water and hydrobiont respectively.

The first two equations of expressions (3.29) reflect the exchange kinetics of the parent radionuclide in the system, while the third and fourth represent those of the daughter radionuclide.

With a negligible value of m_{sp}, the solution to the first equation of system (3.29) has the form:

$$C_{hP}(t) = \left[C_{wP0}CF_{SP} + (C_{hP0} - C_{wP0}CF_{SP})e^{-p_P t}\right]e^{-\beta_P t}, \qquad (3.30)$$

while the second has the form:

$$C_{wP}(t) = C_{wP0}e^{-\beta_P t}, \qquad (3.31)$$

where C_{wP0} and C_{hP0} are the initial values of the concentrations of the parent radionuclide in the water and hydrobiont.

It can be seen from relations (3.30) and (3.31) that, when determining the parameters of the model based on the results of experiments on removing a radionuclide by a hydrobiont, approximated by dependence (3.2), it is necessary to make a correction for decay. This is typically achieved by performing simultaneous radiometry on the samples at the end of the experiment. When determining the parameters of the model from a curve of the form (3.4), the ratio of the polynomials (3.30) and (3.31) is used, following which the term $e^{-\beta_P t}$ is reduced. Consequently, no correction for radioactive decay is required when determining the parameters of model (3.4) from the results of experiments with short-lived isotopes.

From the third and fourth equations of relations (3.29), it can be seen that for a small m_{sp}, the state of the parent–daughter radionuclide system in the aquarium, the hydrobiont becomes stationary when $C_{wP} = C_{wD}$, $C_{hP} = CF_{SP} \cdot C_{wP}$ and $C_{hD} = CF'_{D} \cdot C_{wD}$, where CF'_{D} is the concentration factor of the daughter radionuclide in the environment with the parent isotope. Substituting these values into the third equation of system (3.29), equating its right-hand side to zero, we get:

$$CF'_D = \frac{p_D CF_D + \beta_D CF_P}{p_D + \beta_D}. \tag{3.32}$$

Formula (3.32) shows that $CF'_D = CF_D$ only for $CF_D = CF_P$. When $CF_P = CF_D$, $CF'_D > CF_D$, and vice versa. A similar conclusion was formulated when each radionuclide in a hydrobiont was associated with two compartments (Egorov and Zesenko 1977).

Thus, the data obtained on the mathematical model show that the level of accumulation of the parent isotope significantly distorts the level of accumulation of the daughter in the system of parent–daughter radionuclides in the larger marine environment–hydrobiont system. This indicates that the fate of different isotopes of the same chemical element, which consist of daughter atoms of various parent radionuclides, may also differ in the marine environment.

3.2.3 Dimensional Characteristics of Hydrobionts

In Sect. 2.6, it is noted that the kinetic regularities of the mineral metabolism performed by marine organisms can be largely determined by the size of their individuals and thus depend on the surface-mass ratio. According to our data (Egorov 1978), additionally supported by the results of joint research carried out with V. N. Ivanov, T. G. Usenko and N. A.Filippov (Egorov et al. 1980), obtained from experiments carried out with oceanic zooplankton and isotopes ^{54}Mn, ^{60}Co, ^{65}Zn, it was found that the indicator V_{rel} of the speed of absorption of radionuclides decreased with an increase in the mass of individual hydrobionts m_h (Table 3.2). These data shown on the graph on a logarithmic scale along the coordinate axes fit satisfactorily on a straight line, testifying to the power-law form of the relationship between V_{rel} and m_h. In experiments carried out with ^{65}Zn (Egorov and Ivanov 1981) the ratio between V_{rel} and m_h the description of power function was performed independently from species composition (Fig. 3.10):

$$V_{rel} = a \cdot m_h^{b} \tag{3.33}$$

At $a = 25.06$ and $b = -0.548$.

Species	Radionuclide	Mass, mg	Absorption rate index, per hour
Calanus gracilis	^{65}Zn	1.17	21.50
Clausocalanus mastigophorus	^{54}Mn	0.15	2.72
	^{60}Co	0.15	3.00
	^{65}Zn	0.15	41.30
Undinula vulgaris	^{65}Zn	0.82	19.30
Euchaeta marina	^{65}Zn	0.45	34.60
Pontella sp.	^{54}Mn	2.50	0.70
	^{65}Zn	2.50	35.40
Octracoda sp.	^{54}Mn	0.20	1.50
	^{65}Zn	0.12	127.50
Idotea metallica	^{65}Zn	127.00	1.80
	^{65}Zn	110.00	6.65
	^{65}Zn	42.00	8.50
	^{65}Zn	36.00	4.10
	^{65}Zn	25.00	6.65
	^{65}Zn	7.00	9.80
Platyscerus serratulus	^{65}Zn	6.00	6.50
	^{65}Zn	4.00	15.60
Simorhynehotus antennarius	^{54}Mn	7.00	0.60
	^{65}Zn	7.00	7.30
	^{65}Zn	4.00	11.00
Euphausiidae	^{65}Zn	5.00	27.00
Metapeneus sp.	^{65}Zn	13.00	5.50
Megalopa decapoda	^{54}Mn	5.50	0.30
	^{65}Zn	5.50	7.30
Miracia efferata	^{54}Mn	0.11	5.23
	^{60}Co	0.11	6.67
	^{65}Zn	0.11	29.00
Hyperoche medusarum	^{54}Mn	70.00	0.40
	^{65}Zn	70.00	1.80

Table 3.2 Absorption rates of ^{54}Mn, ^{60}Co, ^{65}Zn by oceanic zooplankton

From the point of view of compartment modeling, the observations corresponded to an integral dependence on the form (3.4). Differentiating it in time and accepting $t = 0$, we obtain the rate of absorption of the radionuclide by the hydrobiont as:

$$V_{rel} = B_1 p_1 + B_2 p_2. \tag{3.34}$$

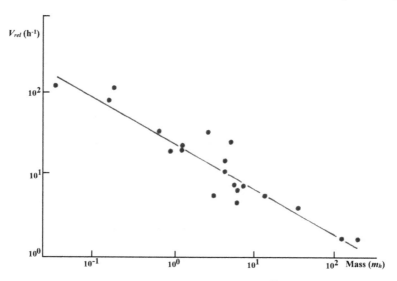

Fig. 3.10 Changes in the speed V_{rel} (per hour) of absorption of ^{65}Zn depending on the change in the mass of individual (m_h) crustaceans (Egorov and Ivanov 1981)

From (3.33) and (3.34) it follows that:

$$B_1 p_1 + B_2 p_2 = a m_h^b. \tag{3.35}$$

Estimates of the absorption rate indices of ^{54}Mn, ^{60}Co and ^{65}Zn by various size groups of oceanic zooplankton are given in Table 3.2.

The change in the rate of absorption of ^{65}Zn depending on a change in the mass of individual (m_h) crustaceans is portrayed graphically in Fig. 9 (3.10).

In Eq. (3.35) the parameters B_1 and B_2 ($B_1 + B_2 = CF_S$) characterise the relative value of exchange resource pools, while p_1 and p_2 are indicators of the rate of their exchange. From the fact of the independence of the parameters B_i and p_i from the value m_h, it follows that dimensional characteristics can simultaneously affect the indicators of both the accumulation and excretion of radionuclides by the hydrobiont.

The dependence of the estimates of the stationary concentration factor (CF_S of crustaceans, made according to the data of short-term experiments, is shown in Fig. 3.11, from which it is seen that the ratio between CF_S and m_h is satisfactorily described by a power function of the form (3.33) at $a = 567.54; b = -567.54$. These data indicate that the volumes of exchangeable resource pools are in a power-law relationship with the masses of individual hydrobionts. The close correspondence of the value of the parameter b, obtained from the results of observations illustrated in Figs. 3.10 and 3.11, indicates that it is parameters B_1 and B_2, and not p_1 and p_2, that make the main contribution to the dependence of the rate of absorption of the radionuclide on the mass of the hydrobiont.

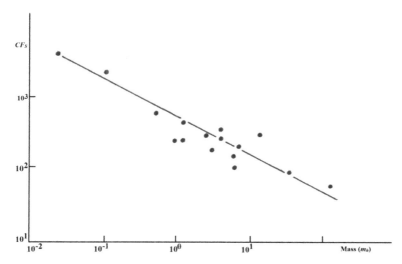

Fig. 3.11 Relationship between the estimates of static coefficients (CF_S) of accumulation of ^{65}Zn and mass (m_h) by crustaceans (Egorov et al. 1975)

The dependence of the change in the parameters p_1 and p_2 on the mass of individuals was studied using the example of ^{65}Zn and sea lice (Egorov and Ivanov 1981). Experiments have shown that *Idotea baltica* individuals of lower mass accumulated higher levels of ^{65}Zn (Fig. 3.12). Transplanted (after three days of preliminary experience of radionuclide accumulation) into water without radioactivity, sea lice of different size groups excreted zinc in such a way that the kinetic regularities, expressed in relative units (C_h/C_0), did not vary (Fig. 3.13). The kinetic regularities were also observed not to be dependent on temperatures ranging from $+ 12$ to $+ 19\ °C$.

Elimination of ^{65}Zn by sea lice was studied under initial conditions

$$C_w(0) = 0; \quad C_{hi}(0) = C_{hi0}.$$

Under the assumption that the model is adequate, the kinetics of the process is described by the dependence (3.16), which is used to determine the parameter p_1 and p_2 from observations. At different m_h, the kinetics of radionuclide elimination does not change (Fig. 3.13), which also indicates that p_i are unchanged as functions of m_h.

Thus, the experiments have shown that when modelling the kinetic regularities of the exchange of radionuclides by marine organisms having different individual masses, the size of the exchange resource pool compartments depends on the power-law form on the individual mass of hydrobionts, while the rate of exchange of an element in the compartments does not depend on the size characteristics of the specimens of this taxonomic group.

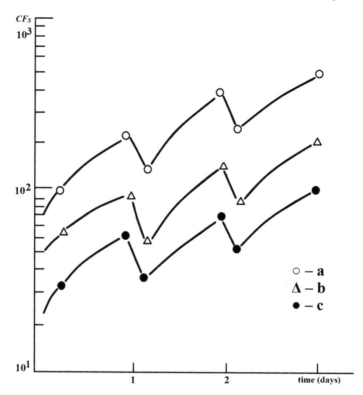

Fig. 3.12 Accumulation kinetics of ^{65}Zn *Idotea baltica* in weight indices of size groups 10–13 (**a**), 43–60 (**b**) and 120–145 (**c**) mg (areas of decrease in concentration factors correspond to periods of feeding in environments without radioactivity); $CF(t)$ is the concentration factor calculated relative to the concentration of the radionuclide in the water of the first aquarium (Egorov and Ivanov 1981)

3.2.4 Growth and Production Processes of Hydrobionts

In experiments with ^{32}P and Black Sea *Ulva*, which were carried out by us in collaboration with G. V. Barinov and reproduced with N. N. Tereshchenko (Tereshchenko and Egorov 1983, 1985), it was established that the thalli of algae concentrate phosphorus more intensively when growing. Experiments with *Ulva* and ^{65}Zn, carried out under conditions of different illumination of the environment (Ivanov et al. 1978), showed that algae grew with the highest intensity at maximum illumination. At the same time, the accumulation of ^{65}Zn reached its maximum values (curve 1 in Fig. 3.14a).

Under conditions of insufficient illumination, the algae had a satisfactory physiological state, but did not grow (curve 3 in Fig. 3.14b), accumulating radioactive zinc weakly (curve 3 in Fig. 3.14a). Algae that stopped growing under conditions of higher illumination (curve 2 in Fig. 3.14b) also stopped concentrating ^{65}Zn (curve 2 in Fig. 3.14a). The data presented in Fig. 3.14 indicate that the increase in the

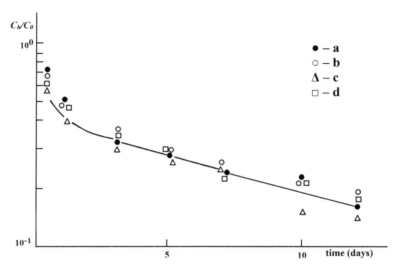

Fig. 3.13 Elimination kinetics of ^{65}Zn *by Idotea baltica* in weight parameters of size groups 10–13 (**a**), 43–60 (**b**) and 120–145 (**c**) mg at + 19 °C and size group 43–60 (**d**) mg at + 12 °C; C_h/C_0 is the ratio of radioactivity in the hydrobiont at the current and initial moments of time, respectively (Egorov and Ivanov 1981)

concentration levels of ^{65}Zn was not caused by a direct increase in the illuminance of the environment, but rather was attributable to an increase in illuminance by the growth of algal thallus samples. After 8 days of accumulation of ^{65}Zn, part of the algae from aquariums with fast (curve 1 in Fig. 3.14b) and insignificant (curve 3 in Fig. 3.14b) thallus growth was transferred to aquariums with non-radioactive water.

The results of experiments on the elimination of ^{65}Zn by samples of *Ulva* are shown in Fig. 3.14b. Here it can be seen that, in algae that accumulated ^{65}Zn under conditions of intensive growth, 70% of radioactivity remained after 6 days (dependence 3 in Fig. 3.14b); in algal thallus samples, growing weakly during the period of radiozinc accumulation, 20% of radioactivity remained after 6 days (dependence 4 in Fig. 3.14b). Thus, the greater the increase in the biomass of algal samples in the preliminary ^{65}Zn accumulation experiment, the less radioactivity was subsequently released into clean water.

Figure 3.15a illustrates the decrease in absolute radioactivity by intensively growing samples of Ulva thallus in the radiozinc of removal experiment, as well as the relative decrease in its radioactivity, expressed in relation to the mass of the samples (curve 2 in Fig. 3.15a). It can be seen from the figure that, if the kinetics of excretion is measured in relative units (C_h/C_0), then there is an "apparent" elimination due to the growth of the hydrobiont samples.

From the standpoint of the applicability of compartment theory, in order for the model to reflect the noted regularities, it was necessary to make the assumption that in hydrobionts, in addition to exchangeable pools, there are also non-exchangeable pools of chemical resources that are fillable only in the process of growth. This

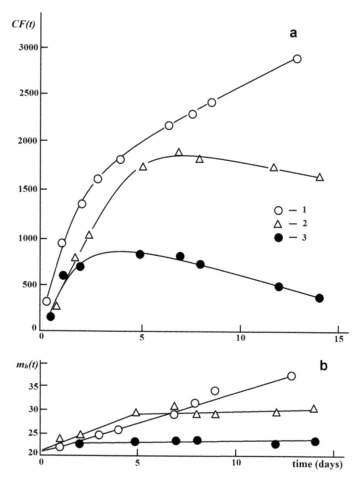

Fig. 3.14 Accumulation of [65]Zn (**a**) and change in the average mass (**b**) of samples of *Ulva rigida* alga at illumination intensities of 2600–4000 lx (1, 2) and 440 lx (3); $CF(t)$ is the concentration factor at time t (Ivanov et al. 1978)

latter category comprises those chemical resource pools whose exchange during the observation period is negligible. From a physiological point of view, the assumption of the presence of resource pools that are filled only in the process of growth is justified, since they can reflect the kinetics of the entry of mineral elements into the main structures of the newly created organic matter.

   The two-compartment structure of the exchange of radioactive and chemical substances by aquatic organisms, supplemented by the compartment of the non-exchangeable chemical resource pool, is described by the equation (Ivanov et al. 1978):

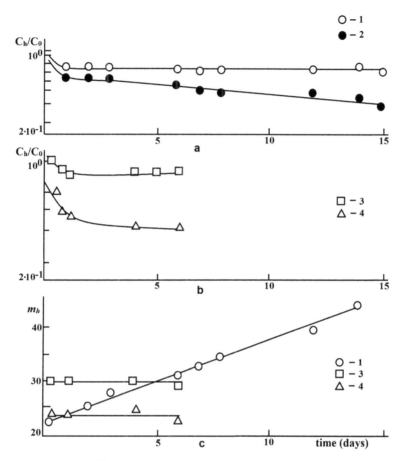

Fig. 3.15 Elimination of ^{65}Zn (**a, b**) and change in the average mass of alga samples *Ulva rigida* (**c**) at an illumination intensity of 4000 lx; 1, 2—after daily exposure in a solution with ^{65}Zn at an illumination intensity of 4000 lx (1—radioactivity is related to the mass of samples at the initial moment of elimination of ^{65}Zn; 2—radioactivity is related to the mass of samples achieved by the time of radioactivity measurement); 3—after an 8-day exposure in an environment with ^{65}Zn at an illumination intensity of 4000 lx and an increase in the mass of the samples from 21 to 29 mg; 4—after an 8-day exposure in a radioactive solution in the dark; C_h/C_0 is the ratio of radioactivity in the aquatic organism at the current and initial time (Ivanov et al. 1978)

$$\frac{dC_h}{dt} = C_w \left(\sum_{i=1}^{2} B_i p_i + r_{ne} \right) - \sum_{i=1}^{2} C_i p_i, \qquad (3.36)$$

where r_{ne} is an indicator of the rate of entry of a radionuclide into the non-exchangeable resources of a hydrobiont.

In order to estimate the parameter r_{ne}, it was hypothesised that the non-exchangeable pool is only filled during periods of a hydrobiont's growth. In this case, the rate of pollutant intake is calculated by the formula:

$$r_{ne} = \frac{CF_{ne}}{m_h} \frac{dm_h}{dt}, \tag{3.37}$$

where CF_{ne} is the stationary concentration factor of an element in a hydrobiont's non-exchangeable chemical resource pool ($CF_{ne} = C_{ne}/CF_w$). Here C_{ne} is the stationary concentration of the element in the non-exchangeable pool.

If the physicochemical forms of radioactive and chemical substances are identical, then, obviously, Eq. (3.36) is applicable to describe the kinetics of absorption by a hydrobiont of both a chemical and its radioactive analogue. The accumulation of a radionuclide by a hydrobiont under initial conditions $C_{10} = C_{20} = 0$; $C_{30} = 0$ and $C_w(0) = C_w0$; $m_h(0) = m_{h0}$ is described by the expression:

$$CF(t) = CF_{ex} - B_1 e^{-p_1 t} - B_2 e^{-p_2 t} + CF_{ex}(m_h - m_{h0})/m_h. \tag{3.38}$$

From relation (3.38) it follows that the concentration factor of the radionuclide $CF(t)$, as determined from the results of observations, is always higher, the greater the increase in the biomass of the hydrobiont. In the limit, with an increase in the ratio $(m_h - m_{h0})/m_h$, the value of $CF(t)$ tends to the value $CF_S = CF_{ex} + CF_{ne}$, which is equal to the concentration factor of a stable element by a hydrobiont at $m_{h0} = 0$. If the hydrobiont does not grow, then, with respect to the radionuclide, Eq. (3.38) is transformed into function (3.4).

Following exposure of the hydrobiont's concentration of a radionuclide to a stationary state in the preliminary experiment, the kinetics of changes in its concentration in a medium without radioactivity is described by the expression:

$$C_h(t)/C_0 = B_1 e^{-p_1 t} + B_2 e^{-p_2 t} + CF_{ne}(m_2 - m_1)/m_h, \tag{3.39}$$

where m_1 and m_2 are the initial and final masses of a hyrobiont in a preliminary experiment carried out in a medium with a radioactive label.

From the relation (3.39) it can be seen that, all other things being equal, the greater the increase in mass ($\Delta m = m_2 - m_1$) of the accumulation of the radioactive label in the preliminary experiment, the less the radionuclide is eliminated from the aquatic organism. Under the experimental conditions of removing a radionuclide in a non-radioactive environment, if the biomass m_h increases, then there is an "apparent" removal of the radionuclide by the hydrobiont. Thus, the development of compartment theory by supplementing the structure of the model with transport communication r_{nb} allowed the kinetic regularities of mineral metabolism to be captured in those cases when an increase in the intensity of accumulation of elements is observed along with a relative decrease in the degree of their elimination when the growth rate of the hydrobiont increases.

Along with the data indicating an increase in the concentrating ability of hydrobionts in relation to radionuclides during production processes, the literature contains materials illustrating a decrease in the degree of concentration of radionuclides in the process of cell division of marine organisms (Davies 1973; Tsytsugina and Lazorenko 1983). In the empirical work carried out by Z. P. Burlakova and her colleagues (Burlakova et al. 1979), it was found that under experimental conditions ^{65}Zn is accumulated by *Stephanopixis palmeriana* diatoms in the dark to higher levels than in the light (Fig. 3.16); moreover, this is primarily due to the intensity of the process of cell division.

In order to take account of the noted effect on the model, the problem of concentrating a radionuclide in one exchange pool of a growing hydrobiont was considered. In this case, the change in the content of the radionuclide in the hydrobiont (A_h) depending on its content in water (A_w), according to (3.11), is described by the equation:

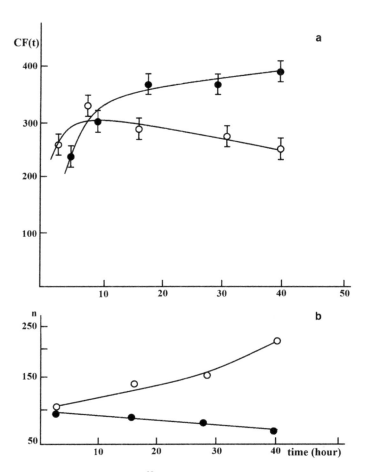

Fig. 3.16 Kinetics of accumulation of ^{65}Zn by *Stephanopixis palmeriana* (**a**) due to their cell division (**b**); $CF(t)$ is the concentration factor at the moment of time t; n is the density of the culture of unicellular algae, cells/ml (Burlakova et al. 1980)

$$\frac{dA_h}{dt} = r_{12}A_w - r_{21}A_h. \tag{3.40}$$

Substituting the values $A_h = m_h C_h A_w = m_w C_w$ and $r_{12} = m_h p C F_S / m_w$; $r_{21} = p$ and differentiating by the variables m_h and C_h, we get:

$$\frac{dC_h}{dt} = C_w C F_S p - \left(p + \frac{1}{m_h} \frac{dm_h}{dt} \right) C_h. \tag{3.41}$$

From Eq. (3.41) it can be seen that the lower the rate of concentration of a radioactive or chemical substance, the higher specific production rate of the hydrobiont $\frac{1}{m_h} \frac{dm_h}{dt}$. At the same time, due to the state of dynamic vacancy of the newly formed exchange pool of an element during the growth of a hydrobiont, both the concentration of this element in the hydrobiont and the concentration factor $CF(t)$ can decrease. The termination of the growth processes $\left(\frac{dm_h}{dt} = 0 \right)$ leads to a relative filling of the exchange pool, while the concentration factor $CF(t)$ tends to its stationary value CF_S.

Thus, the development of compartment theory based on the assumption of new exchangeable elemental resource pools arising during the growth of a hydrobiont helped to account for the decrease in the concentration and concentration factors of radionuclides by hydrobionts during the production of organic matter observed in radiolabel experiments. It was found that the kinetics of the concentration and exchange of radioactive and chemical elements by hydrobionts during the production of organic matter can be described by a model, two compartments of which correspond to the accumulations of an element in hydrobionts during growth, exchanging indicators of the rates of first-order metabolic reactions with the environment. The third compartment reflects the intake of the element into the basic, non-exchangeable or weakly exchanging structures only during the production of organic matter by the hydrobiont.

3.2.5 Specific Biomass and Population Density

The study of the dependences of the influence of the specific biomass of hydrobionts in the aquatic environment on the characteristics of their sorption and mineral metabolism in radiolabel experiments is associated with the risk of experimental conditions occurring in which the limits of applicability of radio indicator methods may be violated ($m_{sp} \cdot CF_S \ll 1$). A failure to take this circumstance into account led to the creation of a "theory of population density", which, however, was not confirmed by the mineral metabolism factor (Khailov and Popov 1983). For this reason, the modelling of such processes requires verification of the parameters, taking into account both the methods of interpreting the results of radio-indicator observations, as well as analytical methods and physiological approaches.

For the purposes of mathematical description of the kinetics of these processes, the characteristics of the absorption and excretion of radionuclides by hydrobionts in connection with a change in their specific biomass were studied under experimental conditions. Studies into the regularities of exchange of ^{32}P by unicellular algae carried out jointly with A. Y. Zesenko, A. V. Parkhomenko and Z. Z. Finenko (Egorov et al. 1982) showed that the greater the change in the specific radioactivity of water in aquariums, the higher the specific biomass of the hydrobiont in the environment. This was due to the processes of leveling the specific radioactivity of phosphorus taking place as a result of metabolism in the hydrobiont and the medium. Since the specific radioactivity of the hydrobiont was zero at the initial moment of time when its stationary level was established, the greater the specific biomass of the hydrobiont in the medium, the more the radionuclide was removed from the water. At the same time, since the amount of the mineral element absorbed from the water was compensated by its excretion from the hydrobiont as a result of metabolism, the concentration of the isotope carrier in the water remained constant. As a result of the processes noted above, an "apparent" decrease in the rate of absorption of the mineral element was observed, while the estimates of V_{ex}, obtained by Eq. (3.5) from the results of experiments with radioactive indicators, with other factors being equal, were more underestimated the higher the specific biomass of the hydrobiont in the environment. According to our measurements, the "apparent" decrease in the rate of absorption of mineral phosphorus, determined by the uptake of ^{32}P by unicellular algae, could differ three times from its true value with an order-of-magnitude change in their stocking density in an aquarium (Egorov et al. 1982).

A study of the absorption kinetics of ^{137}Cs by *Bacteria album* bacteria (Parkhomenko and Egorov 1979) showed that the accumulation rates of this radionuclide decreased with increased bacterial stocking density in experimental aquariums (Fig. 3.17a).

It can be observed that the time for the accumulation curves to reach a plateau was the same in each case. Bacteria placed in an environment without radiocaesium excreted it according to the same kinetic curves (Fig. 3.17b). Data presented in Table 3.3 reflecting the kinetics of exchange of ^{137}Cs by bacteria, which were obtained from a model of the form described in (3.15), gave satisfactory agreement between parameter values derived from theoretical calculations and observations.

Table 3.3 depicts a mathematical model used to describe the kinetics of metabolic processes in bacteria in relation to caesium, in which the elimination rate indicators (p_1 and p_2) did not depend on this factor. According to our estimates, the values of CF_S and the density of bacteria from the above table fell on the straight line plotted on the graph in double logarithmic coordinates. This indicated that the relationship between the stationary concentration factors and the specific biomass of an hydrobiont in the medium can be described by a power function, comprising a typical manifestation of a violation of the conditions for setting up radio indicator experiments $m_{sp} \cdot CF_S \ll 1$ (Egorov 1975). For this reason, when parametrising models (3.11) and (3.15), the relationship between m_{sp} and CF_S should be refined using other analytical methods.

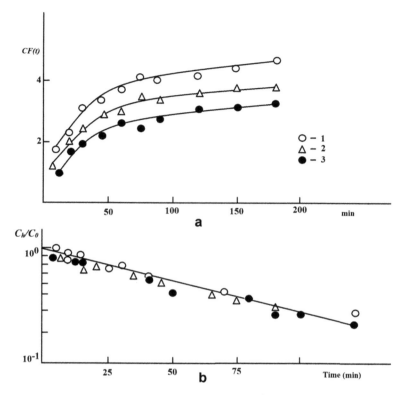

Fig. 3.17 Kinetics of accumulation (**a**) and excretion (**b**) of ^{137}Cs by *Bacteria album* at a specific biomass of bacteria 0.033 (1); 0.066 (2) and 0.133 (3) mg/mL; $CF(t)$ is the accumulation rate at time t (Parkhomenko and Egorov 1979)

Table 3.3 Parameters of concentration and exchange of ^{137}Cs by *Bacteria album*

Specific biomass of bacteria, mg/mL	Relative rate of exchange (V_{rel})	B_1	p_1, min^{-1}	B_2	p_2, min^{-1}	CF_S
0.033	0.306	2.22	0.126	2.68	0.00977	4.90
0.066	0.264	1.93	0.126	2.14	0.00977	4.07
0.133	0.210	1.61	0.126	1.93	0.00977	3.49

3.2.6 Concentration of Chemical Elements in the Aquatic Environment

One of the main problems of model verification consists in their parametrisation under conditions of changes in the concentration of chemical elements and their radionuclides in the aquatic environment. Here, one of the verification tasks is associated with the determination of the stationary values of the concentrating function of

living and inert matter, estimated by the Freundlich and Langmuir equations, while
the second involves the determination of the dynamic characteristics of concentration
in accordance with the Michaelis–Menten equation.

Stationary characteristics of the concentrating function

Figure 3.18 presents the results of long-term observations of changes in the concentration of lead in water (a) and in the surface layer (b) of bottom sediments in the
central part of the Sea of Azov. From Fig. 3.18c, it can be seen that the results are
consistent with a hyperbolic function. Their representation on a logarithmic scale
along the ordinate axes (Fig. 3.18d) demonstrate that this pattern is described with
the coefficient of determination $R^2 = 0.529$ by an equation of the form:

$$C_h = 15.31 \cdot C_w^{-0.299}. \tag{3.42}$$

By representing this series of observations in the form of a dependence of the
change in the concentration factor of bottom sediments on changes in its concentration in water, it is shown that, on a linear scale, this dependence corresponds to a
hyperbola having a smaller scatter of data (Fig. 3.18e), while on a logarithmic scale
(Fig. 3.18f) it is described at $R^2 = 0.955$ by the Langmuir equation:

$$CF = 15.3 \cdot C_w^{-1.299}. \tag{3.43}$$

On the whole, these studies indicated a higher adequacy of data reflection by the
theoretical Freundlich dependence, written in the form of Eq. (3.43). Previously, this
method of approximating observational data was applied in the work of Matishov
et al. (2017). It makes sense to refer to an equation of the form (3.43) as a modified
Freundlich equation.

The results of the determination of the concentration of the organochlorine
compound Arochlor 1254 in the bottom sediments of the Black Sea at different
concentrations in the water (C_w) are shown in Fig. 3.19a. It shows a trend of saturation of bottom sediments with Arochlor 1254 with an increase in the concentration of
this organochlorine compound to values of more than 80 ng/g. Calculations showed
that the concentration factors of Arochlore 1254 by bottom sediments (CF) at low
values of C_w exceeded the levels of 5×10^4 units; at $C_w > 70$ µg /L they decreased
by three orders of magnitude (Fig. 3.19b), while the graphical relationship between
CF and C_w took the form of a hyperbolic function. Its approximation by the modified
Lineweaver–Burk equation showed that the relationship between the concentration
factors of Aroclor 1254 in bottom sediments of the Black Sea and its concentration
in water can be described by a Langmuir equation having the form:

$$C_{BS} = 340 \cdot C_w/(1/k_L + C_w) \tag{3.44}$$

at $k_L = 0.5$ µg/L.

In this case, the limiting sorption capacity of bottom sediments in relation to
Aroclor 1254 was $C_{max} = 340$ µg/kg.

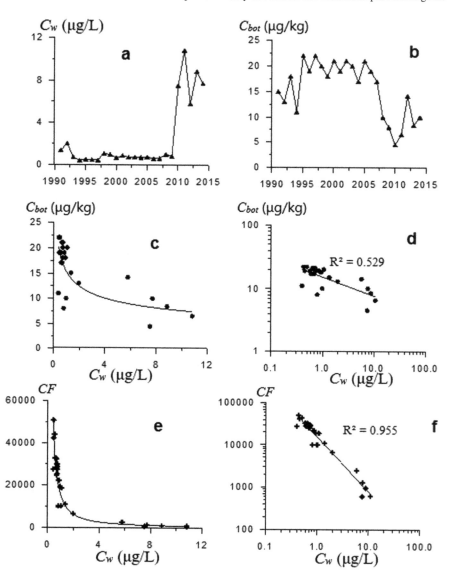

Fig. 3.18 Characteristics of changes in lead content in water and in the surface layer of bottom sediments in the open part of the Sea of Azov (Matishov et al. 2017). Concentration in water (**a**) and in bottom sediments (**b**) in the period 1990–2015. Dependence of the change in the concentration of lead in bottom sediments on the change in concentration in water on a linear (**c**) and on a logarithmic scale (**d**). Dependence of the change in the concentration factors of lead in bottom sediments on the change in concentration in water on a linear (**e**) and on a logarithmic scale (**f**)

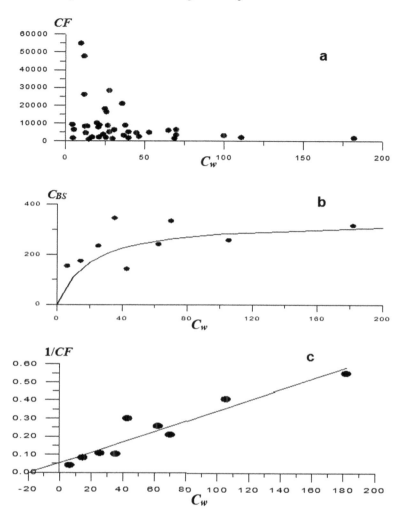

Fig. 3.19 Dynamic characteristics of the concentration of Aroclor 1254 in bottom sediments of the Black Sea: change in the concentration factors CF (**a**); concentration of Aroclor 1254 in the surface layer of bottom sediments (**b**); dependence of the change in quantity $1/CF$ in the coordinates of the modified Lineweaver–Burk equation (**c**) from concentration (C_{BS}) of Aroclor 1254 in the water (C_w) (Malakhova et al. 2019)

Studies have shown that the response of living and inert matter to water pollution in the area of different ranges of their concentrations can take the form of both Langmuir's and Freundlich's equations. Figure 3.20 shows the results of long-term observations of changes in the concentration of cadmium in water (C_w) and in the upper layer of bottom sediments (C_{BS}) in the central part of the Sea of Azov.

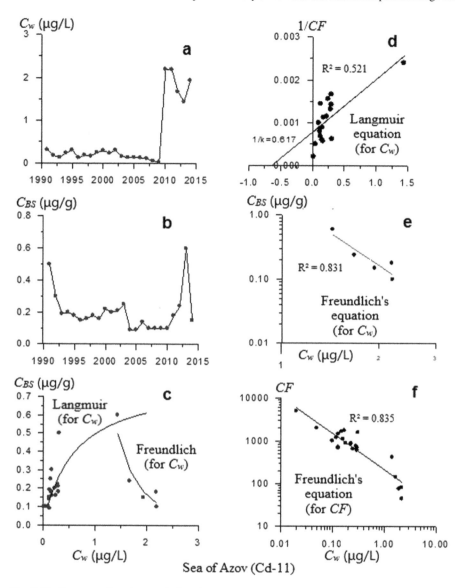

Fig. 3.20 Concentration characteristics of the distribution of cadmium in bottom sediments of the open part of the Sea of Azov (Matishov et al. 2017): **a** in water; **b** in the surface layer of bottom sediments; **c** dependence of the concentration of cadmium in bottom sediments on its content in water (solid line on the left—tabulation according to Langmuir; solid line on the right—tabulation according to Freundlich); **d** graph for the Langmuir equation (the dependence of the change in the value of $1/CF$ in the coordinates of the modified Lineweaver–Burk equation for C_W); **e** graph of the Freundlich equation for C_W; **f** Freundlich equation graph for CF

The dashed line in Fig. 3.20c depict the results of describing the relationship between C_{BS} and C_w with a polynomial of the third degree. They showed: at $C_w < 1.4$ µg/L, the trend of this dependence corresponded to an increase in the levels of C_{BS}, while, with a subsequent increase in water pollution with cadmium, a decrease in the values of C_{BS} was recorded.

Statistical analysis of the results of these observations showed that in the range of C_w values from 0 to 1.5 µg/L, the dependence of the change in its concentration in bottom sediments (at $R^2 = 0.521$) was described by the Langmuir equation, while at $C_w > 1.5$ µg/L it was described (at $R^2 = 0.142$) by the Freundlich equation (Fig. 3.20d). At the same time, the relationship between CF and C_w over the entire range of changes in C_w values having the highest statistical reliability ($R^2 = 0.835$) was described by the modified Freundlich equation (Fig. 3.20e).

It was found that the concentration of mercury by the suspended matter of the Black Sea (CF_{bs}), depending on the change in concentration in water (C_w), was three orders of magnitude higher than the coefficients of its accumulation by bottom sediments of the Sea of Azov (CF_{az}) (Fig. 3.21). The trends of dependences between the parameters CF and C_w with statistical reliability determined by the coefficients of determination $R^2 = 0.713$ and $R^2 = 0.352$, respectively, were described by the modified Freundlich equations with the following parameters:

$$CF_{bs} = 9.421 \cdot C_{bs}^{-0.836} \quad \text{and} \quad CF_{az} = 4.258 \cdot C_{az}^{-0.902}, \tag{3.45}$$

where C_{bs}, CF_{bs} and C_{az}, CF_{az} are the concentrations in water and the concentration factors of mercury in the Black and Azov seas, respectively.

The fairly close values of the exponents of these equations (with a relative error of $(0.902 - 0.836) \times 100/0.902 = 7.3\%$) indicated the coincidence of the general

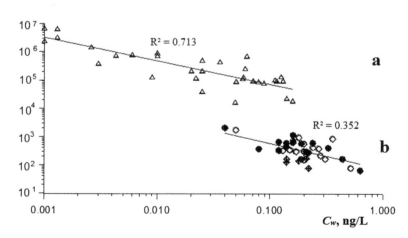

Fig. 3.21 Changes in the accumulation factors (CF) of mercury in suspended matter from the surface layer of the Black Sea (**a**) and bottom sediments of various waters of the Sea of Azov (**b**) with a change in the concentration of mercury in water (C_w, ng/L) (Stetsyuk and Egorov 2018)

trends in the dependencies between the concentration of mercury in suspensions and bottom sediments, in which the concentration levels were determined only by the granulometric characteristics of suspensions and soil particles.

Dynamic concentration characteristics Significant studies were devoted to testing the hypothesis of the applicability of the Michaelis–Menten equation to reflect the kinetics of the absorption of radionuclides of mineral elements of various biological significance by hydrobionts when the concept of isotopic carriers in water is changed. The effect of changes in the concentration of isotopic carriers in the aquatic environment on the rate of elimination of radionuclides by hydrobionts was studied separately. The correspondence of the regularity of the rate of absorption of mineral elements to the Michaelis–Menten equation was estimated by the degree of linearity of the empirical dependences plotted in the coordinates of the linearised Lineweaver–Burk equations (Patton 1968).

The results of our studies of the rate of absorption of mineral phosphorus by suspended matter of the Black and Tyrrhenian Seas, carried out jointly with L.G. Kulebakina and V.N. Popovichev, are illustrated in Fig. 3.22.

The dependence of the rate of phosphorus absorption by Black Sea Ulva (Tereshchenko and Egorov 1985) is shown in Fig. 3.23, while the patterns of phosphorus absorption by representatives of pontellid and Idotea zooplankton, studied by us together with L. G. Kulebakina and V. N.Popovichev, are displayed in Fig. 3.24. The experimental results illustrated by these figures were obtained from observations on the absorption of ^{32}P by hydrobionts under the conditions of adding an isotopic

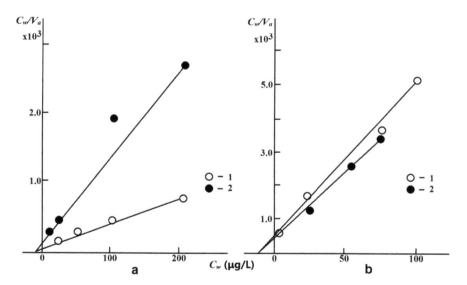

Fig. 3.22 Dependence (in Lineweaver–Burk coordinates) of the absorption rate of mineral phosphorus by a suspension from the Black (**a**) and Tyrrhenian (**b**) seas with a change in the phosphorus concentration (μg/L) in water: a fraction larger than 0.60 μm (1) and 0.23–0.60 μm (2) (according to the data of the experiments of V.N. Egorov, L.G. Kulebakina, and V.N. Popovichev)

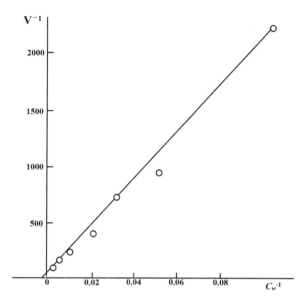

Fig. 3.23 Dependence (in Lineweaver–Burk coordinates) of the absorption rate of mineral phosphorus in the Black Sea *Ulva rigida* at its various concentrations (μg/L) in water (Tereshchenko and Egorov 1985)

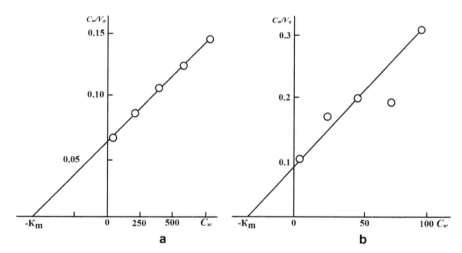

Fig. 3.24 Dependence (in Lineweaver–Burk coordinates) of the absorption rate of mineral phosphorus by *Pontella* sp. (**a**) and *Idotea metallica* (**b**) at different concentrations (μg/L) in water (according to the data of the experiments of V. N. Egorov, L. G. Kulebakina and V. N. Popovicheva)

carrier. They show that in all cases the data lay on a straight line in the coordinates of the variables included in the Lineweaver–Burk equation. This indicated that the hypothesis of the applicability of the Michaelis–Menten equation for describing the absorption of mineral phosphorus by hydrobionts and suspended organic matter can be accepted.

The experimental study of the characteristics of absorption of radionuclides of microelements and elementsrepresenting potential chemical pollutants of the environment by hydrobionts under the conditions of various additions of isotopic carriers revealed two types of regularities.

In one set of experiments, the concentration factors of radionuclides by marine organisms did not depend on C_W across a wide range of concentrations of isotopic carriers in the environment up to levels at which their toxic effects suppressed the vital functions of hydrobionts or changed the pH of seawater. Such regularities were obtained in our experiments with L.G. Kulebakina when studying the absorption of ^{203}Hg and ^{60}Co by Ulva (Fig. 3.25), as well as in experiments performed jointly with V. N. Popovichev with ^{137}Cs on *Ulva* (Fig. 3.26) and *Idotea* (Fig. 3.27). Calculations made on the basis of the results of these observations showed that in all cases the rate of absorption of mineral elements by hydrobionts depended linearly on C_W over the entire investigated range of variation of this value.

In another series of experiments, the rate of absorption of radionuclides was determined by the concentration in water of their isotopic carriers, while the relationship

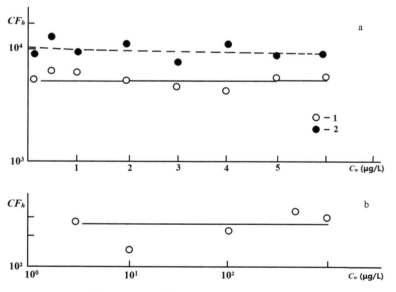

Fig. 3.25 Accumulation ^{203}Hg (**a**) and ^{60}Co (**b**) *Ulva rigida* at different concentrations (C_w) of mercury and cobalt in water: 1—daily; 2–4—day exposure (according to the experiments of V. N. Egorov with L. G. Kulebakina)

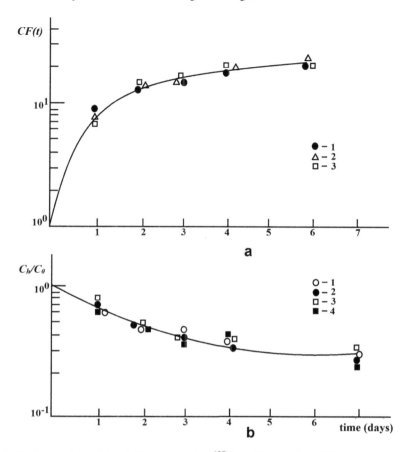

Fig. 3.26 Accumulation (**a**) and elimination (**b**) of ^{137}Cs by *Ulva rigida* at different concentrations of caesium in water; (**a**): 1—water without addition of caesium and with additions of 50 (2) and 100 (3) μg/L; (**b**): removal into water without the addition of caesium (1) and into the medium with the addition of 100 μg/L(2) after preliminary accumulation of ^{137} Cs in water without the addition of caesium; (3) and (4)—elimination into the medium without addition of caesium and with the addition of 100 μg/L accordingly after preliminary accumulation of ^{137} Cs by seaweed in the water with the addition of caesium (according to the data of experiments carried out by V. N. Egorov with V. N. Popovichev)

between V_a and C_w was described by the Michaelis–Menten equation. Such regularities were obtained when studying the absorption of ^{65}Zn by unicellular algae *Chaetoceros curvisetus* (Fig. 3.28a); ^{65}Zn by unicellular algae *Gyrodinium fissum* (Fig. 3.28b); ^{65}Zn and ^{54}Mn by Black Sea *Ulva* (Fig. 3.28c); ^{65}Zn and ^{54}Mn by suspended matter of the open and coastal areas of the Black Sea (Fig. 3.29); ^{109}Cd by brine shrimp in the Black Sea water (Fig. 3.30). In these experiments, estimates of the values of the Michaelis–Menten constant (K_m) were recorded, ranging from 20 to 450 μg/L. For suspended organic matter in open waters of the Ligurian Sea in spring,

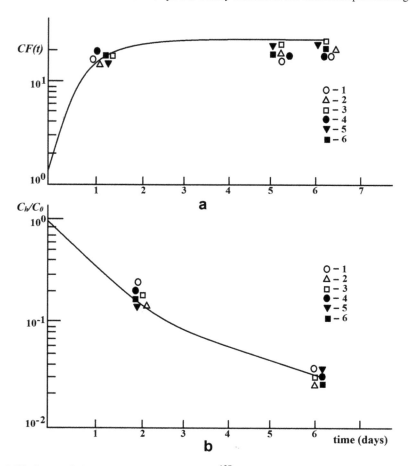

Fig. 3.27 Accumulation (**a**) and elimination (**b**) of ^{137}Cs by*Idotea baltica* at different concentrations of caesium in water: medium without caesium (1) and with additives: 50 (2), 100 (3), 500 (4), 1000 (5) and 5000 (6) μg/L (according to the data of the experiments of V. N. Egorov with V. N. Popovichev)

it was accepted (Fig. 3.31) that the relationship between V_a and C_w is described by the Michaelis–Menten equation at a value of $K_m = 0.5$ μg/L (Egorov et al. 1983).

In order to describe the above regularities within the framework of a unified theory, the compartment model (3.5) and Michaelis–Menten equations (3.20) were considered.

From Eq. (3.15) at $n = 1$ it follows that:

$$V_a = CF_S \cdot p \cdot C_w. \tag{3.46}$$

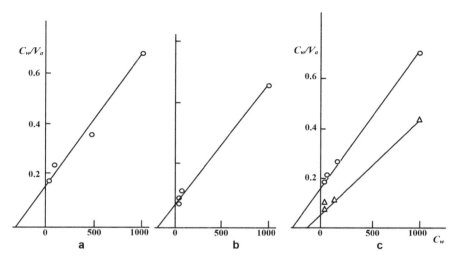

Fig. 3.28 Dependence (in Lineweaver–Burk coordinates) absorption rates of Zn (white circle) and Mn (white triangle) by *Chaetoceros curvisetus* (**a**), *Ulva rigida* (**b**) and *Gyrodinium fissum* (**c**) with a change in the concentration of Zn and Mn (μg/L) in the Black Sea water (according to the experiments of V. N. Egorov with L. G. Kulebakina)

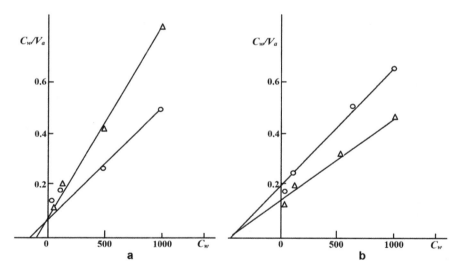

Fig. 3.29 Dependence (in Lineweaver–Burk coordinates) of the absorption rates of Zn (white circle) and Mn (white triangle) by suspended matter in the 10-mile (**a**) and coastal (**b**) zones of the Black Sea with a change in concentration (μg/L) of Zn and Mn in water (according to the experiments of V. N. Egorov with L. G. Kulebakina and V. N. Popovichev)

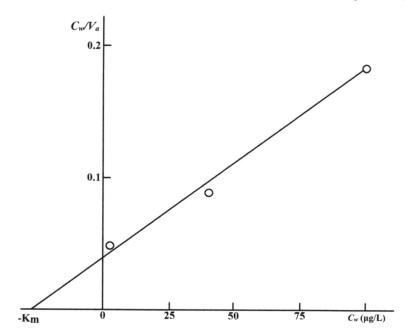

Fig. 3.30 Dependence (in Lineweaver–Burk coordinates) of the absorption rate of ^{109}Cd *by Artemia salina* with a change in its content in water (according to the experiments of V.N. Egorov with L. G. Kulebakina and V. N. Popovichev)

It is evident that the relationship of the form (3.46) reflects the results of all experiments in which the rate of substrate absorption corresponds to the Michaelis–Menten regularity. As already noted, if the condition $C_w \ll K_m$, is met in the Michaelis–Menten Eq. (3.20), then the value of C_w can be neglected, while V_a is proportional to the change in C_w. Therefore, the Michaelis–Menten equation can be used to describe the entire spectrum of regularities in the absorption of mineral elements by aquatic organisms as per the results of the performed experiments (see Figs. 3.22, 3.23, 3.24, 3.25, 3.26, 3.27, 3.28, 3.29, 3.30 and 3.31). For this reason, the Michaelis–Menten equation should be included in model (3.15) instead of the term $C_w \sum B_i p_i$. Relations (3.15) also include an indicator of the rate of elimination of the element from the hydrobiont (p_i); for this reason, it is necessary to study the degree of dependence of the parameters p_i on the change in the value of C_w.

The observation results shown in Fig. 3.25 indicate that, with the accumulation of mercury and cobalt, the value $C F_S \cdot p = $ const. The experiments with *Ulva* (Fig. 3.26) and *Idotea* (Fig. 3.27) showed that the kinetics of accumulation and excretion of ^{137}Cs by hydrobionts does not depend on changes in the concentration of caesium in water. Since in these experiments the concentration factors were constant, the value p_i did not depend on C_w.

Literature data (Ivanov 1979) and observations from experiments carried out with L.G. Kulebakina (see Fig. 3.32) confirm that, in cases where the accumulation factors

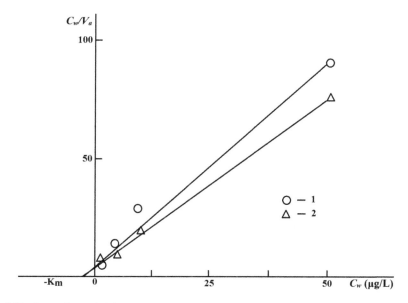

Fig. 3.31 Dependence (in Lineweaver–Burk coordinates) of the absorption rate of Hg by suspended matter in the Ligurian Sea with a change in the concentration (μg/L) of mercury in water: 1 and 2—results of parallel experiments (Egorov et al. 1983)

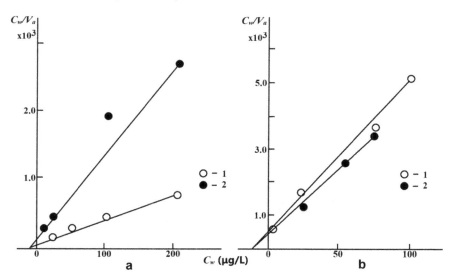

Fig. 3.32 Accumulation of ^{65}Zn by *Ulva rigida* in the water of the Black Sea (1) and in a medium with the addition of zinc at 10 (2); 100 (3) and 1000 (4) μg/L (Ivanov et al. 1978)

of radionuclides by hydrobionts depended on changes in the concentration of isotopic carriers in the medium, the output time plateaus of the radionuclide kinetic accumulation curves were the same. The transient time in the system described by Eq. (3.15) is determined by the parameters p_i, which indicates that the values p_i did not depend on the change in C_w.

A more detailed assessment of the degree of dependence of the parameters p_i on changes in C_w was carried out according to the results of accumulation and elimination of ^{32}P by *Ulva* under natural light conditions and with various additions of mineral phosphorus in water (Tereshchenko and Egorov 1985). In Fig. 3.33, the graphical dependence 2 reflects the kinetics of elimination of ^{32}P by *Ulva* in a medium with a natural content of mineral phosphorus (20 µg/L), while curve 1 shows the kinetics of elimination of ^{32}P in an aquarium supplemented with mineral phosphorus at 700 µg/L. In this series of experiments, radioactive phosphorus was absorbed by the *Ulva* in a preliminary experiment for 23 h under natural conditions of mineral phosphorus content in the medium. In another series of experiments, ^{32}P was absorbed for 23 h by algae in the preliminary stage of the experiment in an aquarium with the addition of 700 µg/L of mineral phosphorus. Following that, the *Ulva* thalli were placed in two aquariums without radioactivity. In one of them, the phosphate content was natural; the other was supplemented with 700 µgr/L. In the first case, the kinetic regularities of elimination of ^{32}P by the *Ulva* changed in accordance with dependence 4 in Fig. 3.33, while in the second—according to dependence 3. In Fig. 3.33 it can

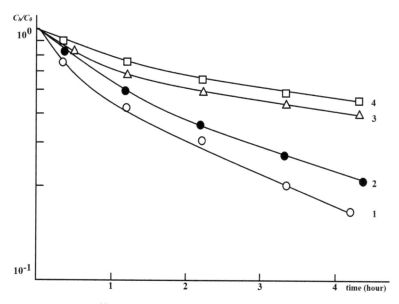

Fig. 3.33 Elimination of ^{32}P *Ulva* in a medium supplemented with 700 µg/L (1, 3) and natural phosphate content (2, 4) following accumulation in a medium with natural phosphate content (1, 2) and with an addition of 700 µg/L (3, 4) (Tereshchenko and Egorov 1985)

be seen that the kinetic curves 1 and 2, as well as curves 3 and 4, varied slightly from each other. This was due to the fact that re-absorption of eliminated ^{32}P by the alga took place during the course of the experiments.

In aquariums 2 and 4, which did not contain mineral phosphorus additives, the probability of re-absorption of ^{32}P was higher; this explains the relatively large amount of radionuclide that remained in the algae thalli. In general, the results of the experiments illustrated in Fig. 3.33 indicated that the kinetic patterns of elimination of ^{32}P by *Ulva* depended only on the intracellular phosphorus content in the alga and were not determined by the concentration of mineral phosphorus in the aquatic environment. For this reason, the experimental results (see Fig. 3.33) directly indicated that the parameters p_i, which reflect the kinetic regularities of the excretion of inorganic substances by hydrobionts, do not depend on the content of these substances in the aquatic environment in the entire studied range of changes in C_w.

Thus, the results of the experimental observations illustrated by Figs. 3.22, 3.23, 3.24, 3.25, 3.26, 3.27, 3.28, 3.29, 3.30, 3.31, 3.32 and 3.33 showed that the parameters p_i of models (3.15) do not depend on C_w. Consequently, the kinetic regularities of the mineral metabolism of hydrobionts under conditions of a changing concentration of mineral elements in the environment can be described on the basis of this model by the following equations:

$$\frac{dC_h}{dt} = \frac{V_{\max} C_w}{K_m + C_w} - \sum_{i=1}^{n} C_i \, p_i;$$

$$\frac{dC_w}{dt} = m_{sp} \left(\sum_{i=1}^{n} C_i \, p_i - \frac{V_{\max} C_w}{K_m + C_w} \right).$$

(3.47)

Although Eqs. (3.47) are a more generalised form of mathematical description of the kinetics of mineral exchange, their use is advisable only if the stationary coefficient CF_S of accumulation of inorganic matter by hydrobionts depends on the change in C_w. CF_S does not depend on C_w under conditions $K_m \gg C_w$. Then only the ratio $V_{\max}/K_m$ can be determined empirically from experimental observations, but not each of these parameters. If the kinetics of mineral exchange of a hydrobiont is described by a model containing a non-exchangeable pool of a particular element, then, in accordance with Eq. (3.36), expression (3.47) should include parameters reflecting the process of intake of an element into the non-exchangeable pool, as well as a term that takes into account the distribution of this substance over exchangeable and non-exchangeable pools.

3.2.7 Concentration of Elements—Chemical Analogues

The problem of studying the regularities of the exchange of radionuclides by hydrobionts in the environment with their non-isotopic chemical analogues was solved by

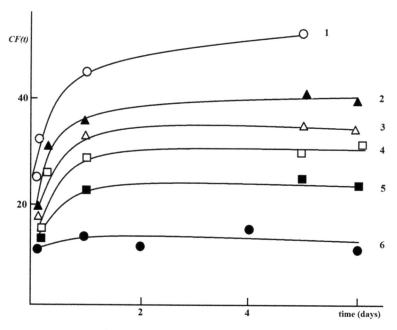

Fig. 3.34 Accumulation of ^{90}Sr by *Cystoseira barbata* in the water of the Black Sea (1) and in a medium with Ca additions 250 (2), 500 (3), 1000 (4) and 2000 (5) mg/L, as well as in a medium supplemented with Sr 1000 (6) mg/L (Egorov et al. 1989)

the example of the analysis of experimental data on the absorption and elimination of ^{90}Sr by *Cystoseira*brown alga at various concentrations of Sr and Ca (Egorov et al. 1989; Polikarpov and Egorov 1986). Figure 3.34 shows the kinetics of changes in the concentration factors of ^{90}Sr cystose in conditions of various contents of Sr and Ca in the water of experimental aquariums.

From Fig. 3.34, it can be seen that the coefficients of accumulation of ^{90}Sr by algae in natural Black Sea water having a concentration of strontium at 4.2–4.4 mg/L and calcium at 267–280 mg/Lexceeded 50 units during the observation period, while, with the addition of Sr into the medium at a concentration of 1000 mg/L, they decreased by four times. A decrease in the values of the concentration factors of Sr was also noted with an increase in the Ca content in the aquatic environment of aquariums. The addition of 1000 mg/L of Sr caused a greater decrease in the concentration factor of ^{90}Sr in cystose than a similar addition of Ca, although the time for the concentration factors of ^{90}Sr to plateau did not change either with the addition of Sr or Ca.

As shown in Fig. 3.34, the effect can be caused both by a decrease in the absorption rate and an increase in the elimination rate of strontium by cystose with a change in the concentration of Sr or Ca in an aqueous medium. In order to investigate these processes, experiments were carried out on the rates of absorption and excretion of *Cystoseira*. The results of the experiment on the elimination of ^{90}Sr by algae (Figs. 3.35 and 3.36) indicate the following: regardless of the content of Ca accumu-

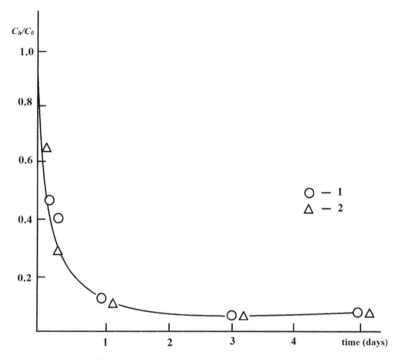

Fig. 3.35 Elimination of ^{90}Sr by *Cystoseira barbata* after 5 days of preliminary accumulation in a medium with radioactive strontium and natural calcium content: 1—in a medium without Ca addition; 2—in a medium supplemented with Ca 2000 mg/L(Egorov et al. 1989)

lated in ^{90}Sr by algae in the preliminary experiment, the kinetics of its elimination by *Cystoseira* into non-radioactive water, expressed in relative units, coincided and did not depend on the Ca concentration in the medium, up to a level exceeding the natural concentration by 2000 mg/L.

Therefore, the observed decrease in the concentration factors of ^{90}Sr with increasing Ca concentration in water (Fig. 3.34) cannot be due to an increase in the rate of elimination of Sr by algae.

The experimental dependences of determining the rate of absorption of strontium by cystose with a change in the concentration of strontium (C_{Sr}) and calcium (C_{Ca}) in water, expressed in Lineweaver–Burk coordinates, are shown in Fig. 3.37a. The continuous straight lines approximate the observation results. The satisfactory location of the experimental points on the straight lines indicates that the rate of absorption of strontium by *Cystoseira* with an increase in the concentration of both Sr and Ca in the medium changes in accordance with the pattern described by the Michaelis–Menten equation. In our experiments (Fig. 3.37a), it was found that with the addition of Sr to the medium, the value of K_m is 1165 mg /L, while, with the addition of Ca, the value reached 2040 mg/L. These data indicate that a halving of the rate of strontium uptake by *Cystoseira* in relation to the maximum (V_{max}) can be

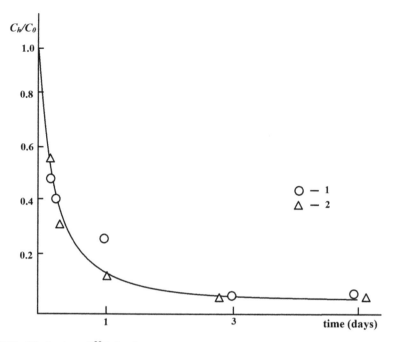

Fig. 3.36 Elimination of ^{90}Sr by *Cystoseira barbata* after 6 days of its preliminary accumulation in a medium with radioactive strontium and with an addition of calcium at 2000 mg/L: 1—in a medium without Ca; 2—in a medium supplemented with Ca at 2000 mg/L (Egorov et al. 1989)

achieved with a lower addition of Sr than Ca. Consequently, calcium influences the absorption of strontium by *Cystoseira* not according to its partial concentration in the medium, but with a certain coefficient (a), which is less than one.

From the results of experiments with additions of Sr, it follows that the rate of its absorption by algae decreased by half at a concentration of strontium in water of $C_{Sr} = 885$ mg/L and calcium $C_{Ca} = 280$ mg/L. It can be seen from experiments with additions of Ca that the rate of Sr uptake by algae decreased by half at $C_{Sr} = 5$ mg/L and $C_{Ca} = 2035$ mg/L. The following can be assumed: if the value of a was taken into account when interpreting the experimental data, the state of the system in which $V_a = V_{max}/2$ would be established at the value of the denominator in expression (3.20):

$$K_m = C_{Sr} + a \cdot C_{Ca}, \tag{3.48}$$

where C_{Sr} and C_{Ca} are such numerical values of the concentration of Sr and Ca in water at which the rate of absorption of strontium by algae is half of the maximum.

Therefore, we can write:

$$C'_{Sr} + aC'_{Ca} = C''_{Sr} + aC''_{Ca}, \tag{3.49}$$

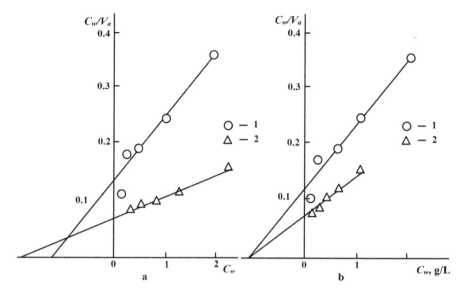

Fig. 3.37 Dependence of the rate of absorption of strontium by *Cystoseira barbata* with a change in the concentration of Sr (1) and Ca (2) in water: the graph of the Lineweaver–Burk equation is constructed without taking into account (**a**) and taking into account (**b**) the degree of chemical similarity of elements (Egorov et al. 1989)

whence it follows that:

$$a = \frac{C'_{Sr} - C''_{Sr}}{C''_{Ca} - C'_{Ca}} \quad \text{or} \quad C'_{Sr} + aC'_{Ca} = C''_{Sr} + aC''_{Ca}. \tag{3.50}$$

The results of the experiments, which are reflected in Fig. 3.37a, interpreted taking the evaluation $a = 0.5$ into account, are shown in Fig. 3.37b. They indicate that the linearity of the location of points in the Lineweaver–Burk coordinates did not change, whille the Michaelis–Menten constant corresponding to both experiments took the value $K_m = 975$ mg/L. An analysis of the results of experiments on the elimination of ^{90}Sr by *Cystoseira* (Figs. 3.35 and 3.36) according to the graphical method of exponential extraction showed that the kinetics of this process is satisfactorily described by a double-exponential function:

$$C_h/C_0 = 0.914e^{-0.1980t} + 0.086e^{-0.0028t}, \tag{3.51}$$

where t is time, h.

These observations confirmed the previously obtained results (Egorov and Kulebakina 1973), indicating that the kinetics of absorption and elimination of ^{90}Sr by *Cystoseira* can be considered as a process of exchange of strontium by two pools having rates of first-order metabolic reactions. They showed that the indicators of the exchange rate of pools p_1 and p_2 are independent of changes in the concentration of Sr and Ca in water within wide limits; therefore, taking into account formulas (3.20) and (3.48), a balance equality can be obtained that describes the kinetics of exchange of strontium with *Cystoseira* in a medium with varying concentrations of strontium and calcium:

$$\frac{dC_h}{dt} = \frac{V_{max}C_{Sr}}{K_m + C_{Sr} + aC_{Ca}} - \sum_{i=1}^{2} C_i p_i, \qquad (3.52)$$

where C_{Sr} is the concentration of strontium, or ^{90}Sr, in *Cystoseira*;

C_i is the concentration of strontium, or ^{90}Sr, in the i-th exchange pool of the alga.

From the standpoint of the theory of mineral metabolism, it can be noted that the difference in the concentration factors of elements/chemical analogues of the hydrobiont can be explained by the fact that their absorption by the hydrobiont from the environment is not partial, but their elimination takes place in accordance with the coefficient a and iscarried out in accordance with the rates of p_1 and p_2, which have different sets of values for each of the elements/chemical analogues. Consequently, if, when absorbed by hydrobionts, some elements can be perceived by them as chemical analogues, then these elements are identified in hydrobionts during the course of biochemical reactions and the patterns of their metabolism may vary.

3.2.8 Physical and Chemical Sorption

The relevance of forecasting the radioactive contamination of bottom sediments under anthropogenic impact is due to the fact that they represent the slowest stage of the transfer of radionuclides in the aquatic environment, that is, the final depot of biogeochemical migration (Polikarpov et al. 1995).

For this, a study of the transformation of physicochemical forms of radionuclides is carried out in connection with a change in the granulometric and mineral composition of soils (Voitsekhovitch et al. 1991; Neiheisel et al. 1992), as well as an analysis of the sorption properties of bottom sediments in experiments using a radioactive tracer.

Experiments have shown (Lyubimova 1973; Polikarpov et al. 1987a; Lazorenko and Polikarpov 1990) that the kinetics of concentration of radionuclides by bottom sediments is always directed to a stationary level characterised by the concentration (distribution) coefficient. It was noted (Timofeeva-Resovskaya 1963)that a number of radionuclides, especially ^{137}Cs following its preliminary accumulation by bottom sediments from water having an increased radiocaesium content, are withdrawn into the aqueous environment, which is cleaner for this radionuclide, either insignificantly or with a greater lag time with respect to the accumulation process. This fact can be explained by the prevalence of processes involved in the chemical component of sorption, which occur due to a chemical reaction (Benson 1964). Desorption of radionuclides from bottom sediments due to such reactions may be absent or proceed much slower than when due to the physical component of these processes. A search for and empirical verification of a theoretical basis for the mathematical description of the kinetic regularities of the migration of radionuclides used the example of the regularities of the distribution of ^{137}Cs at the separation of the phases of the aqueous medium/bottom sediments as a result of the physical and chemical components of sorption. Compartment models (Riziĉ 1972)were used as a methodological basis for the mathematical description of the kinetics of the processes of interrelation between bottom sediments and radionuclides contained in the aquatic environment. Verification of the adequacy of their application and parameterisation was carried out on the basis of experimental results.

When developing the model, it was assumed that the concentration of the radionuclide in the bottom sediments (C_h) at any time is determined by the sum of the contributions of the processes of physical (C_{ph}) and chemical (C_{ch}) sorption:

$$C_h = C_{ph} + C_{ch}. \tag{3.53}$$

The kinetics of physical sorption processes were described by a model having the form:

$$dC_{ch}/dt = \sum r_{phi}(C_w B_i - C_i), \tag{3.54}$$

where C_w is the concentration of the radionuclide in water;

B_i and r_{ph} are the relative volumes of exchangeable pools of a radionuclide in the surface layer of bottom sediments and indicators of their exchange rate, respectively;

C_i is the concentration of the radionuclide in the i-th exchange pool at the time t.

The following equation was used to reflect the kinetic regularities of the chemical sorption of radionuclides by bottom sediments:

$$dC_{ch}/dt = (1 - CF_{ch}(t)/CF_{Sch}) \cdot r_{ach}C_w - r_{dch}C_{ch}, \tag{3.55}$$

where $CF_{ch}(t)$is the concentration factor of the radionuclide by bottom sediments as a result of chemical sorption at the moment in time$t$:

$$CF_{ch}(t) = C_{ch}(t)/C_w, \tag{3.56}$$

where $CF_{Sch}(t)$ is the stationary value of the concentration factor established in bottom sediments under the influence of chemical sorption only;

r_{ach} and r_{dch} are indicators of the rate of chemical adsorption and desorption.

In Eq. (3.55), it is assumed that the value of the stationary concentration factor (CF_{Sch}) depends on the content of the radioisotope and isotopic carrier in water in accordance with the Langmuir equation. In order to verify the model implemented by Eqs. (3.54) and (3.55), the experiments investigated the kinetic characteristics of the accumulation and removal of radionuclides by samples of bottom sediments, which are capable of description by integral solutions of these equations under the appropriate initial conditions. In this case, the parameters of the model were determined by an approximation of the experimental dependences, while the comparison of the results of numerical experiments on the model and the corresponding experiments made it possible to draw a conclusion about the adequacy of the theoretical basis used.

The interpretation of experimental observations was based on the general solution of Eq. (3.54), which for a closed system has the following form:

$$C_{ph}(t) = R_{aph} \cdot C_{w0} + \sum (C_{i0} - C_{wi} \cdot B_i) \cdot e^{-p_i t}, \tag{3.57}$$

where C_{w0} and C_{i0} are the content of the radionuclide at the initial moment of time, respectively, in water and in bottom sediments that have accumulated it under the influence of physical adsorption.

From relation (3.57) it can be seen that, for any value of $C_w \neq 0$ and $t \to \infty$, the concentration of the radionuclide by bottom sediments is described by a double-exponential function tending to a stationary state, at which $C_{ph}(\infty) = CF_{Sph} \cdot C_w$. In accordance with this equation, at zero initial conditions, when $C_{w0} = 0$, the change in the radionuclide content in bottom sediments along a double-exponential relationship tends to $C(\infty) = 0$.

In cases when only chemical sorption is significant in its contribution to the concentration of radionuclides by hydrobionts, or when only the chemical sorption role is considered, under initial conditions when $t = 0$ and $C_a \neq 0$, the kinetic dependence at $t \to \infty$ tends to the level:

$$CF_{ch}(\infty) = CF_{Sch} \cdot r_{ach}/(r_{ach} + r_{dch}). \tag{3.58}$$

With a more significant contribution of chemical sorption in comparison with the contribution of physical sorption $r_{dch} \leq r_{ach}$, supporting the assumption that, under stationary conditions at $t \to \infty$, the value $CF_{ch}(\infty) \to CF_{Sch}$. It can also be seen from relation (3.55) that, in relation to the initial conditions of the experiments, when $C_w = 0$, chemical desorption will be determined only by the value of the constant rate of elimination r_{dch}. Since $r_{dch} \ll r_{ach}$ and $r_{dch} \ll r_{phi}$, chemical desorption can be neglected in the time scale of experimental observations.

From the above reasoning, it can be seen that observation under experimental conditions of the kinetics of radionuclide accumulation by bottom sediments allows the determination of the total stationary concentration factor $CF_h = CF_{ph} + CF_{ch}$, as well as, by monitoring the elimination kinetics, an estimation of the contribution of chemical sorption (CF_{ch}) to the accumulation process. The approximation of the kinetic dependences of the accumulation or removal of the radionuclide by bottom sediments by a function of the form (3.57) makes it possible to estimate the parameters of the model r_{phi} and r_{ach}.

The theoretical prerequisites for describing the kinetic regularities of physical and chemical sorption of radionuclides have been verified on the example of experimental observations (Polikarpov et al. 1995). In June 1986, experiments were carried out on the accumulation and removal of ^{137}Cs by bottom sediments of various granulometric composition, sampled in the lower reaches of the Dnieper River and in the Dnieper-Bug estuary (Table 3.4). The procedure used for setting up experiments with bottom sediments is described in (Polikarpov et al. 1987a; Lazorenko and Polikarpov 1990).

Typical kinetic curves of the accumulation and excretion of ^{137}Cs obtained as a result of an experiment carried out on fine silty dark grey sand sampled from the Dnieper-Bug estuary are shown in Figs. 3.38 and 3.39 and Table 3.4. The high degree of irreversibility of the accumulation-elimination processes is apparently due to the presence of strong chemical binding of the studied radionuclide to this type

Table 3.4 Characteristics of bottom sediments in experiments on accumulation and elimination of ^{137}Cs

Sampling area	Geographical coordinates	Depth, m	Bottom sediment type
Dnieper River (Kherson)	46° 37.5′N, 32° 36.5′E	6.0	Finely dispersed silty dark grey sand
Dnieper River (Kherson)	46° 33.4′N, 32° 27.5′E	7.0	As above
Dnieper-Bug estuary	46° 35.9′N, 31° 43.2′E	5.8	Medium-dispersed sand
Dnieper-Bug estuary	46° 42.4′N, 31° 55.9′E	7.5	Silty light grey sand
Dnieper-Bug estuary	46° 31.7′N, 32° 08.0′E	7.0	Finely dispersed silty dark grey sand

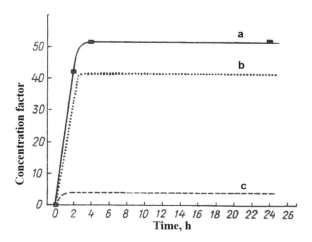

Fig. 3.38 Sorption of ^{137}Cs by fine silty dark grey sand: approximating curve (**a**) and the contribution to it of chemical (**b**) and physical (**c**) sorption processes (Lazorenko and Egorov 1994)

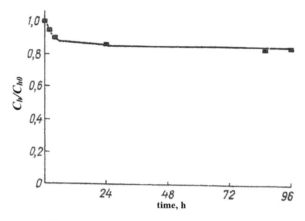

Fig. 3.39 Desorption of ^{137}Cs by finely dispersed silty dark grey sand. C_h/C_{h0} is the ratio of the concentration ^{137}Cs in bottom sediment samples at the observed and initial time points (Lazorenko and Egorov 1994)

Table 3.5 Calculation of the kinetic parameters of sorption of ^{137}Cs for various types of bottom sediments

Type Bottom sediment type	$CF_h(\infty)$	CF_{Sch}	B_1	r_{ph1}	B_2	r_{ph2}	r_{ch}
Finely dispersed silty dark grey sand	345.0	331.1	0.257	0.736	0.743	0.050	59.9
As above	464.0	445.3	0.355	0.447	0.645	0.020	235.6
Medium-dispersed sand	34.8	31.7	0.849	0.526	0.151	0.031	4.1
Silty light grey sand	38.9	32.4	0.837	2.727	0.163	0.039	3.7
Finely dispersed silty dark grey sand	52.5	45.3	0.756	0.445	0.244	0.046	40.1

of bottom sediments. Similar kinetic curves of accumulation and excretion of ^{137}Cs were obtained for the other types of studied bottom sediments (see Table 3.5).

3.2.9 Biotic Transformation of Physicochemical Forms of Radionuclides

The problem of modeling the kinetics of the biological transformation of radionuclides was solved by us on the example of studying the kinetics of accumulation and elimination of various forms of ^{131}I of the Black Sea *Ulva* (Polikarpov et al. 1983).

In experiments with various forms of radioactive iodine, it was found that ^{131}I in the form of iodide (I^{1-}) was accumulated by the green alga *Ulva rigida* more intensively (Fig. 3.40a) than ^{131}I in the pentavalent form IO_3^{5-} (Fig. 3.40b). The study of the elimination of various forms of iodine in clean water after a 5-day exposure to their

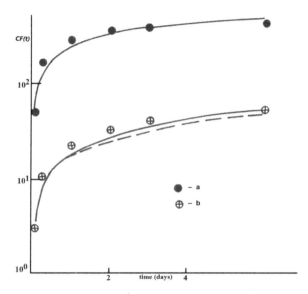

Fig. 3.40 Accumulation of monovalent I^{1-} (**a**) and pentavalent IO_3^{5-} (**b**) forms of radioactive iodine *Ulva rigida*. The solid lines show the approximating curves; the dashed lines depict the theoretical accumulation curve of ^{131}I in the form IO_3^{5-}. $CF(t)$—the concentration factor of this physicochemical form of the element at the moment in time t (Polikarpov et al. 1983)

preliminary accumulation in *Ulva* showed that the kinetics of elimination, expressed in relative units, coincided (Fig. 3.41). This indicates that, in accumulating iodine in monovalent (I^{1-}) and pentavalent (I^{5+}) forms to different levels, *Ulva* exchanges them with the same rates of first-order metabolic reactions. In order to study the biogenic

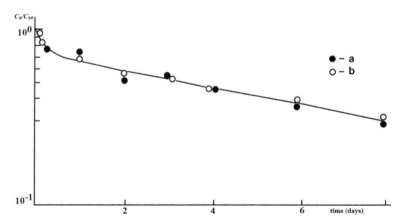

Fig. 3.41 Elimination of ^{131}I, accumulated *Ulva rigida*, in the form I^{1-} (**a**) and IO_3^{5-} (**b**) after a 5-day preliminary exposure of accumulation experiments ^{131}I; C_h/C_0 is the ratio of the remaining radioactivity to the initial one (Polikarpov et al. 1983)

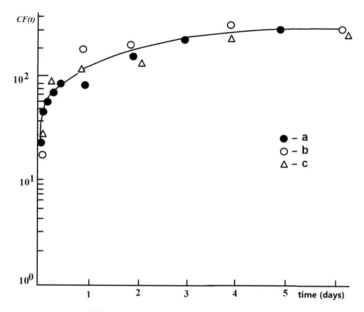

Fig. 3.42 Accumulation of ^{131}I by *Ulva rigida* in the form I^{1-} (**a**), as well as in physicochemical forms transformed by algae upon preliminary absorption of ^{131}I in form I^{1-} (**b**) and IO_3^{5-} (**c**) (Polikarpov et al. 1983)

transformation of the physicochemical forms of iodine, a series of experiments was carried out in which the concentration of active iodine removed by the *Ulva* into clean water was studied. Experiments have shown that the concentrations of the eliminated forms of iodine coincide with each other, as well as with the kinetics of the accumulation by *Ulva* of monovalent [131]I (Fig. 3.42). This permits the conclusion that, as a result of interaction with *Ulva*, the pentavalent form of iodine is transformed into its physicochemical forms, which, when accumulated by algae, behave as monovalent (Polikarpov et al. 1987b).

With this interpretation of the observations, it follows that the concentration of pentavalent iodine in the experiment illustrated in Fig. 3.40b was accompanied by the intake into water of its monovalent form and its subsequent accumulation together with the pentavalent form. As a result, the content of pentavalent ions in water decreased, while the content of monovalent ions increased; that is, chemical forms of iodine were subjected to a biogenic transformation.

In order to be able to develop a mathematical description of the kinetics of this process, it was necessary to determine the parameters of the model for the concentration by *Ulva* of various forms of iodine.

Approximation of the results of observations on the elimination of ^{131}I by *Ulva* allowed us to isolate two exponents with the parameters $p_1 = 0.124$ days^{-1} and $p_2 = 4.62$ days^{-1}, which required the reflection of the exchange kinetics of each *Ulva* form by means of a two-compartment model. The parameters of the exponential function describing the process of accumulation of monovalent ^{131}I by *Ulva* were calculated using the formula (3.4) and the concentration factor of monovalent ^{131}I by *Ulva* after a 5-day exposure:

$$CF(t) = 1033 - 905e^{-0.124t} - 132e^{-4.62t}. \qquad (3.59)$$

Comparison of the theoretical description of the kinetic dependence of the accumulation of monovalent iodine by *Ulva* (depicted by the solid line in Fig. 3.40) with observational data demonstrated their satisfactory agreement and testified to the sufficient adequacy of the model.

The parameters of the metabolism model of pentavalent iodine *Ulva* were obtained using the absorption rate of $^{131}IO_3^-$ by the *Ulva* at time $t = 0$. The validity of such an assessment was determined by the evident circumstance of only the pentavalent form of iodine being present in the experiment illustrated by Fig. 3.40b at the initial moment of time. According to the calculations, formula (3.24) has the form (at $C_w = 1$):

$$V_{abs} = B_1 p_1 + B_2 p_2 = 60.24(day^{-1}). \qquad (3.60)$$

In expression (3.60), the values p_1 and p_2 are determined from observations. Under the assumption that the ratios of B_1/B_2 in the models of the exchange of penta- and univalent iodine are equal, the equation for the concentration of $^{131}IO_3^-$ by *Ulva* is obtained:

$$CF(t) = 87 - 76e^{-0.124t} - 11e^{-4.62t}. \qquad (3.61)$$

The time-tabulated dependence (3.61) is shown by a dashed line in Fig. 3.40. This reflects the contribution of the pentavalent form to the concentration of iodine in *Ulva*. In general, the situation of concentration and exchange of the two forms of iodine by *Ulva* corresponds to the model implemented by the equations:

$$
\begin{aligned}
C_h' &= C_1' + C_2' \\
\frac{dC_1}{dt} &= p_1(C_w B_1 - C_1) \\
\frac{dC_2}{dt} &= p_2(C_w B_2 - C_2) \\
\frac{dC_w'}{dt} &= \frac{m_h}{m_w}\left[(C_1 + C_1')p_1 + (C_2 + C_2')p_2 - C_w'(B_1 p_1 + B_2 p_2)\right] \\
\frac{dC_1'}{dt} &= p_1(C_w' B_1' - C_1') \\
C_h &= C_1 + C_2; \\
C_h' &= C_1' + C_2',
\end{aligned}
\qquad (3.62)
$$

where C_w and C_w' are the concentrations of mono- and pentavalent iodine in water; C_1', C_1 and C_2', C_2 are the concentrations of mono- and pentavalent iodine in the compartments of the hydrobiont.

In relations (3.62), Eqs. 1 and 4 reflect the balance of the penta- and monovalent forms of iodine in water, while Eqs. 2, 3, 5 and 6 represent these forms in the compartments of the hydrobiont.

In the model interpretation, observations (see Fig. 3.40b) correspond to the accumulation parameter calculated by the formula:

$$CF(t) = \frac{C_h + C_h'}{C_w + C_w'}. \qquad (3.63)$$

The results of calculating this parameter by numerically solving differential Eqs. (3.62) are shown in Fig. 3.40. These results indicate a good convergence of theory and observational data. Consequently, compartment models are applicable to describe the kinetics of biogenic transformation of physicochemical forms of pollution.

Thus, using the example of an empirical study and a mathematical description of the concentration kinetics of mono- and pentavalent ^{131}I by *Ulva*, it has been shown that the kinetics of exchange of different physicochemical forms of pollutants by hydrobionts can be considered as a result of its entry into the exchange pools of hydrobionts corresponding to each physicochemical form. The interaction of chemical resource exchange pools with the environment is carried out at rates of first-order metabolic reactions. The numerical values of the rates of intake into the pools are determined by the physical and chemical forms of pollutants. The rates of elimination of different physicochemical forms of inorganic substances from the corresponding exchange pools in hydrobionts are the same.

3.2.9.1 Limitation of Primary Production Processes by Biogenic Elements

Chemical elements are considered in terms of biogens, which can limit the course of primary production processes in ecosystems. Thus, the primary constituents of biogenic inorganic matter are the organogenic elements (N, P, Si, S). A lack of microelements (Zn, Mn, Co, Mo, Fe) in the environment (Dobrolyubskii 1956; Turn et al. 1982) can also limit the growth of aquatic organisms (hydrobionts).

The problem of providing an accurate mathematical description of the regularities of chemical limitation of production processes has aroused wide interest among researchers. In 1962, M. R. Droop (Droop 1962) proposed that dynamics of changes in the specific biomass of unicellular algae (m_{sp}) depending on the concentration of nutrients in a continuous cultivator could be described by the equation:

$$\frac{dm_{sp}}{dt} = \mu m_{sp} - R_v m_{sp};$$

$$\frac{dC_w}{dt} = R_v(C_{w0} - C_w) - C_h \mu m_{sp}, \tag{3.64}$$

where R_v is an indicator of the flow rate of the nutrient medium through the cultivator (per day);

μ is the specific growth rate of the culture (per day);

C_w and C_{w0} are the concentrations of nutrients in the medium and at the inlet of the cultivator, respectively;

C_h is the intracellular concentration of the limiting substrate.

In Eq. (3.64), the change in C_h was calculated from the ratio (Burmaster and Chisholm 1979):

$$\frac{dC_h}{dt} = V_a - \mu C_h, \tag{3.65}$$

in which V_a is the rate of absorption of a nutrient by algae proposed (Dugdale 1967) to describe the Michaelis–Menten equation. Taking this circumstance into account, the equality (3.65) took the form:

$$\frac{dC_h}{dt} = \frac{V_{max} C_w}{K_m + C_w} - \mu C_h. \tag{3.66}$$

Two main formulas for determining the value of μ in the relation (3.66) are known from the literature. One of these, proposed by Jacques Monod (Monod 1942), associated this parameter with the concentration of a nutrient in the environment:

$$\mu = \frac{\mu_{max} C_B}{K_n + C_B}, \tag{3.67}$$

where μ_{max} is the maximum physiologically possible specific growth rate of a cell culture;

K_n is a constant.

Another regularity, established by Droop (1974), determined the relationship between the parameter μ and the specific concentration of a nutrient in algae:

$$\mu = \mu_{max}\left(1 - \frac{q_{min}}{C_h}\right), \tag{3.68}$$

where q_{min} is the minimum intracellular concentration of a biogenic element for ensuring the viability of a cell culture.

The use of Monod's formula (3.67) in Eq. (3.66) reflected the state of such a system in which the stationary state Ch at $K_m = K_n$ was calculated from the relation:

$$C_h = \frac{V_{max}}{\mu_{max}}, \tag{3.69}$$

that is, the intracellular concentration of a nutrient in algae was constant and did not depend on the concentration of this element in the environment.

When using the Droop Eq. (3.68), the kinetics of changes in the intracellular concentration of a biogenic element were described by the equation:

$$\frac{dC_h}{dt} = \frac{V_{\max} C_w}{K_m + C_w} - \mu C_h \left(1 - \frac{q_{\min}}{C_h} \right), \tag{3.70}$$

according to which the stationary value of the intracellular concentration of the biogenic element at $\frac{dC_h}{dt} = 0$ is:

$$C_h = \frac{V_{\max} C_w}{(K_m + C_w)\mu_{\max}} + q_{\min}. \tag{3.71}$$

From (3.70) it can be seen that the value of C_h was dependent on C_w. The minimum value of C_h was $q_{\min}$ at $C_w = 0$, while the maximum value that this value could take with an increase in C_w (at $C_w \gg K_m$ and $C_h \gg q_{\min}$) tended to the level calculated by the formula (3.69) as determined using Monod's equations.

It has been empirically established that unicellular algae can store nutrients under conditions of increased concentrations in the environment and subsequently use them for growth when the environment is depleted in these nutrients (Brunel et al. 1982; Giammatteo et al. 1983). From expressions (3.69) and (3.70), it can be seen that the noted pattern could be reflected in the model only when using the Droop equation.

Thus, the literature data have shown that, in order to describe the regularities of changes in the concentration of nutrients in the medium and primary producers grown in a culture of continuous type (or in open systems), a balance equality can be used, which reflects the processes of entry of nutrients by hydrobionts according to the Michaelis–Menten equation, which also comprises Monod or Droop dependencies that determine the level of chemical limitation of production processes. Monod's equation reflects the dependence of limitation as a function of the content of biogenic elements in the medium, while Droop's equation reflects their intracellular concentration.

The task of a separate study arose from the need to compare the compartment theory with the existing methods of mathematical description of the kinetics of mineral metabolism of hydrobionts under conditions of chemical limitation of production processes. In the studies carried out by us jointly with Z. Z. Finenko, A. Ya. Zesenko and A. V. Parkhomenko (Egorov et al. 1982), a model derived from the relation (3.41) was used to describe the kinetics of the exchange of biogenic elements by producers in relation to a single exchange pool of a biogenic element in a hydrobiont:

$$\frac{dC_h}{dt} = \frac{V_{\max} C_w}{K_m + C_w} - \left(p + \frac{1}{m_{sp}} \frac{dm_{sp}}{dt} \right) C_h. \tag{3.72}$$

The validation of the model (3.72) was made on the basis of experimental observations on the concentration of radioactive phosphorus by the culture of the single-celled algae *Skeletonema costatum*, which grew on a modified Goldberg medium under conditions of a limiting concentration of mineral phosphorus in the medium. According to the results of a series of independent experiments on the absorption and excretion of ^{32}P by algae, the parameters of the model were calculated $q_{min} = 3.1$ μgP/mg; $V_{max} = 7.34\,\mu$gP/hour; $K_m = 11\,\mu$g/L; $p = 0.231\,$h^{-1}. The value of μ_{max} for this culture of algae was 5 days^{-1} (Finenko and Krupatkina-Akinina 1974). A comparison of the results of observations and calculations under the initial conditions of $C_h(0) = 3.1\,\mu$gP/mg, $C_w(0) = 11\,\mu$gP/L and $m_h(0) = 1.09\,$mg/L demonstrated their satisfactory agreement (Fig. 3.43).

The dependence of the parameter p in the model of the form (3.72) on the change in the concentration of mineral phosphorus in the medium (C_w) was investigated by us in a wider context in joint experiments carried out with L. G. Kulebakina and V. N.Popovichev on the example of regularities of elimination of ^{32}P by unicellular algae and suspended matter with variation in the additions of the isotopic carrier. Observations showed (Fig. 3.44) that unicellular algae *Gimnodinium lanskaya* excreted ^{32}P to a greater extent in an environment with a high phosphorus content; moreover, in aquariums with phosphate supplements of 20 and 100 μg/L the elimination kinetics coincided. At the same time, the relative elimination of ^{32}P by suspensions from the pelagic waters of the Tyrrhenian Sea coincided in aquariums both with additives and natural phosphorus content (Fig. 3.45). In the first case (Fig. 3.44), the experiments were carried out with a specific biomass of an aquatic organism in water of 10 mg/L, while in the second (Fig. 3.45), they were undertaken with a biomass of 1 mg/L. According to Tereshchenko's determinations, the concentration of mineral phosphorus in the waters used for these experiments did not exceed 3 μgP/L. To explain the differences in the kinetics of excretion of ^{32}P by unicellular algae at different concentrations of phosphates in the aquatic environment illustrated in Fig. 3.44, we used the modelling method. The dynamics of the model (3.72) were investigated at a specific biomass of the hydrobiont of $m_{sp} = 10$ mg/L and $C_w = 3$ μgP/L, as well as $C_w = 23$ and $C_w = 103\,\mu$gP/L.

It was shown that the difference between the kinetic regularities of the elimination of ^{32}P by algae (curve 1 in Fig. 3.44) in an environment with a natural content of mineral phosphorus from the data of other experiments is due to the reabsorption of previously eliminated radionuclide by the algae. This effect was manifested due to the increased specific biomass of the hydrobiont in the environment of this aquarium, as well as, consequently, the increased specific radioactivity of water arising due to the processes of elimination of ^{32}P by algae. On the basis of the results of experiments with unicellular algae and suspensions, as well as in experiments with ^{32}P and *Ulva* (Fig. 3.45), the hypothesis about the dependence of the kinetics of phosphorus excretion by hydrobionts with a change in its concentration in water was rejected. Since the approximation of the coinciding kinetic curves of the removal of radioactive phosphorus by suspended matter gave the same estimates of the parameters p_i, the results of the experiments illustrated in Figs. 3.44 and 3.45 were interpreted to mean that the rates of phosphorus excretion by hydrobionts do not depend on C_w.

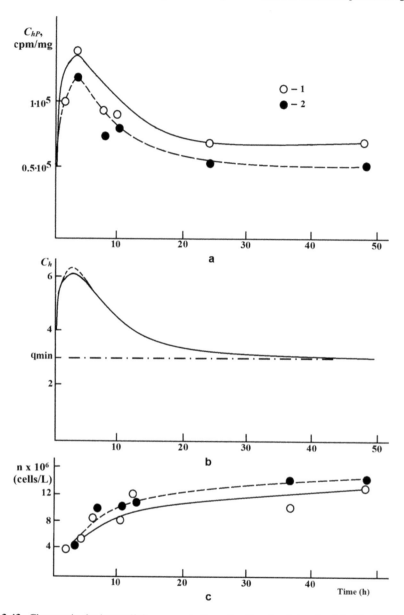

Fig. 3.43 Changes in the intracellular concentration of radioactive (**a**) and stable (**b**) phosphorus and the number of cells *Skeletonema costatum*(c) under light (1) and in the dark (2): C_h and C_{hP}— respectively, the intracellular phosphorus concentration (P—μg/mg; ^{32}P—cpm/mg); n is the number of cells (cells^{-1}); dashed and solid lines represent model data; dash-dotted line—level q_{min} (Egorov et al. 1982)

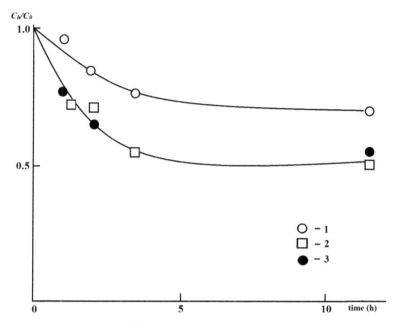

Fig. 3.44 Elimination kinetics of ^{32}P by*Gimnodinium lanskaya* in Mediterranean water (1) and in a medium with additives 20 (2) and 100 (3) μgP/L; C_h/C_0 is the ratio of the remaining radioactivity relative to the initial level (according to the data of V. N. Egorov with V. N. Popovichev and L. G. Kulebakina)

This indicated that the model implemented by Eq. (3.72) can be used to describe the kinetics of the exchange of biogenic elements by hydrobionts in a wide range of changes in their content in the aquatic environment.

From the conditions of application of Eq. (3.72) to a closed system or to a stagnant culture of algae (at $R_v = 1$) it follows that the second term in parentheses on the right side of Eq. (3.72) is equal to μ. Substituting the value μ from the Droop formula (3.68) into Eq. (3.72), we get:

$$\frac{dC_h}{dt} = \frac{V_{max}C_w}{K_m + C_w} - \left[p + \mu_{max}\left(1 - \frac{q_{min}}{C_h}\right) \right] C_h. \tag{3.73}$$

A comparison of Eq. (3.73) obtained on the basis of compartment theory with relation (3.70) showed that they differ only in the parameter p. It is easy to see that Eq. (3.70) is applicable for describing the kinetics of the processes of changes in the intracellular concentration of a biogenic element depending on its content in the medium when this element is limited by the growth process. If production processes are limited by another biogenic element (when $C_h' = q_{min}$) and there is no growth, the second term on the right-hand side of expression (3.70) goes to zero, while the intracellular concentration of element C_h tends to infinity with an

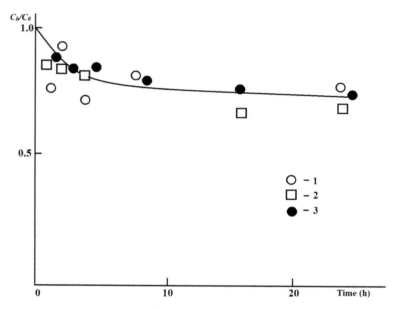

Fig. 3.45 Kinetics of elimination of ^{32}P by suspended matter of the Tyrrhenian Sea in water without mineral phosphorus additives (1) and in a medium with the addition of 25 (2) and 100 (3) μgP/L; C_h/C_0 is the ratio of the remaining radioactivity relative to the initial level (according to the experiments carried out by V. N. Egorov with V. N. Popovichev and L. G. Kulebakina)

increase in t. That is why, in contrast to equality (3.73), Eq. (3.70) is generally inapplicable for reflecting the process of absorption of nutrients by non-growing algal cultures, since growth limitation can be not only chemical, but also, for example, due to photosynthetically active radiation. Equation (3.70) also does not reflect the kinetics of metabolic processes, since it does not take into account the removal of an element from cells as a result of metabolism. In order to eliminate the noted disadvantages, model (3.70) is typically supplemented with a parameter reflecting the respiration process of aquatic organisms (Menshutkin and Finenko 1975; Vinogradov and Menshutkin 1977). If this parameter is numerically equal to the parameter of nutrient removal from the hydrobiont as a result of the cumulative impact of all abiotic and biotic interaction mechanisms, Eqs. (3.70) and (3.73) become identical.

Thus, the studies made it possible to conclude that the development of the compartment theory for describing the kinetic regularities of the exchange of biogenic elements under conditions of chemical limitation of the production of organic matter allowed the relationships to be substantiated consistently with the equations obtained by other methods, as well as to reflect the processes of elimination of biogenic elements via the metabolism of hydrobionts.

3.3   Alimentary Pathway for the Mineral Nutrition of Hydrobionts

3.3.1   Features of the Nutritive Pathway of the Absorption of Radionuclides and Their Isotopic and Non-Isotopic Carriers by Hydrobionts

The food, or alimentary, path of mineral nutrition is inherent to biological consumers. The food they assimilate is used to replenish energy expenditures, for the biosynthesis of organs and tissues, as well as for the catalysis of enzymatic reactions. The undigested part of the food is removed in the form of liquid and solid excretions. As a result of metabolism, mineral elements are removed from aquatic organisms through the gastrointestinal tract and external integuments of the body, as well as in the process of intravital excretion of organic matter.

Studies have shown that, along with chemical environmental pollutants, radioactive substances can enter the bodies of marine organisms with their food. The degree of assimilation of different elements from food varies considerably. Gamma spectra analysis of ^{51}Cr, ^{65}Zn and ^{137}Cs in organisms at three trophic levels—phytoplankton and detritus, euphausiids (krill) and shrimps—supported the hypothesis that fragmented radionuclides sorbed on finely dispersed particles and second-tier plankton are only passed to the third trophic level in insignificant quantities (Osterberg and Pirsi 1971). Atomic adsorption spectrometry was used to established that ciliates, when feeding on bacteria in which Mn, Hi, Zn and Pb were concentrated, accumulated Zn to a greater extent; however, higher levels of Ni and Mn accumulated by the bacteria themselves did not pass into the ciliates (Mansouri-Aliabadi and Sharp 1985). When studying the accumulation of nickel by Scenedesmus algae and Daphnia plankton, it was found that Ni consumed by Daphnia with algae did not accumulate but was instead concentrated in their faeces and released into bottom sediments (Wartas et al. 1985). At the same time, such a highly toxic element as mercury was accumulated along the trophic chain comprised of phytoplankton $\rightarrow$ zooplankton $\rightarrow$ fish fry $\rightarrow$ planktophagous fish $\rightarrow$ squid (Morozov and Petukhov 1979).

A study of the mineral metabolism of hydrobionts with the alimentary intake of radioactive indicators of inorganic substances showed that the kinetic regularities of the processes of accumulation and excretion of radionuclides by marine organisms can be described by exponential functions (Pentreath 1973; Weers 1975a, 1975b; Fowler and Guary 1977). Compartment representations are used in differential models to reflect the regularities of mineral exchange. Kowal (1971) proposed a two-pool modelfor describing the kinetics of cobalt uptake by aquatic organisms. Here, one exchangeable pool of the model reflected the intake of this element into the internal organs, while the other reflected its intake into tissues. Conover and Francis (Conover and Francis 1973) proposed a compartment model to describe the dynamics of nutrient transfer processes along trophic chains. Each compartment of this model corresponds to a trophic link, while the exchange of an element between the compartments is carried out in accordance with first-order metabolic reactions.

When developing a semi-empirical theory, the main task consisted in verifying the adequacy of the compartment model when describing the alimentary pathway of absorption of radionuclides by hydrobionts; the dependencies of its parameters were studied when changing the trophic characteristics of marine organisms, as well as the concentration of mineral elements in their food. An important part of the work consisted in the development of compartment theory for describing the regularities of mineral metabolism in the process of organic matter production by hydrobionts.

3.3.2 Empirical Verification of Compartment Models

The adequacy of compartment models and the influence of environmental factors on the alimentary absorption of inorganic substances by consumers were studied in experiments with ^{65}Zn using the example of the trophic chain consisting of macrophytes—isopods and unicellular algae—copepods. In the first series of experiments, *Idotea* sea lice were fed on green alga Black Sea *Ulva*, which had unchanged radioactivity at ^{65}Zn. It was established (Ivanov et al. 1980), that zinc obtained from food was concentrated to higher relative levels by organisms having lower mass (Fig. 3.46).

This was due to the fact that smaller organisms had higher specific nutrient budgets. The average daily nutrient budget of 13–20 mg idotes was 38% of the body weight, for the size group 90–99 mg—16; for 120–145 mg—7.5. After 9-day accumulation of ^{65}Zn from food, the *Idotea* were transferred to non-radioactive food. In these experiments, a decrease in the concentration of ^{65}Zn in animals was observed, as illustrated by Fig. 3.47. As shown in the figure, although in the preliminary experiment the Idotea accumulated radioactive zinc to different levels (Fig. 3.46), the kinetics of elimination by animals of different size groups, as expressed in relative units, coincided. On the graph plotted on a logarithmic scale along the ordinate axis, the experimental data of observations on the removal of the radionuclide fell on a straight line, indicating the exponential character of the kinetics of the process. The experimental results were satisfactorily approximated by an exponential having a half-life decay period of 12.9 days. The approximating function was:

$$C_h/C_0 = 0.95e^{-0.0537t}, \tag{3.74}$$

where C_h and C_0 are the concentrations of the radionuclide in the hydrobiont at the moments in time t and t_0, taken as initial.

From the experimental observations, approximated by the curve (Fig. 3.47), it followed that the organisms eliminated the food-ingested ^{65}Zn with a rate of $p = 0.0537$ days^{-1}. This made it possible to interpret the kinetic regularities of the behaviour of radiozinc in *Idotea* as a result of alimentary intake and metabolic excretion of the radionuclide from the exchangeable zinc pool in the animals. The rate of intake and metabolic excretion of the radionuclide into the exchange pool

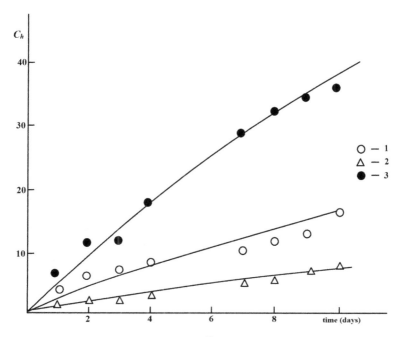

Fig. 3.46 Kinetics of alimentary accumulation of ^{65}Zn by *Idotea baltica* in size groups of different weights: 1: 13–20 mg; 2: 90–99 mg; 3: 120–145 mg; C_h: specific radioactivity of organisms (cpm/mg) (Ivanov et al. 1980)

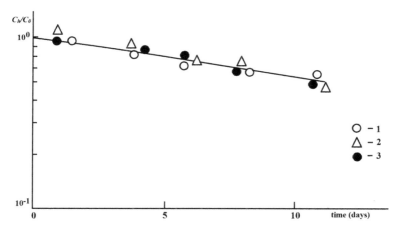

Fig. 3.47 Elimination kinetics of ^{65}Zn by *Idotea baltica* size groups of different weights: 13–20 mg (1); 90–99 mg (2); 120–145 mg (3) (Ivanov et al. 1980)

was determined by the specific nutrient budget (R) and nutrient radioactivity (C_N), as well as the degree of assimilation of the element from food (q), and the excretion proceeded with the rate of the first order metabolic reaction (p). In differential form, this dependence is described by the equation:

$$\frac{dC_h}{dt} = C_N R q - C_h p. \tag{3.75}$$

The solution of the differential Eq. (3.75) having the initial condition $C_h(0) = 0$, corresponding to the experimental conditions of accumulation of radionuclide from food (Fig. 3.46), is obtained in the following form:

$$C_h = \frac{C_N R q}{p} (1 - e^{-pt}). \tag{3.76}$$

Statistical analysis showed that the approximating curves of the form (3.76) did not contradict the results of the experiment at $q = 0.6$. In Fig. 3.46, these curves are shown as solid lines. A comparison of the calculated data with observations attested to the adequacy of the model (3.75).

It is known that not all captured food is absorbed by consumers. In work carried out jointly with V.N. Ivanov, T.G. Usenko and N.A. Filippov (Ivanov et al. 1979), measurements of the radioactivity of food prior to, as well as immediately after feeding, showed that *Idotea metallica* absorbed from 69 to 89% of ^{65}Zn and ^{54}Mn contained in the nutriment. In the experiments, no dependence on a decrease in the concentration of trace elements in Idotea specimens and the amount of food presented was detected. An *Idotea* specimen weighing 94 mg after 20 h of fasting was given food consisting of pontellid copepods weighing 2 mg and containing radioactive zinc in stages for 1 h 25 min. The amount of radioactivity consumed as a percentage of that offered was: 72.5%; 75.0%; 76.0%; 70.0%; 69.2%. This was due to the fact that the chitinous exoskeletons of pontellids were not eaten by the *Idotea* specimens. Due to this aspect of the trophic behaviour of consumers, according to which part of the feed may not be absorbed but is instead transformed into suspended organic matter as waste material, the parameter q of expression (3.75) in the methodological approach used for determining the adequacy of the model took the form of an estimate based on the proportion of the consumed food in relation to that offered being equivalent to that of the radionuclide assimilated from food.

In the second series of observations – in experiments with zooplankton organisms of the Indian Ocean *Euchirella bella*(Piontkovsky et al. 1983) – animals weighing 2.7 mg each ate ^{65}Zn from single-celled algae *Peridinium trochoideum* with a density of 940 c/ml, or 15 mg/L. The radioactivity of the cells, which was at 4230 cpm/mg in terms of the wet weight of algae, was kept constant for 61 h of the experiment. Although, animals accumulated ^{65}Zn when given radioactive nutriment, over time the rate of accumulation approached the stationary level (Fig. 3.48a).

When transferring *E. bella* to non-radioactive food following 61 h of the experiment, the animals began to lose radioactivity (Fig. 3.48b). The kinetics of this process

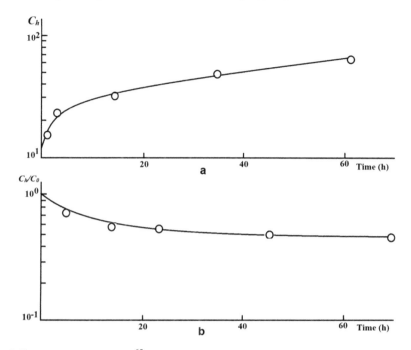

Fig. 3.48 Accumulation (**a**) of ^{65}Zn by *Euchirella bella* with food and **b** excretion into a non-radioactive medium following 61 h of accumulation exposure. C_h – concentration of ^{65}Zn (cpm per mg) in organisms at the current time; C_0–concentration of ^{65}Zn (cpm per mg) at the initial moment of the experiment in a non-radioactive environment (Ivanov et al. 1979)

was approximated with a sufficient degree of accuracy by a double-exponential function:

$$C_h/C_0 = 0.4e^{-4.150t} + 0.6e^{-0.076t}, \tag{3.77}$$

where t is time in hours.

By analogy with the interpretation of experiments with ^{65}Zn and *Idotea*, it was assumed that the entire set of mechanisms determining the exchange kinetics of ^{65}Zn by *E. bella* can be formally reflected by the process of zinc concentration by two pools, each of which exchanges an element at metabolic reaction rates of the first order. The change in the concentration of a substance in exchange pools is described by the equations:

$$\frac{dC_1}{dt} = \lambda C_N Rq - C_1 p_1;$$
$$\frac{dC_2}{dt} = (1 - \lambda)C_N Rq - C_2 p_2, \tag{3.78}$$

where C_1 and C_2 are the concentrations of the radionuclide in the exchangeable pools of the organism $(C_1 + C_2 = C_h)$;

λ is part of the radionuclide flux entering the first exchange pool.

With constant values of the parameters included in Eq. (3.78), the change in the concentration of the radionuclide in the hydrobiont is described by the expression:

$$C_h(t) = K_N C_{N0} + \sum_{i=1}^{n=0} (C_{i0} - C_{N0} B_i) e^{-p_i t}, \tag{3.79}$$

where C_{i0} and C_{N0} are the concentrations of the radionuclide in the exchangeable pools of the hydrobiont and its food source at the moment $t = 0$;

B_1 and B_2- the relative volumes of the exchangeable pools of the hydrobiont:

$$B_1 = \frac{\lambda R q}{p_1}; \; B_2 = \frac{(1 - \lambda) R q}{p_2} \tag{3.80}$$

CF_N is the limiting or static coefficient of food accumulation of the radionuclide by the hydrobiont $CF_N = B_1 + B_2$.

In expressions (3.80), the value of λ reflected the assumption about the proportionality of the radionuclide fluxes obtained from food to the exchangeable pools of the hydrobiont:

$$\lambda = \frac{B_1 p_1}{B_1 p_1 + B_2 p_2}. \tag{3.81}$$

A comparison of expression (3.79) with (3.80) showed that they are analogous in meaning – and that for $C_{w0} = C_{N0}$ they are identical. This made it possible to apply the method of determining B_1 and B_2 when evaluating the parameters and adequacy of models based on the results of experimental observations. According to the experimental data (Fig. 3.48), at $q = 1$ the following was determined: $B_1 = 6 \cdot 10^{-3}$, $p_1 = 4.15$ days^{-1} and $B_2 = 52 \cdot 10^{-3}$, $p_2 = 0.102$ days^{-1}. The results of calculations on the model with the obtained parameters are shown by solid lines in Fig. 3.48. The calculations showed that the model (3.78) describes the kinetics of the exchange of ^{65}Zn by organisms with alimentary intake of the radionuclide with a sufficient degree of adequacy.

Thus, the studies showed that, given the constancy of environmental factors and under the conditions of the alimentary route of intake, the kinetics of the concentration of radionuclides by consumers is adequately described by mathematical compartment models reflecting the exchange of a chemical substance by one or two resource pools of a hydrobiont. The intake of a substance into the pools is proportional to the specific radioactivity of the food, the diet and the degree of assimilation of the radionuclide from food;the introduction from the pools proceeds at rates of first-order metabolic reactions.

3.3.3 Concentration of Isotopic Carriers in Nutriment

In the previous paragraph, since the atoms of the radioactive zinc isotope had practically zero mass, the parameters of Eqs. (3.75) and (3.78) were determined at a constant concentration of the isotopic carrier in food and organisms. It is evident that, when using these models under conditions of varying concentrations of mineral elements in a food source and a hydrobiont, it is necessary to study the dependences of p and q as functions of C_h and C_N. The dependence of the change in the exchange rate index p on the concentration of the isotope carrier in the body of a hydrobiont was studied in experiments with the Black Sea *Idotea baltica* (Popovichev and Egorov 1987). The food for the *Idotea* was the green alga Black Sea *Ulva*, which accumulated ^{137}Cs for several days in a medium with a natural concentration of caesium, as well as in a medium supplemented with caesium in 10^2 and 10^4 μg/L. The accumulation rates of ^{137}Cs *Ulva* were the same in all cases. Three groups of *Idotea* (having an average individual weight of 140 mg) were fed with nutriment prepared as described above. After daily feeding, the *Idotea*, which had assimilated different amounts of caesium at equal nutrient budgets, were transplanted into clean seawater. Subsequent measurements of their radioactivity showed that the kinetics of elimination of ^{137}Cs, expressed in relative units, coincides (see Fig. 3.49). Identical kinetic dependences of the elimination of ^{137}Cs by these groups were also observed in the medium with the addition of the isotopic carrier 10^2 and 10^4 μg/L. The initial conditions $C_{N0} = 0$; $C_1(0) = C_{10}$; $C_2(0) = C_{20}$ of the experiment, whose results are shown in Fig. 3.49, corresponded with the solution of the Eqs. (3.79). From (3.79) it can be seen that the kinetics of elimination, expressed in relative units of $C_h(t)/(C_{10} + C_{20})$, was determined only by the parameters p_1 and p_2. For this reason, the coincidence of the kinetics of elimination of ^{137}Cs by *Idotea*, which had previously accumulated caesium to different levels, showed that the parameters p_1 and p_2 did not depend on changes in the concentration of caesium in the *Idotea*.

The dependence of the rate of exchange indicators (p_1 and p_2) on the form of nutrient were studied in experiments carried out with V. N. Ivanov, T. G. Usenko and N. A. Fillipov on the Southwest Atlantic *Idotea metallica* (Ivanov et al. 1979). In these experiments, eucalanus, pontellid and euphausiid zooplankton organisms, labelled with ^{65}Zn and ^{54}Mn, served as food for Idotea specimens. After being fed with the indicated nutriment types, the radioactivity of the *Idotea* was measured. They were then placed in aquariums with seawater without radioactivity, and periodically subjected to in vivo radiometric testing in weighing bottles with 1–2 ml of water. Following radiometry, the zooplankters were again placed in aquariums. The results of the experiments showed that the kinetics of excretion of ^{65}Zn (Fig. 3.50) and ^{54}Mn (Fig. 3.51) expressed in relative units did not depend on either the type of food or the size of the organisms. These data indicate that ^{65}Zn and ^{54}Mn, assimilated from different food types are exchanged by hydrobionts at the same rates of metabolic reactions for each of the microelements.

The degree of assimilation of mineral elements from food as a function of C_W was studied in experiments with ^{137}Cs and *Idotea*. The green alga Black Sea*Ulva rigida*,

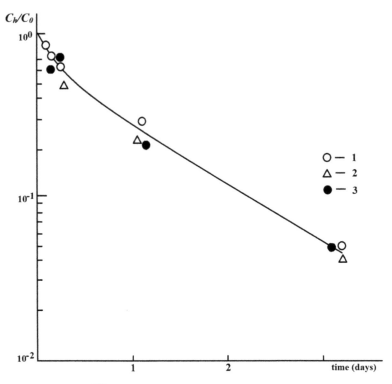

Fig. 3.49 Elimination of ^{137}Cs *Idotea baltica* following daily preliminary absorption from food contained in natural sea water (1) and in a medium with caesium additives 10^2 (2) and 10^4 (3) μg/L (according to experiments carried out by V.N. Egorov and V.N. Popovichev)

labelled with ^{137}Cs, served as nutriment for the *Idotea* specimens (Popovichev and Egorov 1987). The algae accumulated radiocaesium in two aquariums. In the first of them, the concentration of stable caesium in seawater was of the order of 0.4 μg/L, while in the second, the concentration was 50 μg/L. After one week of exposure, the concentration factors of ^{137}Cs in *Ulva* from different aquariums did not differ significantly. This indicated that the concentration of caesium in the *Ulva* from the first and second aquariums began to differ in the same ratio as the concentration of caesium in the water of the same aquariums—that is, by about two orders of magnitude. Following the accumulation of radiocaesium, the algae were placed in aquariums with clean seawater. Groups of *Idotea* having an average individual weight of 32.6 mg were placed in the same aquariums for three hours. During the three-hour feeding period, the pre-starved *Idotea* specimens received an average nutrient intake equal to 17.67% of body weight in the first aquarium, and 17.90% in the second. During this time, they excreted faecal pellets at least once. Radiometric measurements showed that the concentration of ^{137}Cs in the *Ulva* did not change significantly during the feeding period; the *Idotea* in the first aquarium absorbed

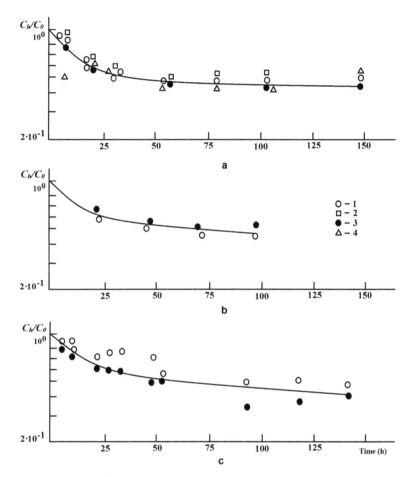

Fig. 3.50 Elimination of ^{65}Zn *Idotea metallica* weighing 32 (1), 52 (2), 85 (3) and 100–200 (4) mg after feeding them with radioactive eucalanus (a), pontellid (b) and euphausiid (c) nutriment sources. C_h/C_0 is the ratio of the remaining radioactivity to the initial value (Ivanov et al. 1979)

22.7%, while the specimens in the second aquarium absorbed 22.9% of the food-carried radioactivity. During the three-hour feeding period, part of the radioactivity in the *Ulva* was returned to the water as "food waste". The process of elimination of ^{137}Cs through the surface integuments of the bodies and faeces of the organisms took place simultaneously with its assimilation from food. It is shown above that the loss of microelement radionuclides with "food waste" and nutriment was, on average, the same at different feeding rates. The kinetics of elimination of ^{137}Cs did not depend on the concentration of caesium in the nutriment of the organisms (Fig. 3.49). In this regard, the fact of close coincidence of the values of the absorbed ^{137}Cs radioactivity made it possible to conclude that the *Idotea* assimilated an equal percentage of caesium from the nutriment. Consequently, when the concentration of

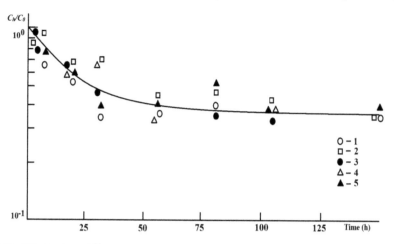

Fig. 3.51 Elimination of ^{54}Mn by *Idotea metallica* with masses of 32 (1), 52 (2), 65 (3), 85 (4) and 116 (5) mg following their preliminary feeding with radioactive eucalanus. C_h/C_0 is the ratio of the remaining radioactivity to the initial value (Ivanov et al. 1979)

caesium in the food source changed by more than an order of magnitude, the relative efficiency of its alimentary absorption was constant, that is, q did not depend on C_N.

A study of the relationship between the feeding rate of marine organisms and indicators of the rate of metabolism of chemicals assimilated from nutriment was carried out on Atlantic planktonic crustaceans (Ivanov et al. 1979). Zooplankton *Scolecithrix danae* organisms, weighing 0.75 mg after 25 h of feeding with single-celled algae*Peridinium trochoideum* and labelled with ^{65}Zn were distributed into two equal-sized groups. One of them was placed in an aquarium with clean sea water, while the the other was placed in an aquarium with phytoplankton. Although, after 41 h of exposure, the organisms of the first aquarium had 82%, while those in the second had 94% radioactivity, it was not possible to obtain a statistically significant difference between the estimates.

In a similar experiment with ^{65}Zn-labelled unicellular *Peridinium trochoideum* algae and *Undinula vulgaris* copepods, the latter demonstrated 55% of the initial radioactivity after 60 h of stay in an aquarium with food, while in an aquarium without food, the corresponding ratio was 53%.

The kinetics of radiozinc elimination, having been previously absorbed by *Calanus gracilis* zooplankters when fed on phytoplankton*Gynmodinium lanskaya* algae labelled with ^{65}Zn, are shown in Fig. 3.52. It can be seen that the excretion of radiozinc by calanoid copepods in a saltwater aquarium without food was slightly higher (Fig. 3.52a) than in an aquarium with added food (Fig. 3.52b). At the same time, no significant differences in the elimination of ^{65}Zn by feeding and starving copepods, as in the experiment with *Scolecithrix danae*, were observed. In our view, the most likely reason for the insignificant observed difference (Fig. 3.52) is the repeated entry of ^{65}Zn into zooplankters along with their unicellular algae food, which had accumulated the radionuclide eliminated by the copepods as part

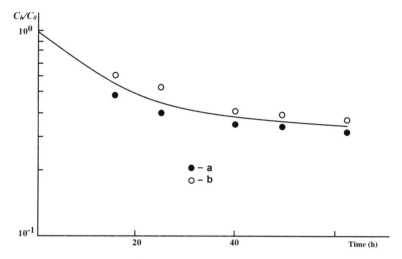

Fig. 3.52 Elimination kinetics of ^{65}Zn *by Calanus gracilis* in an aquarium with filtered sea water (a) and in a medium with unicellular algae *Gymnodinium lanskaya* (b); C_h/C_0 is the ratio of the radioactivity of animals at the current and initial moment of time (according to the experiments of V.N. Egorov and V.N. Ivanov)

of their metabolic processes. In general, the results of experiments with scolecithrix, undinula and calanus showed that there is no significant difference in the kinetics of radiozinc elimination between feeding and starving animals. They demonstrated that the rates of exchange of metals assimilated from food, comprising potential inorganic pollutants of the marine environment, do not depend on the intensity of feeding of zooplankters.

Thus, the study of the absorption of nutrients ^{54}Mn, ^{65}Zn and ^{137}Cs by zooplankton showed that the parameters p_1, p_2 and q are not dependent on an alteration in the magnitude of R, C_N and C_h within a wide range. This indicates that models (3.75) and (3.79) can be used to describe the kinetics of the exchange of consumers of both radionuclides and their isotopic carriers under the conditions of the changing impact of the considered biotic and abiotic environmental factors.

3.3.4 Somatic and Generative Growth

Over the course of a life cycle, the mass of individuals of consumers increases; when they reach sexual maturity, they synthesise reproductive products. As a result of these production processes, the specific biomass of hydrobionts in the environment changes. Equations (3.75) and (3.78) should take into account the kinetic regularities of mineral metabolism, including the growth factors of hydrobionts. We will use relation (3.75) to derive an equation that reflects the kinetics of radionuclide exchange in the process of projection of organic matter by hydrobionts. In Eq. (3.75), the value of C_h is:

$$C_h = A_h/m_h, \tag{3.82}$$

where A_h is the amount of radionuclide or isotopic carrier in the hydrobiont;
m_h is the biomass of an individual hydrobiont.

For $m_h = const$, expression (3.75) is equivalent to:

$$\frac{dA_h}{dt} = C_N Rqm_h - A_h p. \tag{3.83}$$

If the hydrobiont is growing, then m_h is a variable. Substituting in (3.83) instead of A_h its value $A_h = C_h m_h$ and differentiating Eq. (3.83) with respect to the variables C_h and m_h, we obtain an expression describing the kinetics of the exchange of a radionuclide or isotope carrier:

$$m_h \frac{dC_h}{dt} + C_h \frac{dm_h}{dt} = C_N Rqm_h - C_h m_h p. \tag{3.84}$$

From Eq. (3.83) it can be seen that the higher the specific production of individuals of the hydrobiont, the lower the stationary level of the value C_h. A comparison of expressions (3.75) and (3.84) showed that the second of these additionally has a term that is numerically equal to the specific production. This reflects the kinetic mechanism of a decrease in the concentration of a chemical element during the production of organic matter by a hydrobiont due to the appearance of unfilled exchange pools of this element during the growth process.

The following is evident: having made similar transformations with expression (3.78), an equation can be obtained that reflects the kinetic regularities of the mineral metabolism of hydrobionts in the case of using the two-pool model:

$$\begin{aligned}
\frac{dC_1}{dt} &= \lambda C_N Rq - \left(p_1 + \frac{1}{m_h} \frac{dm_h}{dt} \right) C_1; \\
\frac{dC_2}{dt} &= (1 - \lambda) C_N Rq - \left(p_2 + \frac{1}{m_h} \frac{dm_h}{dt} \right) C_2.
\end{aligned} \tag{3.85}$$

It is known that the growth rate of a hydrobiont is determined by the difference between assimilated nutrient and expenditure on respiration (Zaika 1972). Taking this circumstance into account, we can write:

$$\frac{dm_h}{dt} = m_h R(A_N - A_R). \tag{3.86}$$

where A_N is the assimilated part of the nutrient;
A_R is the portion of the nutrient spent on the respiration of a hydrobiont.

The part of the nutrient assimilated for growth, which we denote by q_N, is equal to:

$$q_N = A_N - A_R. \tag{3.87}$$

Substituting the value $\frac{dm_h}{dt}$ from (3.86) into (3.84) taking into account (3.87) we get:

$$\frac{dC_h}{dt} = R(C_N q - C_h q_N) - C_h p. \tag{3.88}$$

From Eq. (3.88) it can be seen that, depending on the sign of the difference $(C_N q - C_h q_N)$, the nutrient intake of a chemical substance or its radionuclide can lead both to an increase and a decrease in its concentration in the hydrobiont. With an equal degree of assimilation of matter and nutriment for growth as a whole ($q = q_N$), its concentration in the hydrobiont will decrease if $C_N < C_h$, which means, in relation to contaminants, that nutriment less contaminated than the hydrobiont will lead to the cleansing of the chemical pollutant from the consumer. If the kinetics of the consumer's mineral exchange is described by a two-compartment model, then from expression (3.86) we obtain:

$$\frac{dC_1}{dt} = \lambda R(C_N q - C_1 q_N) - p_1 C_1;$$
$$\frac{dC_2}{dt} = (1 - \lambda)R(C_N q - C_2 q_N) - p_2 C_2. \tag{3.89}$$

Thus, the development of the compartment theory as applied to the description of mineral metabolism under the conditions of organic matter production by hydrobionts has shown that the kinetics of their exchange by growing hydrobionts can described by one- and two-compartment models during alimentary absorption of inorganic substances or their radionuclides. Each of the compartments corresponds to an exchange pool. The intake of an inorganic substance or its radionuclide into the pools is proportional to the budget or concentration of this substance in the nutrient, the digestibility of the nutrient budget and the efficiency of assimilation of the element from the nutriment. Elimination of substances from exchange pools is determined by the rates of first-order metabolic reactions.

In hydrobiology, production refers to all types of synthesis of organic matter by aquatic organisms (Zaika 1972). Somatic growth refers to an increase of protein reflected in the body weight of individual organisms, as well as an increase in energy – mainly composed of fat – reserves (Shulman 1972). By generative growth is understood the biosynthesis of generative tissues – roe, gonads and eggs. In addition, in the course of vital processes, consumers release organic metabolic products into the environment consisting of exoskeletons and other exuviae. The contribution to production of individual growth components varies across different taxonomic groups of aquatic animals. For example, the growth of marine copepods (Sazhina

1980) is accomplished discretely in the transition from nauplius to copepodite stages. Each stage ends with moulting, with a direct increase in body weight occurring at those moments of time when the organisms lack chitinous integuments. As a rule, adult copepods undergo no somatic growth, but only produce reproductive tissues. In light of the above, it can be seen that model (3.89) reflects only the integral effect of growth processes on the kinetic regularities of the mineral metabolism of hydrobionts, making it possible to calculate the balance of radionuclides and their isotopic carriers in ecosystems during the production of organic matter.

3.4 Generalised Characteristics of the Semi-Empirical Theory

3.4.1 Conjoint Parenteral and Alimentary Absorption of Chemical Matter by Hydrobionts

Until now, we have given separate consideration to the processes of direct absorption of chemical and radioactive substances from the aquatic environment by living and inert components of ecosystems to those processes involving alimentary nutrition. While absorption through sorbing surfaces and outer integuments is the only form of mineral interaction and nutrition for inert matter and producers, consumers also ingest radionuclides and their isotopic and non-isotopic carriers from food. Meanwhile, decomposers can absorb chemical elements both metabolically through cell membranes from water, as well as via sorption during the mineralisation of organic substances and chemical compounds.

Since the absorption of mineral elements occurs simultaneously via parenteral and alimentary pathways, at any time, the equality $C_h = C_{pr} + C_{al}$ is self-evidently true, where C_h is the concentration of a mineral element (radionuclide) in a hydrobiont, established as a result of absorption from the aqueous medium (C_{pr}) and from food (C_{al}). With regard to single-compartment models, reflecting the parenteral and alimentary processes of concentration of an element by one exchange pool, we can write:

$$\frac{dC_h}{dt} = \frac{V_{max}C_w}{K_m + C_w} + C_N Rq - C_{pr}p_{pr} - C_{al}p_{al}. \qquad (3.90)$$

where p_{pr} and p_{al} are the indicators of the exchange rate of the element received directly from the aquatic environment and with food, respectively.

With the independence of the parameters p_{pr}, p_{al} and q of Eq. (3.90) from C_w and C_N, this expression is clearly applicable for describing the kinetics of a hydrobiont's exchange of both radioactive substances and their stable analogues.

Under the conditions of experiments designed for studying the intensity of the removal of mineral elements from hydrobionts using the radioactive tracer method, at

$C_w(0) = 0$; $C_N(0) = C_{N0}$; $C_{pr}(0) = C_{pr0}$; $C_{al}(0) = C_{al0}$ the solution to Eq. (3.90) has the form:

$$C_h(t) = C_{pr0}e^{-p_{pr}t} + C_{al0}e^{-p_{al}t}. \qquad (3.91)$$

It can be seen from this relationship that, at $p_{pr} = p_{al}$, the kinetics of the elimination process can be described by one exponential. It follows from this that, in those cases when the exchange of an inorganic substance directly from the aqueous medium and from food is the same, the kinetics of removing its radioactive label must also coincide. If an element is included in several exchange pools of hydrobionts, then the data on elimination, plotted on a graph with a logarithmic scale along the ordinate axis, can be described by a dependence of the form (3.2). If the rates of exchange of an element by the pools of the hydrobiont are the same along the parenteral and alimentary intake pathways, then the dependences reflecting the kinetics of removing its radionuclide from the exchange pools can clearly both coincide with and differ from each other. The curves will coincide under identical initial conditions of the relative filling of the pools at the moment of transferring the specimens to non-radioactive food or to a medium without a radioactive label. Under different initial conditions of the relative occupancy of the exchange pools, the kinetics of the removal of radionuclides from the nutriment and from the aquatic environment will not coincide. If dependences (3.2) and (3.91), reflecting the kinetics of elimination in the final phase of the experiment, have parallel rectilinear sections, we can conclude that p_{pr} and p_{al} coincide for slowly exchanging pools. If the parallelism of the linear sections of the radionuclide elimination curves on the graph is established at the same time points, it can be hypothesised that $p_{pr} = p_{al}$ also applies to rapidly exchanging pools. A more accurate comparison of the rates of exchange of resource pools requires the isolation of exponential parameters according to the data of empirical observations, reflecting the kinetics of elimination of radionuclides both in the parenteral and alimentary pathways of their intake by the hydrobiont.

Observations have shown that the kinetics of the elimination of radionuclides that enter a hydrobiont from a nutriment and directly from the aquatic environment can both coincide and differ. It was found that the elimination of ^{65}Zn by *Eucalanus elongatus* crustaceans practically coincided irrespective of whether received alimentarily or parenterally (Fig. 3.53a), while the alimentary and parental kinetics of elimination of ^{54}Mn (Fig. 3.53b) differed (Ivanov et al. 1979). In experiments with ^{65}Zn and *Idotea* (Egorov and Ivanov 1981), it was found that the half-life of a radionuclide that entered the pool parenterally at a lower metabolic rate was 9 days (Fig. 3.13). When studying the alimentary pathway of mineral element intake (Ivanov et al. 1980), it was determined that the half-life of ^{65}Zn from the pool of *Idotea* at a lower metabolic rate was of the same order of 12.9 days (Fig. 3.47).

The results of experiments simultaneously studying the excretion of ^{65}Zn by *Idotea* following its preliminary parenteral and alimentary absorption, respectively, (Ivanov et al. 1986), are illustrated in Fig. 3..3.54. The graphical isolation of the exponents based on the data of these observations showed that, in both cases, the half-life of radioactive zinc from the pool with a fast exchange was 0.7 days, while

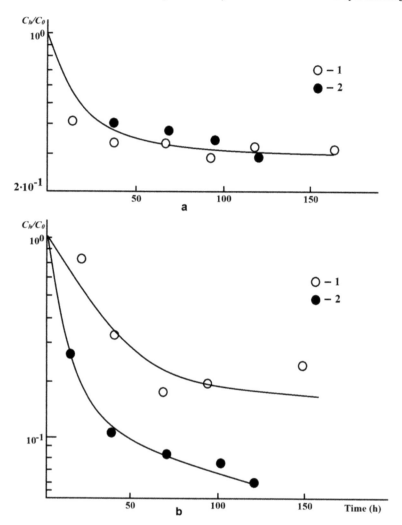

Fig. 3.53 Elimination of ^{65}Zn (a) and ^{54}Mn (b) by*Eucalunus elongatus*obtained from nutriment (1) and accumulated from water (2); C_h/C_0 is the ratio of the remaining radioactivity to the initial level (Ivanov et al. 1979; Egorov et al. 1980)

from the pool with a slow exchange it was 14.8 days. These data indicated that the metabolic rates in the corresponding exchange pools of zinc were the same both in the case of parenteral and alimentary absorption of ^{65}Zn by *Idotea*.

There is reason to believe that the differences in the kinetics of elimination are determined by the difference in the physicochemical forms of radionuclides of mineral elements.

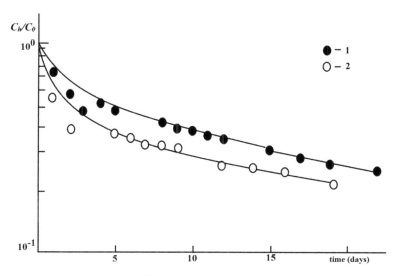

Fig. 3.54 Kinetics of elimination of ^{65}Zn *by Idotea baltica* following absorption of radiozinc from nutriment (1) and directly from water (2). The exposure of preliminary absorption of ^{65}Zn from food was 10 days, while from water, the exposure was 3 days. C_h/C_0 is the ratio of the remaining radioactivity to the initial value (Ivanov et al. 1979; Egorov et al. 1980)

It has been established that ^{60}Co accumulated by hydrobionts from the aquatic environment is eliminated with a greater intensity than ^{58}Co assimilated from food in the form of cobalamin (Nakamura et al. 1982). It is quite natural that differences should exist in the kinetics of exchange by hydrobionts of elements taking different physicochemical forms. The same elements can be a constituent part of inorganic substances of varied biological significance, from chemical substrates that limit production to highly toxic environmental pollutants. The effect of such substances on hydrobionts and their metabolism can also radically differ.

Subsequently, only those cases were considered when mineral elements and the various forms and chemical compounds in which they participate are equally included in the metabolism of hydrobionts. At the same time, the hydrobiont "did not distinguish" the elements that came with food or directly from the aquatic environment but exchanged them at the same rates $p_{pr} = p_{al} = p$. In this case, from (3.90) it follows:

$$\frac{dC_h}{dt} = \frac{V_{\max}C_w}{K_m + C_w} + C_N Rq - C_h p. \tag{3.92}$$

If the hydrobiont grows and accumulates mineral elements both in exchangeable and non-exchangeable pools, taking into account (3.36), we can write:

$$\frac{dC_h}{dt} = \frac{V_{\max}C_w}{K_m + C_w} + C_w r_{ne}(P_h/B_h) + C_n Rq - \sum_{i=1}^{n} C_i p_i - C_h(P_h/B_h), \tag{3.93}$$

where C_i is the concentration of a mineral element (radionuclide) in the i-th exchange pool of the hydrobiont $C_h = \sum C_i - C_{ne}$;

P_h/B_h is the specific production of the hydrobiont;

p_i are the exchange rate constants of the pools;

where r_{ne} is an indicator of the rate of entry of an element or radionuclide into a non-exchangeable pool.

The first term on the right-hand side of expression (3.93) describes the kinetics of the parenteral entry of an element or radionuclide into exchangeable pools according to the Michaelis–Menten equation, the second reflects the filling of non-exchangeable pools during growth, while the third represents the intake of an element with nutriment. The fourth term describes the kinetics of the elimination of inorganic substances from the metabolic pools of a hydrobiont, while the fifth reflects the dynamic state of the relative incompleteness of the exchange pools that arises during the growth of the hydrobiont.

Thus, with the simultaneous absorption of identical forms of mineral elements or their radionuclides from food and directly from the aquatic environment, the kinetic regularities of the exchange of these elements by the hydrobiont can be interpreted as a result of the interaction of the elements contained in water and food with the exchangeable and non-exchangeable pools of the hydrobiont. The parenteral absorption of mineral elements is accomplished in accordance with the Michaelis–Menten equation, while alimentary absorption is proportional to the nutrient budget, the concentration of these elements in food and the efficiency of their assimilation from the nutriment. The elimination of mineral elements or their radionuclides from the exchangeable pools of hydrobionts is carried out in accordance with the rates of first-order metabolic reactions.

3.4.2 Mass Transfer in Open Systems

When describing the kinetics of the interaction of mineral elements and their radionuclides by hydrobionts under the real functional conditions of marine ecosystems, it is necessary to take into account the transfer of these elements with water, as well as with hydrobionts between aquatic areas or water layers on the spatio-temporal scales of the studied processes. In thermodynamically open systems, the average concentration of an element in hydrobionts in a given volume of water can change not only due to interaction with chemical or radioactive components of the environment, but also as a result of the elimination of a part of its biomass from this volume and the entry of hydrobionts into it from a body of water. Taking into account only the processes of migration of hydrobionts into a given volume of water and the elimination of a part of their biomass from the volume, the balance equation of the content of an element or its radionuclide in hydrobionts can be written in the following form:

$$\frac{dA_h}{dt} = p_m A_m - p_{el} A_h, \tag{3.94}$$

where A_h and A_m is the content of the element in the biomass of aquatic organisms from the analysed volume of water and from the adjacent aquatic area;

and p_{el} are indicators of the rates of biomigration and elimination, respectively.

Let us substitute into expression (3.94) instead of A_h and A_m their values $A_h = B_h \cdot C_h$ and $A_m = B_m \cdot C_m$, where B_h, C_h and B_m, C_m are the biomass and concentration of an element (radionuclide) in the biomass from the analysed and adjacent aquatic areas, respectively. Differentiating the Eq. (3.94) with respect to the variables B_h and C_h, we get:

$$\frac{dC_h}{dt} = \frac{B_m}{B_h} C_m p_m - C_h p_{el} - \frac{C_h}{B_h} \frac{dB_h}{dt}. \tag{3.95}$$

If the specific biomass of hydrobionts of the analysed volume of water is determined only by processes of migration and elimination, then we can write:

$$\frac{dB_h}{dt} = B_m p_m - B_h p_{el}. \tag{3.96}$$

Substituting the value $\frac{dB_h}{dt}$ from (3.96) into expression (3.95), we get

$$\frac{dC_h}{dt} = p_b(C_m - C_h), \tag{3.97}$$

where $p_b = B_m p_m / B_h$ is an indicator of the rate of mass transfer of hydrobionts in an open system.

Thus, the change in the average concentration of an element or its radionuclide in hydrobionts in aquatic areas that exchange marine organisms with adjacent waters is determined by the intensity of mass transfer of living matter in these areas and the difference in the element concentrations in migrating hydrobionts. From relation (3.97) it can be seen that the concentration of an element in hydrobionts does not depend on the rate of elimination of biomass from the analysed aquatic area. By analogy with (3.94), it is easy to observe that, in the presence of water exchange of the investigated volume with the external environment, the balance equation of the content of an element in the water of this volume can be written in the following form:

$$\frac{dA_w}{dt} = \Delta A_{wm} - A_w p_w, \tag{3.98}$$

where A_w isthe content of the element (radionuclide) in the analysed volume of water;

ΔA_{wm} is the rate of intake of the element as a result of water exchange;

p_W is an indicator of the rate of water exchange.

In the expression (3.98) $\Delta A_{wm} = C_{wm} \cdot \Delta m_w$ and $A_w = C_w \cdot m_w$, where C_w is the concentration of an element in the water of the analysed volume with a mass of m_w, while C_{wm} is the concentration of an element in water having a mass of Δm_w entering the analysed layer as a result of water exchange.

Substituting the values ΔA_{wm} and A_w into expression (3.98) and taking into account the fact that $m_w = const$, we get:

$$\frac{dC_w}{dt} = \frac{\Delta m_w}{m_w} C_{wm} - C_w p_w. \tag{3.99}$$

In (3.99), since the ratio $\Delta m_w / m_w$ is numerically equal to the rate of water exchange p_w of a given volume of water with adjacent aquatic areas, we finally get:

$$\frac{dC_w}{dt} = p_w (C_{wm} - C_w). \tag{3.100}$$

From (3.97) and (3.100) it can be seen that the ratios reflecting the change in the concentration of mineral elements or their radionuclides in hydrobionts and water as a result of migration and water exchange are analogous. It follows that the rate of change in the concentration of an element in the components of ecosystems as a result of mass transfer is determined by the rate of mass transfer and the difference in the concentrations of the element in the migrating components. Depending on the sign of the difference in expressions (3.97) and (3.100), mass transfer can either increase or decrease the concentration of elements and their radionuclides in a hydrobiont or aquatic environment.

3.4.3 Balance Equalities of the Generalised Model

Under natural conditions, biotic and abiotic components jointly affect the processes of concentration and exchange of radioactive and chemical substances by hydrobionts. Chemical components can be absorbed by hydrobionts both directly from the aquatic environment and from food. At the same time, mass exchange of any considered volume of water is possible, provided by the transfer and migration of hydrobionts and exchange of water with adjacent aquatic areas. In order to describe the kinetics of the interaction of chemical components of the environment with living and inert matter in natural waters, it is necessary to use balance equalities corresponding to open systems comprising a radioactive or chemical substance in the marine environment and a hydrobiont.

Taking into account the sorption, parenteral and alimentary pathways of interactions of living and inert matter with radioactive and chemical components of the marine environment, as well as mass transfer, the equation of radioisotope and mineral balance in open systems has the form:

$$\frac{dC_h}{dt} = \frac{V_{max}C_w}{K_m + C_w} C_w r_{ne}(P/B_h) + C_N Rq + \frac{B_m}{B_h} p_m(C_m - C_h)$$

$$- \sum_{i=1}^{n} C_i p_i - C_h(P/B_h). \tag{3.101}$$

The first term on the right-hand side of expression (3.101) reflects the flow of matter into the exchangeable pools of the hydrobiont directly from the aquatic environment, while the second determines its supply to the non-exchangeable pools only during the process of growth. The third term on the right side describes the alimentary intake of the substance, while the fourth represents the intake of the substance from adjacent aquatic areas as a result of the migration and transport of hydrobionts. The first term on the right side having a negative sign reflects the processes of removal of the chemical component from the exchange pools of the hydrobiont, while the second represents the relative decrease in the concentration of the component in the hydrobionts due to the occurrence of new exchange pools.

The change in the concentration of a substance in an aquatic environment as a result of interaction with a hydrobiont in relation to an open system is described by the expression:

$$\frac{dC_w}{dt} = \frac{B_h}{m_w}\left[\sum_{i=1}^{n} C_i p_i - C_w\left(\frac{V_{max}}{K_m + C_w} + r_{ne}\frac{P}{B_h}\right)\right] + p_w(C_{wm} - C_w) + \Delta C, \tag{3.102}$$

where B_h is the biomass of the hydrobiont in the analysed volume of water having a mass of m_w;

p_w is an indicator of the rate of water exchange of the considered volume of water with adjacent aquatic areas;

C_{wm} is the concentration of a substance in the water of adjacent aquatic areas;

ΔC is the rate of intake of matter from the external environment.

The first term on the right-hand side of Eq. (3.102) is equal to the rate of entry of the chemical component removed into the water from the exchange pools of the hydrobiont, while the second term (in brackets) describes the kinetics of the processes of absorption of matter by the exchange and non-exchange pools of the hydrobiont. The third term reflects the change in the concentration of matter as a result of water exchange in the aquatic area, while the fourth characterises the input from external sources, for example, of anthropogenic origin.

Above, in one-factor experiments, the adequacy of the ratios reflecting the kinetic characteristics of the effect of each of the indicated mechanisms of intake and elimination of substances by hydrobionts was shown. It was demonstrated by comparing the results of modeling and observations, where the parameters of the models, determined from the data of experiments, were for the most part independent of those used to establish the adequacy of the models. For this reason, we can conclude that Eqs. (3.101) and (3.102) implement an empirical model reflecting the kinetics of the exchange of a chemical element or its radionuclide, taking into account their parenteral and alimentary absorption by a hydrobiont. They also describe the influence of production processes and mass transfer of living matter and waters between aquatic areas on the kinetic characteristics of the interaction of hydrobionts with radioactive and chemical substances in the marine environment.

3.4.4 Scope of the Generalised Model

As already mentioned, when solving dynamic problems of radiochemoecology using the methods of the theory of "complex" geosystems (Polikarpov and Egorov 1986), the empirical blocks of semi-empirical models should reflect the kinetic patterns of the interaction of living and inert matter with radioactive and chemical components of the marine environment under the influence of biotic and abiotic factors. The list of factors included in the consideration should be determined by the specific formulation of the problem, the space–time scale of the study, as well as the accuracy and smoothness in frequency of the predicted parameters.

The set of parameters optimal for the problem is not typically established in advance (Belyaev 1978). The established practice for solving dynamic problems relies on a two-stage definition. At the first stage, when setting the problem, a set of a priori variables is established that determine the interaction of factors. It is at the second stage, when studying the dynamic characteristics of the model and checking its adequacy, that the significance of the variables is determined. In this case, both the possibility of excluding a certain number of variables from consideration due to their low significance, as well as the need to take into account new factors, can arise. In this regard, it may be necessary not only to take new parameters into account, but also to detail the impact characteristics of factors that have already been considered. The specified features of the systematic approach to the study of the dynamics of the interaction of living and inert matter with the radioactive and chemical components of the marine environment require that the variables of empirical models reflecting the effects of various factors be independent and that their structure be such as to allow for subsequent modification and development.

In a generalised empirical model of the form (3.101) and (3.102), the kinetics of metabolic processes are interpreted as a result of the absorption and excretion of substances by the metabolic pools of a hydrobiont. The basis for this interpretation was the theory of compartment models (Sheppard and Householder 1951), while the empirical basis for determining its structure and exchange parameters consisted in the exponential approximation of the kinetic characteristics of the accumulation and elimination of chemical elements and their radionuclides by hydrobionts. In this model, the possibility of nucleation and plenishment of the material resource pool during the growth of a hydrobiont was necessarily assumed. Due to this, the total flow of material resources into the hydrobiont can be differentiated into exchange and non-exchange flows. Meanwhile, the methods of interpretation of observations that determine the structure of the model do not provide grounds for identifying the compartments – and therefore exchange and non-exchange pools – with any territorially separated organs or tissues of marine organisms.

It should be noted that the balance Eqs. (3.101) and (3.102), given by a set of empirical parameters, can be applied to both chemical elements and their radionuclides. This is due to the fact that, with the exception of tritium (^{3}H), the isotope effect does not significantly affect the kinetic characteristics of the exchange of chemical elements by hydrobionts. Section 3.1.3 of Chapter 3 shows that the dependence of the rate of absorption by aquatic organisms of substrates of different biological significance can be described by the Michaelis–Menten equation, which is identical to the Langmuir equation. However, the first of these was obtained in the course of studying metabolic processes, while the second is related to sorption. V.I. Belyaev (1964) studied the interaction of clouds in meteorology, as well as the kinetics of the exchange of radionuclides by hydrobionts, using the Lagrangian marker method. Under the same initial conditions, the solutions of his obtained isotopic exchange equations coincide with those of equations that implement compartment models. Consequently, the processes of mixing of particles with physicochemical properties that do not differ in this system can be described by mathematical models whose parameters reflect first-order metabolic reactions. However, these studies exclude the grounds for associating the parameters of the model with any one mechanism responsible for the parenteral absorption of substances by hydrobionts. Rather, they evidently reflect the combined effects of physical, chemical and biological factors.

We may recall that the versatility of the application of the Michaelis–Menten equation lies in its capability of describing absorption kinetics as a result of various interaction mechanisms across the entire range of changes in the content of substances of various biological significance in a medium. If the problem of modelling is solved in the range of microconcentrations of a substance in the medium at $C_w \ll K_m$, then at $r_{ne} = 0$, the solution of Eq. (3.101) in relation to producers can be conveniently analysed in the form of the dependence of the change in the concentration factor $CF(t)$ of a substance by a hydrobiont over time. In this case, there is no need to determine C_w from observations, since $CF(t)$ does not depend on C_w. When the problem is solved in relation to a radioactive pollutant under conditions of a significant change in the concentration of its stable analogue in water, Eqs. (3.101) and (3.102) must be considered simultaneously for both the radionuclide and the isotopic carrier. This is due to the fact that, for the same values of the parameters, the initial conditions of Eqs. (3.101) and (3.102), considered in relation to a chemical element and its radionuclide, may not coincide.

If the sets of initial conditions do not coincide, situations may arise when, simultaneously with the accumulation of the radionuclide by the hydrobiont, the removal of the corresponding stable isotopic carrier from it can be observed. When a chemical analogue is also present with the analysed element in the marine environment, in place of C_w, the value $a_{ch}C_w$ can be indicated in Eqs. (3.101) and (3.102). If a substance is removed from a hydrobiont as a result of metabolism in physicochemical forms that differ from the initial ones, it is necessary to use parameters that take into account the biotransformation of the substance in equations taking the form (3.101) and (3.102). In the generalised empirical model, the role of hydrobiont can be fulfilled not only by individual marine organisms, but also by groups of organisms having approximately the same mass, as well as animals of the same trophic level and the products of their vital activity. If the hydrobiont is not characterised by all the mechanisms of its inter-action with matter, reflected by Eqs. (3.101) and (3.102), the influence of individual absorption or elimination processes can be excluded by assigning zero values to the corresponding parameters of these equations. The remaining model parameters can be set by nonzero constants or functions. When, according to the conditions of the problem, the dimensional characteristics of hydrobionts are not averaged in relation to all constituent age groups, the parameters of the model are determined by a power function of the dimensional characteristics of the hydrobionts. In this case, in rela-tion to consumers, it is necessary to take into account the dependence of the specific nutrient budget R on the physiological size-mass characteristics.

The patterns described above indicate that the rate of absorption of chemical or radioactive substances by hydrobionts is generally determined by the concentration of the substance and its chemical analogues in the medium, the relevant size-mass characteristics, as well as the conditions of their nutrition and growth. The rate of their elimination depends on the level of accumulation in the exchange pools of the hydrobiont and on its metabolic activity in relation to the given substance. In the range of microconcentrations of a substance in the medium, the processes of its absorption proceed in accordance with the rates of first-order metabolic reactions. In the region of macroconcentrations, the order of metabolic reactions changes to zero. Elimination from the hydrobiont does not depend on the content of the substance in the medium but proceeds in accordance with the rate of first-order metabolic reactions across the entire range of changes in its concentration. It should be noted that these regularities apply to the interaction of hydrobionts with chemical substances of different biological significance, from biogenic elements limiting growth processes to chemical toxicants. Meanwhile, when using Eqs. (3.101) and (3.102) for solving production problems, it is evident that that the model parameters p_i and q are likely to depend a certain extent on the physiological state of organisms, nutrient budgets and nutriment energy equivalents.

It should also be noted that, since the generalised empirical model is intended primarily for environmental studies, its parameters reflect the kinetic regularities of the interaction of chemical substances of the marine environment with hydrobionts on the time scale of sorption and metabolic processes. While the parameters of the generalised empirical model are independent, the structure of the model allows its modification and development due to the detailed consideration of interaction char-acteristics and need to take new factors into account. For this reason, Eqs. (3.101) and (3.102) satisfy the requirements for semi-empirical models studied with a system-atic approach for solving problems of the interaction of living and inert matter with radioactive and chemical components of the marine environment, primarily in order tassess and predict its self-cleaning ability in relation to pollutants.

Conclusion

The regularities of the absorption of chemicals having varying biological signif-icance, along with their radionuclides, by the living and inert matter of the marine environment are described in terms of linear functions and the Lang-muir or Freundlich equations. Meanwhile, the rates of mineral and sorption exchange, which are described by the Michaelis–Menten and Droop laws, correspond to the rates of metabolic reactions of the first or zero orders. Regard-less of the biological significance of mineral elements, the kinetics of the exchange of their radionuclides by hydrobionts of different trophic levels can be interpreted in terms of a process of concentration of elements in one or two exchange pools (or compartments) of aquatic organisms, exchanged at rates of first-order or zero-order metabolic reactions. The size of the compartments of exchange pools is in a power-law form of dependence on the individual mass of hydrobionts, while the rate of exchange of an element in the compartments does not depend on the size characteristics of individual specimens in a partic-ular taxonomic group. The kinetics of production processes are described by a model, two compartments of which correspond to the accumulations of an element in a hydrobiont arising during growth, which are exchanged with the environment according to indicators of the rates of first-order metabolic reac-tions. The third compartment only reflects the entry of the element into the main, non-exchangeable or weakly exchanging structures during the somatic growth of the marine organism. The numerical values of the rates of intake into the resource pools are determined by the physical and chemical forms of pollutants. The rates of elimination of different physicochemical forms of inorganic substances from the corresponding exchange pools in hydrobionts are the same. The developed semi-empirical models satisfy the requirements for describing the material, energy, radioisotope and mineral balance in marine ecosystems, taking into account sorption, trophic and chemical interactions.

References

Artsimovich LA (ed) (1963) Spravochnik po yadernoi fizike (translation from English). Gosu-darstvennoe izdatel'stvo fiziko-matematicheskoi literatury, Moscow, 631 p (in Russian)

Atkins GL (1969) Multicompartment models for biological systems. Methuen, London, p 153

Bachurin AA (1968) Matematicheskoe opisanie dinamiki protsessov radioaktivnogo zagryazneniya morskikh organizmov iz vodnoi sredy. Atomizdat, Moscow, 28 p (in Russian)

Barinov GV (1965) Obmen ^{45}Ca, ^{137}Cs, ^{144}Ce mezhdu vodoroslyami i morskoi vodoi. Okeanologiya 5(1):111–116 (in Russian)

Beasley TM, Lorr HV, Conor JJ (1982) Biokinetic behavior of technetium in the red abalone *Haliotis rufescens: a reassessment*. Health Phys 43(4):501–507. https://doi.org/10.1097/00004032-198 210000-00004

Belyaev VI (1964) Metod Lagranzha v kinetike oblachnykh protsessov. Gidrometeoizdat, Leningrad, 187 p (in Russian)

Belyaev VI (1972) Uravneniya obmena radionuklida mezhdu morskimi organizmami i sredoi. In: Radiatsionnaya i khimicheskaya ekologiya gidrobiontov. Naukova dumka, Kiev, pp 62–71 (in Russian)

Belyaev VI (1978) Teoriya slozhnykh geosistem. Naukova dumka, Kiev, 155 p (in Russian)

Benson S (1964) Osnovy khimicheskoi kinetiki. Mir, Moscow, 603 p (in Russian)

Berman M (1965) Compartmental analysis in kinetics. In: Stacy RW, Waxman B (eds) Computer in biomedical research, vol 2. Academic, New York, pp 173–201

Bernhard M (1971) The utilization of simple models in radioecology. In: Marine radioecology symposium, Hamburg, F. R. Germany, 20–24 Sept 1971

Bernhard M, Bruschi A, Möller F (1975) Use of compartmental models in radioecological laboratory studies. Design of radiotracer experiments in marine biological systems. IAEA, Vienna, pp 241–289

Bloom SG, Raines DL (1971) Mathematical models for predicting the transport of radionuclides in a marine environment. Bioscience 21(12):691–696. https://doi.org/10.2307/1295750

Botov NG (1975) O nekorrektnosti zadach statisticheskoi dinamiki protsessov migratsii radionuklidov. Soobshchenie 3, dep. VINITI, no. 2698-75. Chelyabinsk, 15 p

Brownell GL, Berman M, Robertson J (1968) Nomenclature for tracer kinetics. Int J Appl Radiat Isot 19(3):249–262. https://doi.org/10.1016/0020-708X(68)90022-7

Brunel L, Dauta A, Guerri MM (1982) Croissance algale: Validation d'un modèle à stock à l'aide de données expérimentales. Ann Limnol 18(2):91–99. https://doi.org/10.1051/limn/1982016

Burlakova ZP, Krupatkina DK, Lanskaya LA, Yafarova DL (1979) Vliyanie plotnosti populyatsii morskikh odnokletochnykh vodoroslei na potreblenie fosfora i osnovnye fiziologicheskie pokazateli kletok. In: Vzaimodeistvie mezhdu vodoi i zhivym veshchestvom: proceedings of the international symposium, Odessa, 6–10 Oct 1975, vol 1 (in Russian)

Burlakova ZP, Serdyukov OM, Egorov VN, Ivanov VN, Usenko TG, Markova LS (1980) Nakoplenie i vyvedenie tsinka-65 vodorosl'yu *Stephanopixis palmeriana* v eksperimental'nykh usloviyakh. Ekologiya Morya 2:41–44 (in Russian)

Burmaster DE, Chisholm SW (1979) A comparison of two methods for measuring phosphate uptake by *Monochrysis lutheri* Droop grown in continuous culture. J Exp Mar Biol Ecol 39(2):187–202. https://doi.org/10.1016/0022-0981(79)90013-3

Conover RJ, Francis V (1973) The use of radioactive isotopes to measure the transfer of materials in aquatic food chains. Mar Biol 18(4):272–283

Cranmore G, Harrison FL (1975) Loss of ^{137}Cs and ^{60}Co from the oyster *Crassostrea gigas*. Health Phys 28(4):319–333

Davies AG (1973) The kinetics of and a preliminary model for the uptake of radio-zinc by *Phaeodactylum tricornutum* in culture. In: Radioactive contamination of the marine environment: proceeding of a symposium, Seattle, 10–14 July 1972

Dobrolyubskii OK (1956) Mikroelementy i zhizn'. Molodaya gvardiya, Moscow, 124 p (in Russian)

Droop MR (1962) Organic micronutrients. In: Levin RA (ed) Physiology and biochemistry of algae. Academic, New York, pp 141–159

Droop MR (1974) The nutrient status of algal cells in continuous culture. J Mar Biol Assoc UK 54(4):825–855. https://doi.org/10.1017/S002531540005760X

Dugdale RC (1967) Nutrient limitation in the sea: dynamics, identification and significance. Limnol Oceanogr 12(4):685–695. https://doi.org/10.4319/lo.1967.12.4.0685

Egorov VN (1975) Modelirovanie i analiz nablyudenii na EVM v issledovaniyakh vzaimodeistviya radioaktivnosti morskoi sredy s gidrobiontami. Dissertation abstract, Sevastopol, 28 p (in Russian)

Egorov VN (1978) Matematicheskoe modelirovanie kinetiki mineral'nogo obmena morskikh gidrobiontov. In: II Vsesoyuznaya konferentsiya po biologii shel'fa, pt 1. Kiev, pp 37–38 (in Russian)

Egorov VN, Ivanov VN (1981) Matematicheskoe opisanie kinetiki obmena tsinka-65 i margantsa-54 u morskikh rakoobraznykh pri nepishchevom puti postupleniya radionuklidov. Ekologiya Morya 6:37–43 (in Russian)

Egorov VN, Kulebakina LG (1973) Matematicheskaya model' obmena ^{90}Sr mezhdu tsistoziroi i morskoi vodoi. In: Radioekologiya vodnykh organizmov, vol 2. Zinatne, Riga, p 305 (in Russian)

Egorov VN, Kulebakina LG (1974) Matematicheskaya model' obmena strontsiya-90 mezhdu tsistoziroi i morskoi vodoi. In: Polikarpov GG (ed) Khemoradioekologiya pelagiali i bentali (metally i ikh radionuklidy v gidrobiontakh i srede). Naukova dumka, Kiev, pp 30–39 (in Russian)

Egorov VN, Zesenko AYa (1977) Matematicheskaya model' kinetiki obmena izotopov v sisteme materinskii i dochernii radionuklidy v morskoi srede – gidrobiont. In: Polikarpov GG, Risik NS (eds) Radiokhemoekologiya Chernogo morya. Naukova dumka, Kiev, pp 17–20 (in Russian)

Egorov VN, Rozhanskaya LI, Ivanov VN (1975) Matematicheskoe opisanie protsessa nakopleniya tsinka-65 chernomorskoi vodorosl'yu Ulva rigida. Biologiya Morya 6:63–68 (in Russian)

Egorov VN, Ivanov VN, Usenko TG, Filippov NA (1980) Eksperimental'noe izuchenie obmena mikroelementov u zooplanktonnykh organizmov. Ekologiya Morya 2:44–48 (in Russian)

Egorov VN, Zesenko AYa, Parkhomenko AV, Finenko ZZ (1982) Matematicheskoe opisanie kinetiki obmena mineral'nogo fosfora odnokletochnymi vodoroslyami. Gidrobiologicheskii zhurnal 18(4):45–50 (in Russian)

Egorov VN, Kozlova SI, Kulebakina LG (1983) Kineticheskie zakonomernosti kontsentrirovaniya i obmena rtuti vzveshennym veshchestvom. DAN SSSR 271(6):1488–1491 (in Russian)

Egorov VN, Demina NV, Kulebakina LG (1989) Matematicheskoe opisanie kinetiki obmena elementov – khimicheskikh analogov morskimi makrofitami. Izvestiya AN SSSR. Seriya Biologicheskaya 1:79–87 (in Russian)

Finenko ZZ, Krupatkina-Akinina DK (1974) Vliyanie neorganicheskogo fosfora na skorost' rosta diatomovykh vodoroslei. In: Biologicheskaya produktivnost' yuzhnykh morei. Naukova dumka, Kiev, pp 120–135 (in Russian)

Fowler SW, Guary JC (1977) High absorption efficiency for ingested plutonium in crabs. Nature 266(5605):827–828. https://doi.org/10.1038/266827a0

Fried J (1968) Compartmental analysis of kinetic processes in multicellular systems: a necessary condition. Phys Med Biol 13(1):31–43. https://doi.org/10.1088/0031-9155/13/1/304

Giammatteo PA, Schindler JE, Waldron MC, Freedman ML, Speziale BJ, Zimmerman MJ (1983) Use of equilibrium programs in predicting phosphorus availability. In: Halberg R (ed) Environmental biogeochemistry. Ecol Bull 35:491–501

Glass HJ, Garetta AC (1967) Quantitative analysis of exponential curve fitting for biological applications. Phys Med Biol 12(3):379–388

Gromov VV, Spitsyn VI, Tolkach VV (1979) Pogloshchenie produktov deleniya planktonnymi organizmami. In: Vzaimodeistvie mezhdu vodoi i zhivym veshchestvom: proceedings of the international symposium, Odessa, 06–10 Oct 1975, vol 2, pp 119–126 (in Russian)

Guary JC, Fowler SW (1979) Elimination et répartition du [241]Am et du [237]Pu chez le moule Mytilus galloprovincialis dans son environnement naturel. Rapports Et Procès-Verbaux Des Réunions Commission Internationale Pour L'exploration Scientifique De La Mer Méditerranée 25–26(5):53–55

Hakonson TE, Gallegos AF, Whicker FW (1975) Caesium kinetics data for estimating food consumption rates of trout. Health Phys 29(2):301–306. https://doi.org/10.1097/00004032-197508000-00009

Ivanov VN (1979) Vliyanie razlichnykh kontsentratsii stabil'nogo tsinka v morskoi vode na obmen tsinka-65 i rost vodorosli Ulva rigida. Biologiya morya 50:69–71 (iss "Antropogennoe vozdeistvie na morskie organizmy" (in Russian)

Ivanov VN, Egorov VN, Rozhanskaya LI (1978) Izuchenie nakopleniya i vyvedeniya [65]Zn chernomorskoi vodorosl'yu Ulva rigida v svyazi s ee rostom. Biologiya Morya 44:46–55 (in Russian)

Ivanov VN, Egorov VN, Usenko TG, Filippov NA (1979) Izuchenie pishchevogo puti potrebleniya vyvedeniya radionuklidov u zooplanktonnykh organizmov. Biologiya Morya 50:71–74 (in Russian)

Ivanov VN, Egorov VN, Shevchenko MM (1980) Postuplenie i vyvedenie tsinka-65 u chernomorskogo *Idotea baltica basteri* Aud. Gidrobiologicheskii Zhurnal 16(1):69–72 (in Russian)

Ivanov VN, Egorov VN, Popovichev VN, Shevchenko MM (1986) Matematicheskoe modelirovanie kinetiki obmena mikroelementov u morskikh rakoobraznykh pri pishchevom i parenteral'nom putyakh ikh postupleniya. Ekologiya Morya 23:68–77 (in Russian)

Khailov KM, Popov AE (1983) Kontsentratsiya zhivoi massy kak regulyator funktsionirovaniya vodnykh organizmov. Ekologiya Morya 15:3–16 (in Russian)

Kopchenova NV, Maron IA (1972) Vychislitel'naya matematika v primerakh i zadachakh. Nauka, Moscow, 366 p (in Russian)

Kowal NE (1971) Model of elemental assimilation by invertebrates. J Theor Biol 31(3):469–474. https://doi.org/10.1016/0022-5193(71)90022-1

Krug GK, Sosulin YuA, Fatuev VA (1977) Planirovanie eksperimenta v zadachakh identifikatsii i ekstrapolyatsii. Nauka, Moscow, 208p (in Russian)

Lánczos K (1961) Prakticheskie metody prikladnogo analiza. Fizmatgiz, Moscow, 284 p (in Russian)

Lazorenko GE, Egorov VN (1994) Rol' donnykh otlozhenii v izvlechenii radiotseziya iz vodnoi sredy. In: Radioekologiya: uspekhi i perspektivy: materials of the scientific seminar, Sevastopol, 03–07 Oct 1994 (in Russian)

Lazorenko GE, Polikarpov GG (1990) Sposobnost' donnykh otlozhenii Kakhovskogo vodokhranilishcha k svyazyvaniyu radionuklidov strontsiya i tseziya. Doklady AN USSR. Ser. B 8:64–67 (in Russian)

Lotka AJ (1925) Elements of physical biology. Williams & Wilkins Comp., Baltimore, 495 p

Lowman FG, Rice TR, Richards FA (1971) Accumulation and redistribution of radionuclides by marine organisms. In: Radioactivity in the marine environment. The National Academies Press, Washington, DC, pp 162–199. https://doi.org/10.17226/18745

Lyubimova SA (1973) Sorbtsiya ^{90}Sr i ^{137}Cs donnymi otlozheniyami presnovodnogo ozera. In: Radioekologiya vodnykh organizmov. Zinatne, Riga, pt 2, pp 98–101 (in Russian)

Malakhova LV, Egorov VN, Malakhova TV (2019) Khlororganicheskie soedineniya v komponentakh ekosistem sevastopol'skikh bukht, morskoi akvatorii prirodnogo zapovednika "Mys Mart'yan" i Yaltinskogo porta. Voda: khimiya i ekologiya 1–2:57–62 (in Russian)

Mansouri-Aliabadi M, Sharp R (1985) Passage of selected heavy metals from *Sphaerotilis* (Bacteria: Chlamydobacteriales) to *Paramecium caudatum* (Protozoa: Ciliata). Water Res 19(6):697–699. https://doi.org/10.1016/0043-1354(85)90115-0

Matishov GG, Bufetova MV, Egorov VN (2017) Normirovanie potokov postupleniya tyazhelykh metallov v Azovskoe more. Nauka Yuga Rossii 13(1):44–58 (in Russian)

Menshutkin VV, Finenko ZZ (1975) Matematicheskoe modelirovanie protsessa razvitiya fitoplanktona v usloviyakh okeanicheskogo apvellinga. Trudy Instituta Okeanologii AN SSSR 102:175–183 (in Russian)

Monod J (1942) Recherches sur la croissance des cultures bactériennes. Université de Paris, Thèse de doctorat

Morozov NP, Petukhov SA (1979) Trace elements in hydrobionts and biotope of the surface layer of seawater in the North Atlantic and the Mediterranean. In: ICES council meeting, 7 p

Nakamura R, Nakahara M, Suzuki Y, Ueda T (1982) Effects of chemical forms and intake pathways on the accumulation of radioactive cobalt by the abalone Haliotis discus. Bull Jpn Soc Sci Fish 48(11):1639–1644. https://doi.org/10.2331/suisan.48.1639

Nalimov VV (1971) Teoriya eksperimenta. Nauka, Moscow, 207 p (in Russian)

Neiheisel J, Mcdaniel WL, Panteleyev GP (1992) Sediment parameters of northwest Black Sea shelf and slope: Implications for transport of heavy metals and radionuclides. Chem Ecol 6(1–4):117–131. https://doi.org/10.1080/02757549208035267

Nelepo BA (1970) Yadernaya gidrofizika. Atomizdat, Moscow, 224 p (in Russian)

Nesmeyanov AN (1978) Radiokhimiya. Khimiya, Moscow, 560 p (in Russian)

Osterberg C, Pirsi W (1971) Radioaktivnost' i pishchevye tsepi v okeane. In: Voprosy radioekologii. Atomizdat, Moscow, pp 240–252 (in Russian)

Parkhomenko AV, Egorov VN (1979) Kinetika obmena ^{86}Rb i ^{137}Cs u morskikh bakterii. Gidrobiologicheskii Zhurnal 15(5):94–100 (in Russian)

Patin SA (1979) Vliyanie zagryazneniya na biologicheskie resursy i produktivnost' Mirovogo okeana. Pishchevaya promyshlennost', Moscow, 304 p (in Russian)

Patton A (1968) Energetika i kinetika biokhimicheskikh protsessov. Mir, Moscow, 159 p (in Russian)

Pentreath RJ (1973) The roles of food and water in the accumulation of radionuclides by marine teleost and elasmobranch fish. In: Radioactive contamination of the marine environment: proceeding of a symposium, Seattle, 10–14 July 1972

Petipa TS (1981) Trofodinamika kopepod v morskikh planktonnykh soobshchestvakh: zakonomernosti potrebleniya pishchi i prevrashcheniya energii u osobi. Naukova dumka, Kiev, 242 p (in Russian)

Piontkovsky SA, Egorov VN, Ivanov VN (1983) Opyt ispol'zovaniya tsinka-65 dlya izucheniya ritma pitaniya planktonnykh rakoobraznykh. Ekologiya Morya 15:84–88 (in Russian)

Polikarpov GG (1964) Radioekologiya morskikh organizmov. Atomizdat, Moscow, 295 p (in Russian)

Polikarpov GG (1966) Radioecology of aquatic organisms. Reinhold Publ. Co., New York, p 314

Polikarpov GG, Bachurin AA (1970) Nakoplenie strontsiya i kal'tsiya rodstvennymi organizmami v razlichnykh fiziko-geograficheskikh usloviyakh Mirovogo okeana. In: Polikarpov GG (ed) Morskaya radioekologiya. Naukova dumka, Kiev, pp 244–248 (in Russian)

Polikarpov GG, Egorov VN (1986) Morskaya dinamicheskaya radiokhemoekologiya. Energoatomizdat, Moscow, 176 p (in Russian)

Polikarpov GG, Egorov VN (eds) (2008) Radioekologicheskii otklik Chernogo morya na chernobyl'skuyu avariyu. EKOSI-Gidrofizika, Sevastopol, 667 p (in Russian)

Polikarpov GG, Egorov VN, Zesenko AYa, Svetasheva SK (1983) Prevrashchenie khimicheskikh form ioda chernomorskimi vodoroslyami-makrofitami i biotransformatsionnaya sposobnost' morskikh fitotsenozov. In: Sostoyanie, perspektivy uluchsheniya i ispol'zovaniya morskoi ekologicheskoi sistemy pribrezhnoi chasti Kryma: abstracts of a scientific and practical conference dedicated to the 200th anniversary of the hero city of Sevastopol. Sevastopol, pp 127–128 (in Russian)

Polikarpov GG, Lazorenko GE, Demina NV, Tereshchenko NN (1987a) Radioekologicheskie parametry nakopleniya i vyvedeniya ^{137}Cs donnymi otlozheniyami chernomorskogo limana (v laboratornykh usloviyakh). Doklady AN USSR. Ser. B 5:74–76 (in Russian)

Polikarpov GG, Svetasheva SK, Egorov VN (1987b) Role of chemical species of iodine in its accumulation by seaweed. Actes du 8ème Colloque d'océanographie médicale, 9–12 Oct 1985, vol 85–86. Nice, pp 95–96

Polikarpov GG, Egorov VN, Lazorenko GE, Kulev YuD (1995) Matematicheskoe opisanie kinetiki vzaimodeistviya poverkhnostnogo sloya donnykh otlozhenii s radionuklidami v vodnoi srede. Doklady NAN Ukrainy 5:148–152 (in Russian)

Popovichev VN, Egorov VN (1987) Kineticheskie kharakteristiki parenteral'nogo i alimentarnogo pogloshcheniya ^{137}Cs chernomorskimi idoteyami. Ekologiya Morya 27:68–72 (in Russian)

Riziĉ I (1972) Two-compartment model of radionuclides accumulation into marine organisms. I. Accumulation from a medium of constant activity. Marine Bio 15(2):105–113. https://doi.org/10.1007/BF00353638

Romankevich EA (1977) Geokhimiya organicheskogo veshchestva v okeane. Nauka, Moscow, 256 p (in Russian)

Sazhina LI (1980) Plodovitost', skorost' rosta nekotorykh kopepod Atlanticheskogo okeana. Biologiya Morya 3:56–61 (in Russian)

Sheppard CW (1948) The theory of the study of transfers within a multi-compartment system using isotopic tracers. J Appl Phys 19(1):70–76. https://doi.org/10.1063/1.1697874

Sheppard CW, Householder AS (1951) The mathematical basis of the interpretation of tracer experiments in closed steady-state systems. J Appl Phys 22(4):510–520. https://doi.org/10.1063/1.169 9992

Shulman GE (1972) Belkovyi rost i nakoplenie energeticheskikh rezervov – dve storony problemy rosta ryb. In: Energeticheskie aspekty rosta i obmena vodnykh zhivotnykh: abstracts of reports of the All-Union symposium, Sevastopol, 9–11 Oct 1972, pp 259–260 (in Russian)

Solomon AK (1960) Compartmental methods of kinetic analysis. In: Comar CL, Bronner F (eds) Mineral metabolism vol 1, pt A. Academic, New York, pp 119–168

Stetsyuk AP, Egorov VN (2018) Sposobnost' morskikh vzvesei kontsentrirovat' rtut' v zavisimosti ot ee soderzhaniya v akvatoriyakh shel'fa. Sistemy Kontrolya Okruzhayushchei Sredy 13(33):123–132 (in Russian)

Sukal'skaya SYa, Likhtarev IA (1976) Ob eksperimental'nom modelirovanii i matematicheskom opisanii obmena radionuklidov v sisteme "vodnaya sreda – gidrobiont". Gidrobiologicheskii zhurnal 12(4):55–62 (in Russian)

Sushchenya LM (1972) Intensivnost' dykhaniya rakoobraznykh. Naukova dumka, Kiev, 193 p (in Russian)

Tereshchenko NN, Egorov VN (1983) Kineticheskie kharakteristiki pogloshcheniya i vyvedeniya fosfora zelenoi vodorosl'yu *Ulva rigida* Ag. In: Sostoyanie, perspektivy uluchsheniya i ispol'zovaniya morskoi ekologicheskoi sistemy pribrezhnoi chasti Kryma: abstracts of a scientific and practical conference dedicated to the 200th anniversary of the hero city of Sevastopol. Sevastopol, pp 185–186 (in Russian)

Tereshchenko NN, Egorov VN (1985) Kineticheskie zakonomernosti pogloshcheniya i vyvedeniya fosfora chernomorskoi zelenoi vodorosl'yu *Ulva rigida* Ag. DAN USSR. Seriya B 1:79–82 (in Russian)

Timofeev-Resovskii NV (1957) Primenenie izluchenii i izluchatelei v eksperimental'noi biogeotsenologii. Botanicheskii Zhurnal 42(2):161–194 (in Russian)

Timofeeva-Resovskaya EA (1963) Raspredelenie radioizotopov po osnovnym komponentam presnovodnykh vodoemov. Sverdlovsk, 78 p (in Russian)

Troshin AS (1957) O svyazannom i svobodnom natrii v skeletnykh myshtsakh lyagushki. Biofizika 2(5):617–627 (in Russian)

Tsytsugina VG, Lazorenko GE (1983) Rol' mitoticheskogo deleniya v pogloshchenii biogennykh elementov prirodnymi populyatsiyami fitoplanktona. Ekologiya Morya 12:30–34 (in Russian)

Turn GA, Joshi GHV, Chauhan VD, Rao PS (1982) Effect of metal ions on the growth of *Sargassum swartzii*. Ind J Marine Sci 11(4):338

Urbakh VYu (1964) Biometricheskie metody. Nauka, Moscow, 362 p (in Russian)

Vernadsky VI (1929) O kontsentratsii radiya zhivymi organizmami. Doklady Akademii Nauk SSSR. Seriya A 2:33–34 (in Russian)

Vinberg GG, Anisimov SI (1966) Matematicheskaya model' vodnoi ekosistemy. In: Fotosintez sistem vysokoi produktivnosti. Nauka, Moscow, pp 213–223 (in Russian)

Vinogradov ME, Menshutkin VV (1977) Portretnye determinsistkie modeli funktsionirovaniya ekosistem pelagiali. In: Biologiya okeana, vol 2. Nauka, Moscow, pp 261–276 (in Russian)

Voitsekhovitch OV, Borzilov VA, Konoplev AV (1991) Hydrological aspects of radionuclide migration in water bodies following the Chernobyl accident. In: Proceedings of the seminar on comparative assessment of the environmental impact of radionuclides released during three major nuclear accidents: Kyshtym, Windscale, Chernobyl, vol 2. Luxembourg, 1–5 Oct 1990

Volterra V (1972) Matematicheskaya teoriya bor'by za sushchestvovanie. Nauka, Moscow, 216 p (in Russian)

Wartas CJ, MacFarlane J, Francois MM (1985) Nickel accumulation by *Scenedesmus* and *Daphnia*: Food-chain transport and geochemical implications. Can J Fish Aquat Sci 42(4):724–730. https://doi.org/10.1139/f85-093

Weers AW (1973) Uptake and loss of ^{65}Zn and ^{60}Co by the mussel *Mytilus edulis* L. In: Radioactive contamination of the marine environment: proceeding of a symposium, Seattle, 10–14 July 1972

Weers AW (1975a) The effects of temperature on the uptake and retantion of ^{60}Co and ^{65}Zn by the common shrimp *Crandon crandon* (L.). In: Combined effects of radioactive, chemical and thermal releases to the environment, Stockholm, 2–5 June 1975

Weers AW (1975b) Uptake of cobalt-60 from sea water and from labeled food by the common shrimp *Crangon crangon* (L). In: International symposium on impacts of nuclear releases into the aquatic environment. IAEA, Vienna, pp 359–361

Zaika VE (1972) Udel'naya produktsiya bespozvonochnykh. Naukova dumka, Kiev, 145 p (in Russian)

Zlobin VS (1968) Dinamika nakopleniya radiostrontsiya nekotorymi burymi vodoroslyami i vliyanie solenosti morskoi vody na koeffitsienty nakopleniya. Okeanologiya 8(1):78–85 (in Russian)

Chapter 4
Theory of Radioisotope and Chemical Homeostasis of Marine Ecosystems

Introduction

This chapter presents the parametric bases and practical applications of the theory of radioisotopic and chemical homeostasis of marine ecosystems.

In Sect. 4.1 of the chapter, the characteristics of homeostasis are studied using differential models of the balance of material, energetic, mineral and radioisotopic exchange in the ecosystem of the photic layer, as well as a semiempirical model of the sedimentation function of the pelagic ecosystem of the western halistatic region of the Black Sea as a result of the influence of the natural functional mechanisms of marine ecosystems. It is shown that mathematical models having a structure that takes into account contemporary theoretical concepts and the results of observations on the mechanisms of biotic and abiotic interactions in the marine environment are promising for solving problems of assessing and predicting the state of ecosystems under conditions of changing environmental factors. Based on the results of numerical experiments, it is shown that the dynamic characteristics of the biotic parameters of the models can be determined in terms of damping oscillations having periods corresponding to succession time scales in ecosystems of the photic layer. The stability of the modelled ecosystems with respect to the factors of pollution of the marine environment is shown to depend on the ratio of fluxes of anthropogenic input and the biogeochemical self-purification of waters. This stability is maintained as a result of the influence of natural mechanisms of homeostasis due to the implementation of negative feedbacks in ecosystems, regulated by the functioning of their trophic production, consumer and decomposer levels, as well as the concentrating functions of living and inert matter.

In Sect. 4.2 of the chapter, elements of the theory are considered and a structural diagram of the management of marine ecosystems is presented. In Sect. 4.3, the biogeochemical characteristics of natural homeostasis are demonstrated on the examples of the concentration function of living and inert matter, the adaptation characteristics of hydrobionts, the trophic factor, population

V. Egorov, *Theory of Radioisotopic and Chemical Homeostasis of Marine Ecosystems*, Springer Oceanography, https://doi.org/10.1007/978-3-030-80579-1_4

characteristics of biotopes, as well as the influence of estuarine zones and coastal waters on biogenic elements. In the last paragraph of this section, the biogeochemical mechanisms for implementing compensatory homeostasis in ecosystems are considered in terms of their possible functional and structural rearrangement as a result of anthropogenic impact. It is shown that negative feedbacks in ecosystems are in all cases carried out in accordance with the Le Chatelier–Braun principle.

Section 4.4 of the chapter considers stationary states of dynamic models of marine ecosystems presented in the form of algebraic equations describing the relationships between their parameters. This allowed us to determine functional dependencies (referred to as biogeochemical criteria) between the indicators of the functional intensity of production, decomposer and consumer levels, as well as the characteristics of homeostasis and biogeochemical self-purification of water from radioactive and chemical contamination. It is shown that, for each ecosystem, the biogeochemical self-purification of waters from pollution involves the presence of limiting fluxes. These limiting—or marginal—fluxes have been referred to in terms of the "ecological carrying capacity" or "assimilation capacity" of the marine environment; as such, quantitative ratios have been obtained for assessing the metabolic factor of the carrying capacity and the sedimentation factor of the assimilation capacity of the marine environment, as well as for assessing the radiocapacity, normalised by the maximum allowable concentration (MAC) and criteria of the maximum carrying capacity of aquatic areas.

In Sect. 4.5, problems in marine nature management arising in connection with the implementation of an ecocentric strategy for the sustainable development of aquatic areas by maintaining a balance between the consumption and reproduction of marine resources are considered. It is proposed that the measure of the restoration of water quality resources be assessed according to biogeochemical criteria that refer to their ability to self-purify, while the measure of their consumption should be measured by the ratio of the concentration of pollutants to water in comparison with the MAC.

Section 4.6 of the chapter discusses the practical application of homeostasis theory to marine ecosystems on the examples of environmental regulation of pollution by post-Chernobyl radionuclides, organochlorides and heavy metals in Sevastopol Bay, heavy metals in various waters of the Sea of Azov, radioactive and chemical pollution from critical zones of the Black Sea and biogenic elements in coastal aquatic areas having liquid outer boundaries.

4.1 Dynamic Characteristics of Systems of Interaction of Living and Inert Matter with the Radiochemical Components of the Marine Environment

Under natural conditions, the processes of interaction of living and inert matter with radioactive and chemical components of the marine environment occur on different spatio-temporal scales. As a rule, these processes are not stationary; their implementation is due to biotic and abiotic ecosystemic mechanisms functioning under the influence of natural and anthropogenic trends ranging in scale from diurnal variation to synoptic, seasonal, annual and climatic changes. In order to understand these regularities, it is therefore necessary to study the dynamic modes of formation of their material, energetic, mineral and radioisotopic balance. As already indicated (Belyaev 1978), one of the most promising methodological approaches to solving such problems consists in the use of mathematical models to study the functioning of natural marine ecosystems. An analysis of numerical experiments based on theoretical mathematical models and semi-empirical models of individual regions or aquatic areas reveals the essentially dynamic characteristics of marine ecosystems.

4.1.1 Model of the Photic Layer Ecosystem

The main biogeochemical mechanisms of the interaction of living and inert matter with radioactive and chemical components of the marine environment include primary aquatic production and its limitation by biogenic elements, the sorption and concentrating capacity of hydrobionts and inert matter in relation to chemical substances having varying biological significance, metabolic and trophic interactions in the ecosystem, as well as sedimentation processes that determine the vertical transfer and deposition of conservative chemical substrates in bottom sediments. It can be clearly seen that the influence of the abovementioned biogeochemical mechanisms of interactions is most characteristic of ecosystems of the photic layer of the marine environment. This section presents the results of studies into the dynamic patterns of interaction of the photic layer ecosystem with radioactive and chemical substances based on a numerical analysis of the theoretical model (Polikarpov and Egorov 1986). The problem is considered on the example of the interaction of a pollutant entering the water surface with a six-component ecosystem of the photic layer (Fig. 4.1), whose primary production is limited by phosphates (PO_4) and photosynthetically active radiation (PAR) under conditions in which the sedimentation rate of suspended organic matter (SOM) particles is determined by their size fraction. Trophic and chemical interactions between ecosystem components in each of the ten identified blocks are linked by communications indicated by arrows on the structural diagram (Fig. 4.1). Phosphates are distributed between the layers according to the intensity of water mixing, which also determines the vertical transfer of bacteria and dissolved organic matter (DOM). Both DOM and SOM are subject to non-biological

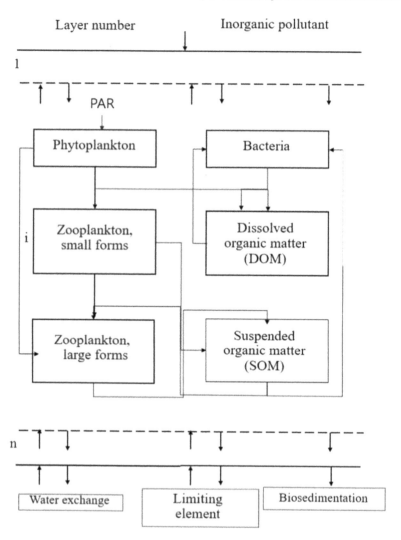

Fig. 4.1 Structure of the model of the interaction of the ecosystem of the photic layer with inorganic pollutants of the marine environment (Polikarpov and Egorov 1986)

dissolution (lysis); moreover, the migration of zooplankton between water layers takes place at a given intensity. Phytoplankton can either be distributed in proportion to the intensity of mixing of waters or, like zooplankton, migrate vertically.

The dynamics of the fluxes of biogenic elements and radionuclides is described separately for each of the trophic groups under the assumptions that hydrobionts have separate exchangeable pools of a pollutant and a limiting substrate, that there are no non-exchangeable pools for the absorption of these substances, but that the chemical substrate or pollutant not assimilated by zooplankton from food enters the SOM.

The balance equations for the biogenic element and substance in relation to the ith layer had the following form.

Phytoplankton:

$$\frac{dB_1}{dt} = P_1 - B_1(a_{10} + a_{13} + a_{14} + a_{15} + a_{16}) + d_{i-1}B_{1i-1} + d_{i+1}B_{1i+1} - 2d_i B_1;$$

$$\frac{dC_1}{dt} = \frac{V_{\max 1} C_w}{K_{m1} + C_w} - (p_1 + P_1/B_1)C_1 + d_{i-1}\frac{B_{1i-1}}{B_1}(C_{1i-1} - C_1)$$

$$\qquad + d_{i+1}\frac{B_{1i-1}}{B_1}(C_{1i-1} - C_1), \tag{4.1}$$

where

> B_1, P_1 and C_1 are the biomass, primary production and concentration of the limiting element in phytoplankton, respectively;
>
> a_{10} is the indicator of the rate of mineralisation due to the respiration of producers;
>
> a_{13} and a_{14} are the rates of trophic absorption of phytoplankton by small and large forms of zooplankton;
>
> a_{15} and a_{16} are indicators of the rate of intravital release of DOM and phytoplankton mortality, respectively;
>
> d with indices are indicators of water mixing rates;
>
> p_1 is an indicator of the rate of exchange of a mineral element that limits production.

Hereinafter, the indices in the parameters indicate that their values refer to the layer numbers indicated by these numbers. If there is no index for the parameter, then it refers to the i-th layer.

In Eq. (4.1), primary production is described by the expression:

$$P_1 = \beta B_1 P_{1\max}(1 - C_{1\min}/C_1), \tag{4.2}$$

where

> β is the coefficient of relative illuminance;
>
> $P_{1\max}$ is the maximum potential specific production of phytoplankton;
>
> $C_{1\min}$ is the minimum conceivable intracellular concentration of the limiting mineral element.

The value β is exponentially dependent on the depth (Menshutkin and Finenko 1975), as well as on the degree of influence of living matter on the transparency of waters (Sorokina 1979). In the model, the value of β was specified by the ratio:

$$\beta = (1 - B_\Sigma/B_{\max})e^{-\gamma_z z}, \tag{4.3}$$

where

B_Σ is the total biomass of ecosystem components, excluding DOM, above the analysed layer;

B_{max} is the critical biomass above the layer at which the sun's energy is completely absorbed within the i-th and overlying layers (at $B_\Sigma = B_{max}$, PAR does not penetrate below the i-th layer);

z is the depth;

γ_z is the indicator of the rate of change of PAR with depth.

The value γ_z is calculated from the condition that, at the lower boundary of the photic zone, the illumination is 1% of that striking the ocean surface. The use of expression (4.3) allowed the universal correlation of a decrease in the depth of the photosynthesis layer with an increase in the biological productivity of waters to be reflected in the model.

It is also well known from the literature that, under the same illumination conditions and with an equal concentration of biogenic elements, the biological productivity of waters is affected by upwelling phenomena that occur in various oceanic regions. However, this fact has not yet been fully explained in theoretical terms. The introduction of the parameter B_{max} into Eq. (4.3) made it possible to empirically account for the different bioproduction capacities of aquatic areas.

The last three terms of the first equation of expression (4.1) describe the transfer of phytoplankton as a result of water exchange between layers, while the two terms on the right of the second equation reflect the change in the concentration of the limiting mineral element in phytoplankton due to the migration of algae along with waters.

Bacteria:

$$\frac{dB_2}{dt} = P_2 - B_2(a_{23} + a_{24} + a_{25} + a_{26}) + \Delta B_2;$$

$$\frac{dC_2}{dt} = \frac{V_{max\,2}C_w}{K_{m2} + C_w} + [a_{52}B_5(C_5q - C_2q_N) + a_{62}B_6(C_6q - C_2q_N)]/B_2$$

$$- p_2C_2 + \Delta C_2, \tag{4.4}$$

where

B_2, B_5 and B_6 are the biomasses of bacteria, DOM and SOM;

P_2 is bacterial production;

C_2, C_5 and C_6 are the concentrations of the production-limiting element in bacteria, DOM and SOM;

q is the efficiency of assimilation by bacteria of the limiting element from a food source;

ΔB_2 and ΔC_2 are the parameters that characterise the transfer of bacteria and the limiting element in bacteria under the influence of water exchange;

ΔB_2 is equal to the sum of the last three terms of the first equation, while ΔC_2 is the sum of the last two terms of the second equation of expressions of the form (4.1) written out in relation to B_2 and C_2.

Relations (4.4) reflect the assumption that both dissolved and suspended organic matter can serve as food for saprophytic bacteria. Bacterial production was calculated using the formula:

$$P_2 = R_2(1 - a_{20}),\qquad(4.5)$$

where

a_{20} is a coefficient reflecting the expenditure on respiration;
R_2 is the daily nutrient budget, calculated from the ratio:

$$R_2 = a_{52}B_5 + a_{62}B_6.\qquad(4.6)$$

Zooplankton:

Small forms:

$$\frac{dB_3}{dt} = P_3 - \left(a_{34} + a'_{36}\right)B_3 + \Delta B_3;$$

$$\frac{dC_3}{dt} = \frac{V_{\max 3}C_w}{K_{m3} + C_w} + [a_{13}B_1(C_1q - C_3q_N) + a_{23}B_2(C_2q - C_3q_N)$$
$$+ a_{23}B_6(C_6q - C_3q_N)]B_3 - p_3C_3 + \Delta C_3.\qquad(4.7)$$

Large forms:

$$\frac{dB_4}{dt} = P_4 - a'_{46}B_4 + \Delta B_4;$$

$$\frac{dC_4}{dt} = \frac{V_{\max 4}C_B}{K_{m4} + C_B} + [a_{14}B_1(C_1q - C_4q_N) + a_{24}B_2(C_2q - C_4q_N)$$
$$+ a_{34}B_3(C_3q - C_4q_N) + a_{64}B_6(C_6q - C_4q_N)]/B_4 - p_4C_4 + \Delta C_4.\quad(4.8)$$

In Eqs. (4.7) and (4.8), the following designations are adopted:

- B, P and C with indices are the biomass, production and concentration, respectively, of the limiting nutrient in the ecosystem components corresponding to the index number;
- a with indices refers to rates of trophic interactions;
- q is the efficiency of assimilation of a nutrient from food;
- a_{36} and a_{46} are the indicators of the die-off rate of zooplankton;
- p_3 and p_4 are the indicators of the rate of exchange of the limiting element by small and large forms of zooplankton, respectively;

- ΔB_3, ΔB_4 and ΔC_3, ΔC_4 are indicators of the transfer rate of zooplankton biomass and the limiting element as a result of zooplankton migration.

The form of the formulas coincided with the expressions for calculating ΔB_2 and ΔC_2; however, instead of the water mixing rate d_i with indices, a coefficient having a constant value for all layers was substituted in them. By this means, the assumption was realised in the model that the intensity of zooplankton migration is constant with respect to depth.

The productivity of small forms of zooplankton was calculated using the formula:

$$P_3 = R_3(1 - a_{30} - a_{35} - a_{36}), \tag{4.9}$$

where a_{30}, a_{35} and a_{36} are coefficients that characterise the expenditure on respiration and undigested food isolated in the form of liquid and solid extracts.

Small forms of zooplankton can feed on phytoplankton, bacteria and SOM. The amount of their nutrient budget was calculated using the formula:

$$R_3 = a_{13}B_1 + a_{23}B_2 + a_{63}B_6. \tag{4.10}$$

where a with indices refers to rates of trophic interactions.

The formulas for calculating P_4 and R_4 were similar to expressions (4.5) and (4.7); however, they additionally included terms reflecting the possibility of large forms of zooplankton feeding on their small forms.

Indicators of the rate of trophic interactions of bacteria (a_{52}, a_{62}) with small (a_{13}, a_{23}, a_{63}) and large (a_{14}, a_{24}, a_{34}, a_{64}) forms of zooplankton were calculated according to the formula of Ivlev (1955), distributed in the case of feeding of hydrobionts with mixed nutriment. In relation to bacteria, the formulas for calculating a_{52} and a_{62} took the form:

$$a_{52} = B_2 S_{52} R_2 / y_2; \quad a_{62} = B_2 S_{62} R_2 / y_2, \tag{4.11}$$

where:

$$y_2 = S_{52}B_5 + S_{62}B_6; \quad R_2 = R_{2\,max}\{1 - \exp[-\zeta(y_2 - B_{20})]\}. \tag{4.12}$$

In relations (4.11) and (4.12):

- S_{52} and S_{62} are coefficients characterising the availability of DOM and SOM for bacteria;
- y_2 and B_{20}—availability for food and the minimum available amount of nutriment for bacteria;
- ζ—coefficient.

The parameters of trophic interactions for small and large forms of zooplankton were calculated using similar formulas (4.11) and (4.12), taking into account the trophic relationships of zooplankton.

Dissolved organic matter:

$$\frac{d B_5}{dt} = a_{15} B_1 + a_{25} B_2 + a_{35} R_3 + a_{45} R_4 - B_5 (a_{50} + a_{51} + a_{52}) + \Delta B_5;$$

$$\frac{d C_5}{dt} = \frac{V_{\max 5} C_w}{K_{m5} + C_w} + [a_{15} B_1 (C_1 - C_5) + a_{25} B_5 (C_2 - C_5)$$
$$+ a_{35} R_3 (C_3 - C_5) + a_{45} R_4 (C_4 - C_5)]/B_5 - p_5 C_5 + \Delta C_5. \qquad (4.13)$$

In (4.13) a_{50} is an indicator of the rate of non-biological mineralisation (lysis) of DOM.

In expression (4.13), ΔB_5 and ΔC_5 were calculated using formulas similar to those used to determine ΔB_2 and ΔC_2, with the appropriate substitution of the values of the mass of DOM and the concentration in it of the limiting element entering the analysed layer as a result of water exchange.

Suspended organic matter (detritus):

Sedimentation was taken into account in the balance equation for SOM. The equations corresponding to the change in the biomass of suspended matter, whose sources consisted of phytoplankton (B_{61}), bacteria (B_{62}), small (B_{63}) and large (B_{64}) forms of zooplankton, were solved separately:

$$\frac{d B_{61}}{dt} = a_{16} B_1 + b_1 (B_{61_{i-1}} - B_{61});$$

$$\frac{d B_{62}}{dt} = a_{26} B_2 + b_2 (B_{62_{i-1}} - B_{62});$$

$$\frac{d B_{63}}{dt} = a_{36} R_3 + a'_{36} B_3 + b_3 (B_{63_{i-1}} - B_{63});$$

$$\frac{d B_{64}}{dt} = a_{46} R_4 + a'_{46} B_4 + b_4 (B_{64_{i-1}} - B_{64}), \qquad (4.14)$$

where B with indices are the indicators of the rate of gravitational settling of the size fractions of the SOM.

In Eq. (4.14), the values of B with indices and subscripts $i - 1$ correspond to the biomasses of different fractions of SOM in the upper water layer relative to the analysed layer.

In general, for SOM, the balance equalities of changes in biomass and concentration of the biogenic element limiting primary production were as follows:

$$\frac{d C_6}{dt} = \frac{V_{\max 6} C_w}{K_{m6} + C_w} + [a_{16} B_1 (C_1 - C_6) + a_{26} B_2 (C_2 - C_6)$$
$$+ (a_{36} R_3 + a'_{36} B_3)(C_3 - C_6)$$
$$+ (a_{46} R_4 + a'_{46} B_4)(C_4 - C_6) + \partial_6 B_{6_{i-1}} (C_{6_{i-1}} - C_6)]/B_6$$
$$+ (1 - q)(\partial_3 + \partial_4) - p_6 C_6;$$

$$\frac{d B_6}{dt} = \sum_{i=1}^{4} \frac{d B_{6j}}{dt} - (a_{60} + a_{62} + a_{63} + a_{64}) + B_6, \tag{4.15}$$

where

a_{50} is an indicator of the SOM lysis rate;
δ_3 and δ_4 are parameters that take into account the intake of the undigested part of the element from food of small and large forms of zooplankton into the SOM;
δ_6 is a coefficient that takes into account the intake of an element with suspensions, descending gravitationally from the upper layer.

The value δ_6 was calculated by the formula:

$$\delta_6 = \frac{\sum_{i=1}^{4} b_j B_{6ji-1}}{B_{6_{i-1}}}. \tag{4.16}$$

The balance equation of changes in the concentration of a nutrient in the water of the analysed layer as a result of interaction with biological components of the ecosystem is presented in the form:

$$\frac{dC_w}{dt} = \frac{10^{-3}}{h} + \sum_{j=1}^{6} C_{j_{jw}} \sum_{j=1}^{6} \frac{V_{\max j}}{K_{mj} + C_B} \frac{\sum_{j=1}^{6} B_j}{m_w}$$
$$+ d_{i-1}\left(C_{w_{i-1}} - C_{w_i}\right) + d_{i+1}\left(C_{w_{i-1}} - C_w\right) + \Delta C_{ws} + \Delta C_{wb} \tag{4.17}$$

where ΔC_{ws} and ΔC_{wb} is the rate of intake of the element, respectively, from the surface (to the first layer) and from the bottom (to the tenth layer);
h is the depth of each of the ten identified surface water layers.

In Eq. (4.17), the first term on the right-hand side in brackets describes the processes of entry of the element excreted by hydrobionts into the water as a result of desorption and metabolism, while the second reflects the kinetics of parenteral absorption of the element by the components of the ecosystem. The factor 10^{-3} before the polynomial in parentheses is used to match the dimensions when measuring the values of B_j in mg/m^3 and when selecting the layer depth h in metres. The balance equations reflecting the interaction of ecosystem components with a radioactive or chemical pollutant of the environment were identical to the ratios described for the primary production-limiting biogenic element. In general, the state of the ecosystem within any layer can be represented by 24 differential equations. Ten of these are comprised of equalities that describe the balance of the ecosystem (phytoplankton, bacteria, small and large forms of zooplankton, DOM, four size groups of SOM and suspensions in general) in terms of matter. Seven additional equations close the system for the biogenic element in water and hydrobionts, while the last seven do the same for the radioactive pollutant. At each step of integration, 240 differential equations were solved sequentially from the 1st to the 10th layer, which corresponded to 240 values of the initial conditions. The parameters reflecting the effects of biotic components had the same values for each layer. The parameters d_i, characterising the intensity of vertical mixing of waters, could be set in the form of tabular functions, allowing the model to take into account layers corresponding to a jump in the density of waters, upwellings, "biological spring" events, as well as oligotrophic, neritic and eutrophic oceanic zones. The model equations reflected the balance of matter under 1 m^2 of the surface of the photic layer. The interaction of the analysed column with adjacent waters was assumed to be negligible. This provided the opportunity to relate the time scale of the implementation of numerical experiments in the model with the succession of biological communities in drifting waters. Since the effects of the pollutant on the morphological, production and genetic characteristics of the ecosystem components were not considered, the task of studying the dynamics was reduced to creating a representative model of the ecosystem of the photic layer and a research model for the concentration and transfer of pollution by hydrobionts against the background of the effect of physical processes of vertical mixing of waters. The model was analysed numerically. The differential equations were solved using the Euler method. The integration step, which was selected experimentally, averaged 0.01–0.05 days.

At the first stage of the study of dynamic characteristics, the stability of solutions was investigated with respect to the values of the parameters and to the initial conditions of the model. In this context, by stable is meant those numerical experiments in which none of the components of the ecosystem over the entire time interval of the consideration of the process, up to the state of stationarity, did not take a zero value or tend to infinity.

When the initial conditions were set for states of the ecosystem of the photic layer observed in nature, the solutions were stable in all cases when the parameters of the model reflecting the production characteristics of the components were determined in such a way that, in the initial periods of simulating the behaviour of the ecosystem, the production capabilities of the previous links of the trophic chain were sufficient to satisfy the physiological needs of the subsequent links. In the first instance, stability was influenced by parameters $P_{1\max} \div P_{4\max}$, coefficients of nutrient availability (S with indices) and the minimum levels of nutrient in the environment.

The intervals of parameter values, according to which the stability of the model is observed, are "outlined" in more than 60 numerical experiments. It became apparent that almost all of the trophic characteristics of hydrobionts determined by ecologists experimentally and in field observations fall into these intervals (Petipa 1981). The only exception was those numerical experiments in which the ecosystems of oligotrophic oceanic regions were modelled. Models of oligotrophic regions with parameters characterising the rates of primary production determined from observations either turned out to be unstable or the structure of the ecosystem in relation to consumers did not correspond to the natural one. This was due to the fact that there was not enough organic matter being produced by phytoplankton to satisfy the energy expenditures of consumers. The noted modelling results are consistent with the estimates of the energy balance in the marine pelagic ecosystems studied through field observations (Greze 1982). When selecting parameters from the interval of "parametric stability", a violation of stability was observed depending on the initial conditions. Loss of stability occurred in cases when the initial conditions set a very low nutrient content in the water, which did not provide the primary production of organic matter sufficient to meet the energy needs of the subsequent links of the trophic chain. At the same time, if the flux of a biogenic element into the photic layer was set at a sufficiently high level, then sometimes "trigger" stationary states of the biological components of the ecosystem appeared; however, simultaneous "control" for the limiting biogenic element and PAR was not observed. The assignment of high biomasses of consumers in relation to producers with the initial conditions sometimes led to the same type of resolutions. In these cases, stability was easily achieved by changing the coefficient values of the availability of various nutrients for consumers. If the initial conditions corresponded to the conditions of a significant difference between the components of the ecosystem and those observed in nature, then the model displayed a mode of relaxation oscillations, during which the structure of interactions was rearranged. After the lapse of the relaxation oscillations, the components of the ecosystem took on values close to those occurring naturally.

The analysis showed that the instability of the model occurred in two main cases. Firstly, when its parameters significantly differed from values determined experimentally. Secondly, when the initial conditions did not substantially correspond to the state of ecosystems observed in nature. Thus, the indicated reasons for instability rather testify in favour of the model's adequacy than vice versa. This is further confirmed by the fact that, across the range of parameter values and initial conditions inherent in ecosystems of the photic layer, the solutions obtained on the model are stable. The instability of the models can be seen to manifest itself in an area that, to a certain extent, corresponds to anomalies in ecosystems. In this case, biological catastrophes can also be observed in nature, which can be considered in terms of a violation of the stability of ecosystems. In all cases, regardless of the initial conditions, the model was reduced to a stable and unambiguously "controlled" state, in terms of PAR and the limiting chemical element, if the ratios of the form (4.11) not only took into account the minimum available amount of nutriment for higher trophic levels, but also limited the availability of food provided to a higher trophic level from the previous trophic level. In this case, the value y_2 in the equality (4.11) was calculated by the formula:

$$y_2 = S_{52}B_5(1 - B_5/B_{5\,min}) + S_{62}B_6(1 - B_6/B_{6\,min}), \qquad (4.18)$$

where B_{5min} and B_{6min} are the minimum specific weights of SOM and DOM available for feeding bacteria.

The available nutrient budgets for small (y_3) and large (y_4) forms of zooplankton were determined similarly.

The change in the total biomass of the photic layer depending on the initial biomasses of the ecosystem components when using a relation of the form (4.18) is illustrated in Fig. 4.2. The figure shows that the stationary level of biomass in the ecosystem did not depend on the initial total biomasses of its components. Model calculations showed that the ratio of the biomasses of the components in the stationary state of the ecosystem was constant, while the value of the stationary level (at constant

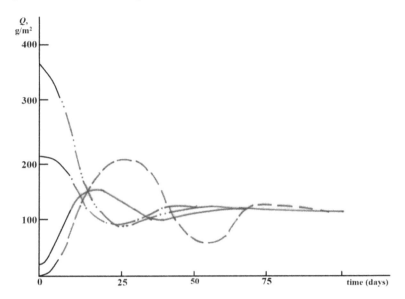

Fig. 4.2 Change in the biomass of the ecosystem of the photic layer (g/m²) with different initial total biomasses of its components (Polikarpov and Egorov 1986)

PAR) was determined only by the rate of input of the nutrient limiting primary production and did not depend on changes in the initial concentrations of the nutrient in hydrobionts and the environment.

In relation to the problem of self-cleaning of the environment, the dynamics of the model of the ecosystem of the photic layer (Fig. 4.1) was studied in numerical experiments with the main parameters given in Table 4.1. The calculations were carried out under the assumption that there is no pycnocline density gradient layer. Daily agitation of mixed layers was taken as 10%. The average rates of gravitational descent were taken as follows: dead bacterial cells—0.05 m/day; phytoplankton—0.2 m/day; faeces, exoskeletons and corpses of small forms of zooplankton—50 m/day; large forms—100 m/day. The influx of the limiting element (phosphates) from below into the 10th layer was 2 µg/L per day. None of the initial conditions of any of the ecosystem components changed with depth. The means of setting the initial conditions determined the relaxation mode in the model. During the relaxation period (60–70 days), there was an almost synchronous surge and subsequent decrease in the biomass and production of all biological components of the ecosystem (Fig. 4.3); during this time, their vertical distribution changed from uniform (Fig. 4.4a) to that qualitatively corresponding to the distribution of biological components in nature (Fig. 4.4c).

The integral values of the biological components of the ecosystem in the photosynthetic layer were subject to damped oscillations. Following the relaxation time, an outbreak of phytoplankton biomass occurred (Fig. 4.3b), after which first small (Fig. 4.3d) and then large (Fig. 4.3e) forms of zooplankton arrived. Subsequently,

Table 4.1 Model parameters of the photic layer ecosystem (Polikarpov and Egorov 1986)

Parameter	Parameter value	Dimension	Parameter	Parameter value	Dimension
P_{1max}	5.00	Per day	a_{35}	0.15	Per day
V_{max1}	10.00	μg/mg per day	a_{36}	0.10	Per day
K_{m1}	11.00	μg/L	d	0.10	Per day
C_{1min}	0.70	μg/mg	S_{13}	1.00	Dimensionless
P_1	0.50	Per day	S_{23}	1.00	Dimensionless
a_{10}	0.10	Per day	S_{63}	0.30	Dimensionless
a_{15}	0.05	Per day	P_{4m}	0.65	Dimensionless
a_{16}	0.10	Per day	a_{40}	0.30	Per day
P_{2max}	2.00	Per day	a_{45}	0.15	Per day
a_{20}	0.30	Per day	a_{46}	0.10	Per day
a_{25}	0.05	Per day	h	10.00	m
a_{26}	0.10	Per day	S_{34}	0.50	Dimensionless
S_{52}	1.00	Dimensionless	S_{24}	0.10	Dimensionless
S_{62}	1.00	Dimensionless	S_{34}	0.80	Dimensionless
P_{3max}	0.90	Per day	S_{64}	0.20	Dimensionless
a_{30}	0.30	Per day	B_{max}	200.00	g/m^2

when the stock of the limiting biogen in the photic layer stabilised (Fig. 4.3a), the ratio of biological components of the ecosystem, both integrally and in terms of depth, equalised for their distribution in the halistatic regions of the ocean (Fig. 4.3). Curve 9 in Fig. 4.4b shows that, in the upper layers of waters, primary production was limited by the intracellular concentration of the nutrient, while in deeper waters, it was limited by photosynthetically active radiation. Thus, after 70 days of a numerical experiment associated with relaxation oscillations in the system, the mathematical model Fig. 4.1 coincided satisfactorily with real observations of pelagic ecosystems.

Considering the functioning of the pelagic ecosystem on the model starting from the 70th day of the numerical experiment, it can be noted that a similar picture is usually observed in planktonic communities that originate in upwellings and develop as the waters drift to halistatic regions. In these communities, succession also begins with an outbreak of phytoplankton biomass: following that, small, and then large forms of zooplankton develop. The depth of the photic layer also increases with distance from the upwelling regions. A similar picture is observed in seasonally vegetative pelagic ecosystems. Analysing these data, we can conclude that the first peak of primary production (on the 80th day of the numerical experiment) corresponded to the "biological spring" (Fig. 4.3), while the second, smaller peak (on the 130th day) corresponded to the "biological autumn".

It should be noted that in numerical experiments, oscillatory processes took place with periods closely comparable with the time scale of the functioning of natural ecosystems of the photic layer. This determined the possibility of using the model

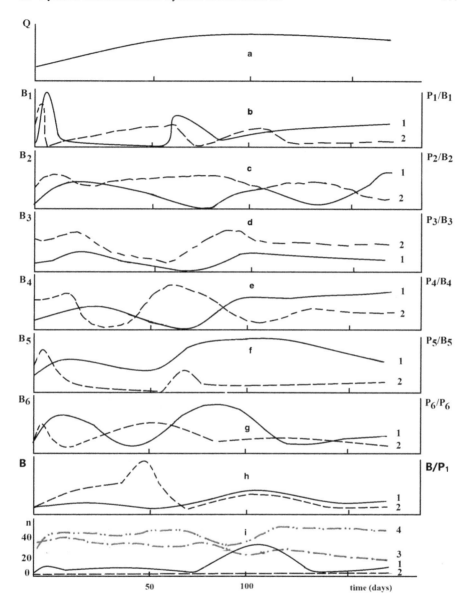

Fig. 4.3 Integral characteristics of the surface water ecosystem 0–100 m: nutrient (**a**); biomass (1) and production (2) of phytoplankton (**b**); bacteria (**c**); small (**d**) and large (**e**) forms of zooplankton; DOM (**f**) and SOM (**g**); biosedimentation (**h**) and relative biosedimentation (**i**) of sources of SOM—phytoplankton (1), bacteria (2), small (3) and large (4) forms of zooplankton. B and P with indices—biomass and specific production of the ecosystem component; *BS* and *BS/P*—biosedimentation and biosedimentation reduced in relation to primary production; *n*—component contribution to biosedimentation (%) (Polikarpov and Egorov 1986)

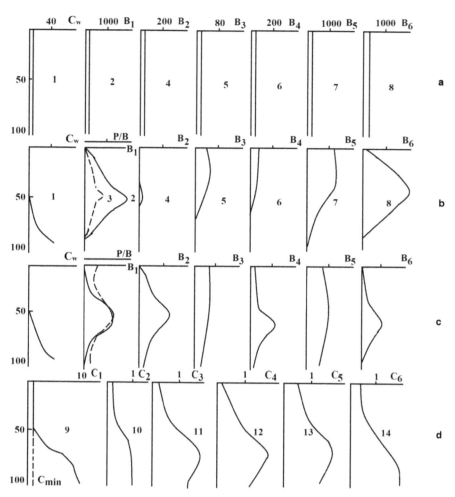

Fig. 4.4 Vertical distribution of ecosystem components: concentration (μg/L) of the limiting element in water (1); biomass (2) and specific production (3) of phytoplankton (μg/m^3), bacteria (4), small (5) and large (6) forms of zooplankton; mass (mg/m^3) DOM (7) and DOM (8) at the initial moment of time (**a**), after 80 (**b**) and 170 (**c**) days; the concentration of the limiting element (μg/mg) of phytoplankton (9), bacteria (10), small (11) and large (12) forms of zooplankton, as well as in DOM (13) and DOM (14). Y-axis—depth, m (Polikarpov and Egorov 1986)

to study the dynamic characteristics of the interaction of the ecosystem of the photic layer with radioactive or chemical pollutants of the marine environment.

Interactions in the ecosystem of the photic layer were studied at low, medium, and high ability of hydrobionts to concentrate radionuclides. The efficiency of assimilation of the contaminant from food by small and large forms of zooplankton was taken as equal to 0.8. Numerical experiments showed that at, low levels of concentration of the pollutant by hydrobionts, its amount in the water of the photic layer increased

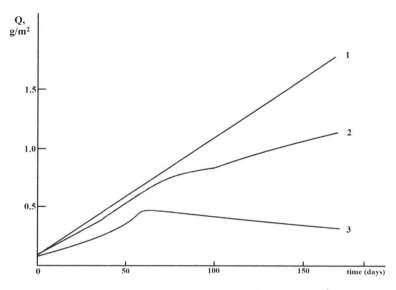

Fig. 4.5   Change in the amount of pollutant in the water of the photic layer (g/m^2) at low (1), medium (2) and high (3) levels of concentration of the pollutant by ecosystem components; Q—quantity of pollutant in the photic layer (g/m^2) (Polikarpov and Egorov 1986)

linearly with time (curve 1 in Fig. 4.5). At the same time, due to the effect of water mixing beinglimited to physical processes, the vertical distribution of the pollutant in the water had an exponential form (Fig. 4.6). At medium concentration levels, the amount of radionuclide in the waters of the photic layer (with equal anthropogenic impact) grew at a lower rate (curve 2 in Fig. 4.5), but the biological components of the ecosystem have not yet coped with water purification.

Dashed lines correspond to the results of a numerical experiment obtained at low levels of concentration of a pollutant by ecosystem components, while solid lines represent high levels of pollutant in the environment in the considered range. When the concentrating ability of hydrobionts in relation to radioactive substances was high as a result of the functioning of biological components of the ecosystem, the quantity of pollutant in the water of the photic layer stabilised at a certain level (curve 3 in Fig. 4.5). This was due to the fact that a portion of the primary production of organic matter of the photic layer as a result of the biosedimentation of suspensions sank towards the bottom and removed the environmental pollutant having accumulated to high levels (Fig. 4.6) and to values of the concentration factors close to their stationary values (Fig. 4.6b). At the same time, the environmental pollutant was concentrated in the SOM to higher accumulation factors (curve 14 in Fig. 4.4d) than in other components of the ecosystem. This was due to the manifestation of the observed (Bogdanov and Kopelevich 1974) natural effect of enrichment of faecal pellets with a pollutant due to its incomplete assimilation from food by consumers.

Numerical experiments have shown that, under halistatic conditions, the main components of SOM, which are 90% responsible for ecological self-purification,

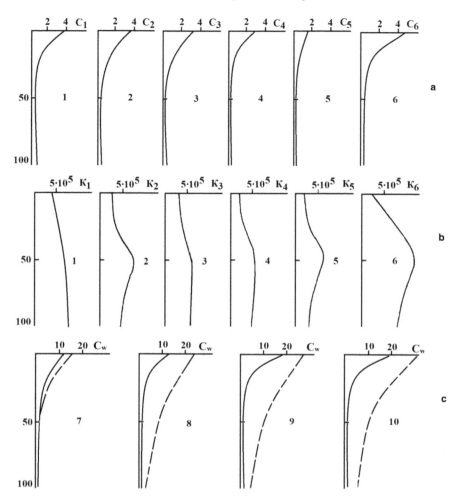

Fig. 4.6 Vertical distribution of the concentration (μg/mg) of the pollutant (**a**) accumulation factors (**b**) in phytoplankton (1), bacteria (2), small (3) and large (4) forms of zooplankton, DOM (5) and SOM (6) after 170 days and vertical distribution of the pollutant (μg/mg) in water (**c**) after 40 (7), 80 (8), 120 (9) and 170 (10) days of numerical experiment (Polikarpov and Egorov 1986)

comprise the waste products of small and large forms of zooplankton (Fig. 4.3i). Under conditions of upwelling or "biological spring", dying phytoplankton cells made the greatest contribution to cleaning the environment.

Figure 4.5 illustrates the predicted characteristics of the contamination of the photic layer with a constant polluting flux and varying degrees of concentration of the radionuclide by hydrobionts. It is evident that, in numerical experiments with a constant concentrating ability of the pollutant by hydrobionts, but at different rates of its entry into the environment, a qualitatively similar result could be obtained. In other words, the maximum conceivable rate of pollutant intake would be determined, at which, as a result of ecosystem functioning, the concentration of the pollutant in the water of the photic layer would not exceed a certain value. This indicates that ecosystems have a certain self-cleaning ability in relation to radioactive and chemical pollutants of the marine environment.

Thus, the analysis of the model (Fig. 4.1) revealed the following properties of the system comprised of marine environment pollutant and photic layer ecosystem:

1. The system is stable in terms of biotic characteristics if the primary productive properties of the ecosystem are limited by PAR and a biogenic element.
2. The dynamic states of the system are determined by damped oscillatory processes whose periods correspond to the time scale of succession in the ecosystems of the photic layer. If the initial conditions deviate too much from the state of the observed ecosystems, a mode of relaxation oscillations arises in the system, during which time period the characteristics of the model do not reflect the state of planktonic communities in the pelagic zone. Following the completion of relaxation oscillations, the states of the simulated system qualitatively corresponded to the real characteristics of the distribution of ecosystem components in the photic layer of waters.
3. The stability of the system with respect to the concentrated characteristics of the distribution of pollutants in the environment is determined by the flux of the pollutant as a result of anthropogenic impact, the biosedimentation productivity of the ecosystem of the photic layer, as well as the level of accumulation and metabolic rate of the pollutant by hydrobionts. Each biosedimentation capacity of an ecosystem and each level of concentration of a pollutant by hydrobionts corresponds to the maximum rate of anthropogenic impact at which the system is stationary.

4.1.2 Model of the Sedimentation Function of the Pelagic Ecosystem of the Western Halistatic Region of the Black Sea

The purpose of the study was to verify the mathematical model of the sedimentation function of the pelagic ecosystem based on the results of observations of its biogeochemical characteristics, as well as to study the dynamic patterns of generation of sedimentation fluxes in the western halistatic region of the Black Sea (Egorov et al. 1992). The theoretical basis for the research consisted in the structure of the mathematical model presented in Fig. 4.1, closed by the balance of biogenic elements, as well as that pertaining to living and inert matter. The mathematical implementation of the model is described in Sect. 4.1.1.

For empirical loading and verification of the model, a test site having the lowest dynamic water activity was selected. This was located in the western halistatic region of the Black Sea having central coordinates 43° 00.0′ N, 31° 30.0′ E. On it on the

25th cruise of the R/V Professor Vodyanitsky in December 1987 and January 1988, two hydrological surveys were made at coordinates 42° 30.0′ N–43° 15.0′ N and 31° 00.0′ E–32° 00.0′ E; the first on 20th–22nd December 1987 and the second on 5th–6th January, 1988. The observations of abiotic and biotic environmental factors after the first survey were carried out at station 3593 at the centre of the test site on 22nd–27th December 1987, while the observations following the second survey took place in the same coordinates at station 3651 on 5th–8th January 1988.

During the first hydrological survey, moderate-to-strong northerly winds were prevailing. The distribution of surface water temperature in the area of the test site from December 22 to 24, 1987 is shown in Fig. 4.7.

In Fig. 4.7 it can be seen that the main gradient of temperature change, which ran in a north-west to south-east direction, did not exceed 1 °C. The aquatic area of station 3593 was characterised by minimal changes in the temperature field. From the profiles of the vertical distribution of temperature, salinity and water density conditions at station 3593 (Fig. 4.8a), as well as from the profiles of the distribution of oxygen and mineral phosphorus (Fig. 4.8b), it can be seen that at this time the winter cooling has

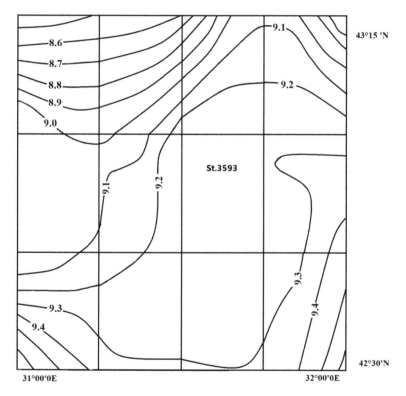

Fig. 4.7 Distribution of surface water temperature at the test site in the western halistatic region of the Black Sea (Egorov et al. 1992)

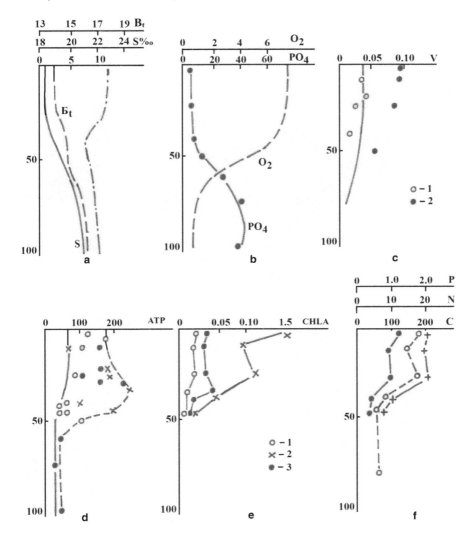

Fig. 4.8 Profiles of the vertical distribution of hydrological, hydrochemical and hydrobiological characteristics of waters at station 3593: **a** temperature (T, °C), salinity (S, ‰), conditional density (σ_t); **b** oxygen (O_2, ml/L), mineral phosphorus (μgP/L); **c** rate of absorption of mineral phosphorus by microplankton (μgP/L per h) at station 3593 (1) and station 3651 (2) (solid curves at positions b and c show the results of numerical experiments obtained on the model); **d** ATP (dashed curves limit the range of values obtained with three replicates of observations); **e** chlorophyll a (1: size fraction 10 μm, 2: 2.5–10, 3: 0.4–2.5 μm); **f** profiles of changes in the content of P (1), N (2), and C_{org} (3) in suspended matter (μg/L) from station 3593 (Egorov et al. 1992)

not yet been completed, the thickness of the upper quasi-homogeneous layer (UHL) was 26 m; under it, there were significant gradients of changes in temperature, salinity and density of waters, as well as the content of oxygen and mineral phosphorus, confined to the lower boundary of the thermocline. The core of the cold intermediate layer (CIL) was located in a depth interval of 36–43 m. The velocity of horizontal currents in the area of the test site calculated by the dynamic method is 0.7–1.7 cm/s; the vertical component of the current velocity was estimated at between 5×10^{-5} and 6×10^{-5} cm/s. According to the results of the second hydrological survey, it was found that by 5th January 1988, the surface water temperature had decreased by an average of 1.0–1.5 °C, with the warmest waters located in the northwestern part of the test site and the coldest ones in the northeastern part. The UHL value at the test site increased to 37 m; the gradients of the vertical profiles of density and hydrochemical characteristics decreased; the core of the CIL is located in a depth interval of 32–77 m.

The analysis of hydrological and hydrochemical data in general showed that the aquatic area of the test site was affected by small-scale cyclonic gyres generated by the Black Sea Rim Current, having a cross section of about 20–40 miles. The results of the first survey testified to the dynamic passivity of the area, while the second was indicative of some increase in the dynamic activity of its waters.

Studies carried out according to the method of radioactive indicators using ^{32}P showed that the change in the dynamic characteristics of the waters of the area corresponded to the change in the indicators of its biological activity. During the first survey (1 in Fig. 4.8c), measured by ^{32}P, the rate of absorption of mineral phosphorus by microplankton was three times lower than during the second stage of the survey (2 in Fig. 4.8c). Analysis of the profiles of vertical distribution of adenosine triphosphate (ATP, Fig. 4.8d) and chlorophyll*a* (CHLA, Fig. 4.8e) in various size fractions of living matter showed that during the observed period a significant quantity of phytoplankton was present in the photic layer, but its biological productivity, measured by the radiocarbon method, was low, since the daily P/B ratio did not exceed 0.3 units. The water transparency measured from the Secchi disk installed at station 3593 was 11–14 m, which corresponded to the occurrence of the lower boundary of photosynthesis on average at a depth of 36 m. The light saturation of phytoplankton at this station ranged from 2.5 to 10 thousand lux, the maximum photosynthesis at light saturation was 5.5–6.1 mgC/m^3 per day, while the daily assimilation number was in the range 3.5–4.3 mgC/mg CHLA per daylight hours (data on ATP and CHLA were obtained and kindly provided by V. G. Shaida and S. V. Zhorov).

The taxonomic and numerical composition of zooplankton in the area of the test site at station 3651 are given in Table 4.2. With the empirical loading of the model, fine filter feeders—nanophages—were conditionally included in the first group (small forms) of zooplankton, while coarse filter feeders—euryphages and predators—made up the second (large forms) group.

The data on the content of phosphorus, nitrogen and organic carbon in suspended matter at station 3593 (Fig. 4.8f), as well as those on ATP and CHLA, were used to obtain estimates of the vertical distribution of detritus and bacterioplankton.

Table 4.2 Abundance (A, spec./m^3) and biomass (B, mg/m^3) of zooplankton at station 3651 (Egorov et al. 1992)

Species	Layers (m)							
	0–10		10–25		25–50		50–93	
	A	B	A	B	A	B	A	B
Nanophages[a]								
Acartia clausi	100	0.1	33	0.0	0	0	10	0.0
Pseudocalanus elongatus	750	3.8	19	2.7	480	2.8	30	0.1
Paracalanus parvus	0	0	33	0.1	1	0	0	0.0
Calanus helgolandicus	50	0.1	67	0.2	60	1.6	0	0.0
Oithona minuta	7860	12.1	3700	5.7	1260	1.9	30	0.1
O. similis	100	0.1	167	0.3	120	0.2	80	0.1
Oikophura dioica	250	1.5	200	1.2	40	0.2	0	0
Polychaeta	4	0	0	0	0	0	170	0.2
Euryphages[b]								
Acartia clausi	100	2.5	33	1.6	0	0	0	0
Pseudocalanus elongatus	1100	27.1	600	14.2	340	8.3	90	3.5
Paracalanus parvus	350	2.0	33	0.3	100	0.5	0	0
Calanus helgolandicus	50	31.5	38	13.6	0.4	0.3	55.8	43.0
Noctiluca miliaris	200	22.0	567	56.4	260	36.0	50	5.9
Predators[c]								
Oithona minuta	1500	5.9	233	0.9	160	0.6	30	0.1
O. similis	350	2.1	67	0.4	240	1.4	140	0.9

[a]Nauplii and I–II copepodites
[b]III–V copepodites + ♀and ♂
[c]♀ ♂

In general, the data on the hydrological and hydrochemical structures of the test site waters, as well as the distribution of living and inert matter, indicated that at the time under consideration, the ecosystem of the photic layer of the western halistatic region of the Black Sea was at the stage of the completion of the summer-autumn blooms, when the production processes were being balanced by destructive ones. For this reason, when performing numerical experiments to verify the model according to the results of observations (Figs. 4.7 and 4.8), the stationary states of the calculated distribution profiles of abiotic and biotic components of the ecosystem and production characteristics were taken into account.

In order to determine the model parameters, the trophodynamic, production and physiological indicators of the functioning of marine ecosystems presented various in publications were used (Finenko and Krupatkina-Akinina 1974; Finenko and Lanskaya 1971; Granstrom et al. 1984; Greze 1977, 1979; Ivlev 1955; Keck and Raffenot 1979; Petipa 1981; Provasoli and Shiraishi 1959; Roots and Kukk 1988; Sorokin 1982; Sushchenya 1972; Tsunogai and Henmi 1971; Vandamme and

Baeteman 1982; Vandamme and Maertens 1983; Vanhaek et al. 1984; Vinogradov 1967; Vinogradov and Lisitsyn 1983; Zaika 1972). Estimates of the indicators of the intensity of phosphate metabolism by hydrobionts were obtained from the results of radiolabelling experiments, presented in the coordinates of a modified Lineweaver–Burk equation (Fig. 4.9). In these experiments, the values of the Michaelis–Menten constant (K_m) and the maximum absorption rate (V_{max}) of phosphates were determined by adding an isotope carrier and its radioactive label to the aquatic environment with hydrobionts sampled in the study area. When calculating the stationary levels of concentration and exchange of phosphates by hydrobionts, in addition to the values of K_m and V_{max}, the stoichiometric ratio $C:N:P$ taken as equal to $100:15:1$ (Romankevich 1977), was used.

The parameters characterising the intensity of the cumulative vertical water exchange, including through the pycnocline, were estimated by comparing the vertical distribution profiles of the limiting nutrient in water obtained on the model with those observed in nature. The thickness of each of the ten layers (Fig. 4.1) was selected as equal to 10 m. The rate of water exchange between the layers was 0.700; 0.500; 0.300; 0.012; 0.025; 0.025; 0.030; 0.050; 0.050 per day. This corresponded

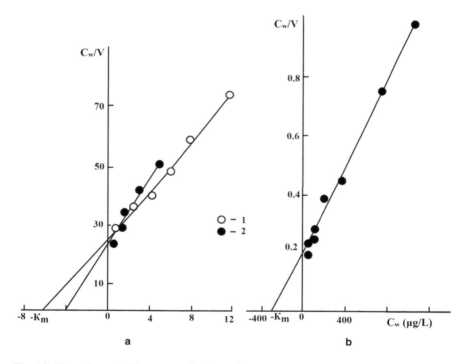

Fig. 4.9 Dependence (in Lineweaver–Burk coordinates) of the rate of absorption of mineral phosphorus by microplankton (**a**) and *Calanus helgolandicus* (**b**) with a change in its concentration in water: 1 and 2 are the results of parallel experiments (Egorov et al. 1992)

to the average effective rate of vertical mixing of waters in the thermocline region, equal to 5×10^{-5} cm/s.

According to the results of experimental observations, the value of the Michaelis–Menten constant for phosphates for microplankton in the area of the test site was 0.5–8.0 μgP/L. In numerical experiments, the calculations were carried out at the values of K_m for phytoplankton and bacteria, equivalent to 4.44 μgP/L. For zooplankton, K_m was 230 μgP/L. The value of V_{max} for phytoplankton in terms of wet weight is 6.45 μgP/mg per day. The rates of phosphate exchange with living matter were within 0.1–0.4 per day. The maximum possible rates of cell division of phytoplankton and bacteria were three generations per day. The minimum intracellular concentration of phosphorus in phytoplankton is $q_{min} - 0.43$ μg/mg wet weight.

In numerical experiments, the expenditure on respiration of living matter was 20–40% of the daily nutrient budget. Food undigested by zooplankton, excreted in liquid form, was taken as equal to 20, while that taking the form of solid excreta was 30% of the daily nutrient budget, whose maximum level was 60–100% of the body weight of the zooplankton organisms. The daily death of living matter amounted to 10%, while the lysis of inert matter was 5%. The rates of gravitational descent of dead bacterial cells were in the range of 0.1–2.5 m/day; phytoplankton: 0.2–5 m/day; carcasses, exoskeletons, faecal lumps and pellets: 50–100 m/day for small forms of zooplankton and 100–150 m/day for large ones. For small forms of zooplankton, it was assumed that the absorption of bacterial and phytoplankton cells is equally probable, while the absorption of detritus particles could decrease. For large forms of zooplankton, the probability of feeding on phytoplankton and bacteria was lower than the possibility of forming a diet due to small forms of zooplankton and detritus. The minimum concentration of food available for nutrition was taken equal to 5 mg/m^3, while the maximum biomass of living components of the ecosystem and detritus, excluding the input of PAR into the layer under consideration, was 150–200 g/m^3 wet weight.

Numerical experiments on the model were carried out when changing the parameters within the variation intervals of their empirical estimates. In this regard, the studied dynamics of the ecosystem model and its stationary states were compared with the results of observations of the investigated test site.

The objectives of such numerical experiments were to determine the possibility of achieving portrait similarity between the model and the object under study in terms of the largest number of stationary states, as well as to evaluate those parameters that are currently not directly observable, but which characterise the dynamics and intensity of sedimentation processes.

A comparison of the results of determining the content of mineral phosphorus in water and the rate of its absorption by microplankton with the curves obtained on the model is illustrated in Fig. 4.8b, c. An analogous comparison of the calculated dependences and profiles of the vertical distribution of the raw biomass of phytoplankton and its daily specific production, estimated according to CHLA and the radiocarbon method, is shown in Fig. 4.10a, b. The profiles of the vertical distribution of wet biomass and daily specific production of bacteria were estimated on the model (Fig. 4.10c) according to the data on the content of C_{org} in suspensions

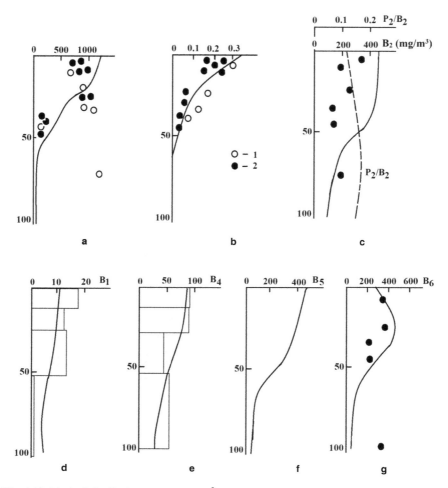

Fig. 4.10 Vertical distribution profiles (mg/m^3) at station 3593: **a** biomass (wet weight) of phytoplankton; **b** daily specific production; **c** estimate of the raw biomass of bacteria; **d** small forms of zooplankton; **e** large forms of zooplankton; **f** concentration of DOM; **g** detritus in the test area. Solid and dashed curves represent the results of numerical simulations, respectively (1 and 2 are the results of duplicate observations) (Egorov et al. 1992)

(Fig. 4.10f), given that bacterioplankton makes up 40% of the organic matter of detritus. The theoretical profiles of the vertical distribution of zooplankton biomass in the test area (according to Table 4.2), as well as DOM and detritus (according to Fig. 4.8d, f) are presented in Fig. 4.10d, g.

The data presented in Fig. 4.8b, c, as well as Fig. 4.10, show that the differences between the values of the factors calculated on the model and determined from the results of the work on the landfill are either small, or have levels of values that lie within the variation range of estimates known from the literature. This indicates that the logical ramification of the model structure and the degree of adequacy of the balance description of the mechanisms of interactions between biotic and abiotic environmental factors are sufficient to obtain a satisfactory portrait similarity in describing the behaviour of a model of such a natural object, which is comprised of the ecosystem of the photic layer. For this reason, it can be expected that the model (Fig. 4.1) will be of practical use in studying the dynamics of both observed and unobserved ecosystems of the photic layer under conditions of changing environmental factors in order to solve problems of forecasting and assessment.

As part of the established tasks, we also used the model (Fig. 4.1) to assess the flux of sediments from the surface waters of the western halistatic region of the Black Sea and to study the dynamic patterns of sedimentation processes as a result of the primary production of organic matter in the ecosystem of the photic layer. The results of studying the dynamics of the ecosystem of the photic layer with a change in the rate of water exchange through the thermocline are shown in Fig. 4.11. They showed that the lower this rate was, the more fluctuations appeared in the supply of biogenic element Q in the photic layer (Fig. 4.11a), although the average value of the nutrient content in surface waters varied slightly under different water exchange equations. At low rates of vertical water exchange, the total mass of living and inert components of the ecosystem B_{Σ} varied with a large amplitude and having an oscillation period of about 50 days (3 in Fig. 4.11b). With an increase in the rate of vertical water exchange, the amplitude of oscillations B_{Σ} decreased (2 in Fig. 4.11b); starting from a certain limit, the oscillatory processes became aperiodic (1 in Fig. 4.11b). Similar patterns manifested in relation to the biomass of phytoplankton B_{ph} (Fig. 4.11c), detritus B_d (Fig. 4.11d) and biosedimentation flux from the photic layer (Fig. 4.11e) were attributed to the primary production of bio sedimentation flux B/P_{ph} (Fig. 4.11f). At the same time, a higher rate of nutrient flux to the photic layer provided both a higher level of water trophicity and greater biosedimentation from the photic layer; however, the relative biosedimentation flux of B/P_{ph} into the waters underlying the photic layer was higher with increased oscillation of the biological components of the ecosystem.

Thus, numerical experiments on the model demonstrated that the supply of nutrients to the photic layer can be the primary generative source of oscillations of ecosystem components; moreover, the efficiency of the "biosedimentation productivity" of the photic layer increases any increase in these oscillations. At the same time, the analysis of the model showed that the relationship between the flux of a nutrient and oscillatory modes in the ecosystem is not unambiguous, but is determined by a number of factors, primarily the minimum concentration of the nutrient in phytoplankton.

Model calculations confirmed that the degree of regeneration of nutrients within the photic layer is numerically equal to the difference $(1 - B/P_{ph})$. For the ecosystem of the photic layer of the western halistatic region of the Black Sea, whose dynamics are reflected in Fig. 4.11 by curves 2, biosedimentation averaged 20% of primary production, which corresponds to the estimates given in the literature (Romankevich 1977; Rudyakov and Tseitlin 1980). Accordingly, 80% of aquatic productivity was

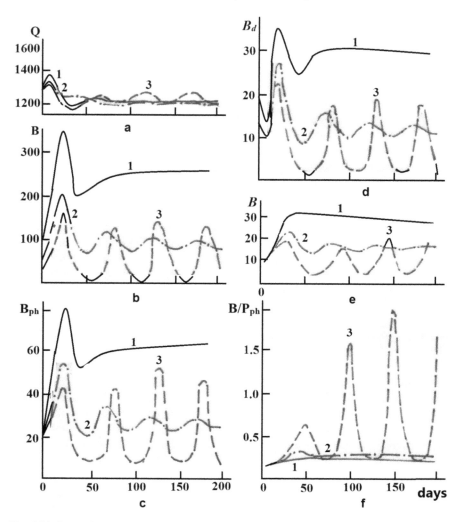

Fig. 4.11 Dynamics of the model of the ecosystem of the photic layer: **a** biogenic element stock (mg/m^2); **b** total biomass of living and inert matter (mg/m^2); **c** phytoplankton biomass (g/m^2); **d** detritus biomass (g/m^2); **e** biosedimentation from the photosynthetic layer (g/m^2 per day), **f** ratio of biosedimentation from the photosynthetic layer to primary production; 1–3: results of numerical experiments on water exchange through the pycnocline of waters 10 m deep at a velocity of 0.050; 0.012; 0.005 days^{-1}, respectively (Egorov et al. 1992)

provided not by influx from below, but by the regeneration of biogenic elements within the photic layer.

Thus, the research has shown that a mathematical model, closed by the balance of matter and biogenic elements, with parameters determined from observations through the indicators of mineral metabolism of hydrobionts, trophodynamic characteristics and energy equivalents of metabolism, reflects the stationary distributions of living and inert components of matter in the ecosystem with a sufficient degree of adequacy and is consequently of use in solving problems of assessment and forecasting under conditions of changing environmental factors. Using the model, it was established that the limited flux of nutrients into the photic layer can be a source of oscillatory processes in the associated ecosystem; moreover, that oscillatory ecosystems will have a higher biosedimentation capacity than ecosystems balanced in terms of production and destruction. The mathematical model predicts that about 20% of the biosedimentation from the photic layer of the western halistatic region of the Black Sea is attributable to the level of its primary productivity, while 80% is provided by the processes of regeneration of biogenic elements.

4.2   Elements of the Theory of Marine Ecosystem Management

According to Vladimir Vernadsky's concept of the global noosphere (1965), marine ecosystems are evolutionarily formed natural objects in which interactions between biotic and abiotic environmental factors occur within a unified spatio-temporal scale. The energy source underpinning their existence consists of solar radiation, while the material source is comprised of chemical substances and their compounds of various biological significance. From a thermodynamic point of view, the functioning of any marine ecosystem is associated with the assimilation and dissipation of part of the solar energy that is transformed into photosynthetically active radiation (PAR). From a biogeochemical point of view, it involves the consumption, transformation of physicochemical forms and return to geological depots of chemical elements having various biological significance, while from an ecological point of view, it consists in the reproduction of organic, mineral and recreational resources.

Under natural conditions, the functioning of marine ecosystems is regulated by the patterns of PAR consumption and the production of organic matter as a result of photo- and chemosynthesis, metabolic reactions and trophodynamics, reproduction of species composition, abundance and biomass under conditions of different-scale successions, as well as larger trends that are caused by evolution, climatic changes and anthropogenic impact.

As already noted (Belyaev 1978), marine ecosystems belong to the category of "complex geosystems"; as such, their responses to impacts and transformations are not linear, but depend on their internal state. At the same time, observations indicate the stable existence of marine ecosystems at various spatial and temporal scales. This indicates the presence of natural mechanisms for the accomplishment of resistance to a variety of biotic, abiotic and biogeochemical factors, implying that methods of control theory and automatic regulation will be applicablefor studying the functional regularities of marine ecosystems. Control systems are currently widely used in engineering, biology, economics and social sciences. The area of these studies is referred to as automatic control theory or control engineering, which in the non-Russian literature is perceived as synonymous with control theory.

Any control system minimally consists of an object of control (the controlled system itself) and a controlling object (controller), closed by a feedback system (Popov 1978). A structural diagram illustrating the management of marine ecosystems, constructed according to the principles of management theory, is presented in Fig. 4.12. Block 1 represents biogeochemical mechanisms of ecosystem function. Block 2 corresponds to its input energetic and chemical parameters limiting production processes. Block 3 reflects the whole complex of influencing biotic and abiotic factors, which include, for example, changes in meteorological, hydrodynamic, hydrochemical and ecotoxicological conditions of ecosystem function, as well as the characteristics of reproduction of its material, energy and chemical resources. Block 4 takes into account the tolerance ranges for the effective action of biogeochemical mechanisms of ecosystem function. Block 5 characterises the extent to which influencing factors from the tolerance ranges are overcome. Finally, block 6 corresponds to feedbacks regulating ecosystem homeostasis.

The functioning of the control system in accordance with the structure presented in Fig. 4.12 can be verbally described as follows. Block 1 reflects the processes of reproduction of the biological structure and functions of ecosystems with the limitation of their material, energy and chemical resources (block 2) under the influence of natural, climatic and anthropogenic factors operating at different scales (block 3). External factors (block 3) can cause the system to move out of the ranges of tolerance. In this case, in block 5, the process of comparing the tolerance ranges obtained at the exit from block 1 and inherent to the system (block 4), while in block 6, the effects

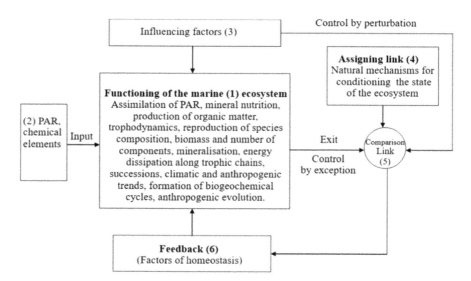

Fig. 4.12 Structural diagram of marine ecosystem management

caused by natural mechanisms of interactions or anthropogenic influence (block 6) are generated in the form of feedbacks, which conduce to the return of mechanisms of ecosystem functionality within the range of tolerance.

Feedbacks can be either positive or negative. The action of positive feedbacks is always oriented towards increasing the degree of mismatch between the output and set parameters of the system control. Feedbacks whose action leads to a reduction in the degree of mismatch between the set and controlled parameters are referred to as negative. Although both positive and negative feedbacks can occur in complex geosystems, the main factor in ensuring their stability is negative feedback. Its analogue in systems related to chemical regulation is expressed in the Le Chatelier-Braun principle (Reimers 1994), which states that: "If a system is in an equilibrium state, then, under the action of forces that perturb this equilibrium, the system will go into a state that results in an attenuation of the disturbing effect".

Thus, for a regulatory mechanism to start acting, it is necessary for the control variable to deviate from the set value. The function of mismatch in control systems is performed by a "sensitive link" or "comparison link", which can be constructed according to various physical, chemical or informational principles. The generation of a control signal, usually taking the form of negative feedback, depends on the structure and principles of the control system. If the feedback control signal is generated from a mismatch between the output from block 1 and the parameters set in block 4, the system can be referred to as controlled "by exception".

There are also systems in which the control signal is generated "by perturbation", taking into account not only the mismatch of parameters, but also the internal state of the system. The principle of governance "by perturbation" is easy to understand by analogy with the system of state regulation of a country's grain supply. Government entities are well aware of the grain consumption requirements on the part of the population, which in turn determines annual plans for its production. If, as a result of meteorological conditions, the yield forecast decreases, then it is by no means necessary to wait for the difference between the planned targets and the yield to manifest itself. Rather, is sufficient to take necessary measures for its timely purchase. From the perspective of control methods, this example represents an implementation of the "perturbation" control principle. In reality, when purchasing grain, it is necessary to take into account not only the current volume of production, but also the accumulated state reserves. In this case, the "perturbation" control principle takes into account knowledge of both the dynamic characteristics of the control system and its internal state.

This principle is especially widely used in the development of digital automatic devices known as finite-state machines. A finite-state machine whose output sequence of signals depends both on its state and on input signals (for example, a combination lock) is called a Mealy machine, while a finite-state machine that does not take into account internal states (e.g., a combination lock) is called a Moore machine. It should be anticipated that the knowledge of the internal state of systems can be a highly significant factor in the management of marine organic resources, since it is under identical conditions of spawning, spring warming of waters and mineral nutrition of the primary trophic links that, almost every year, various fish species appear at the top of the catch pyramid.

The main goal of ecological management, then, is to regulate the stability of ecosystems, manifested at the level of their stationary states or periodic fluctuations within tolerance ranges, referred to in terms of homeostasis. In contemporary reference books, the term "homeostasis" denotes self-regulation, i.e., the ability of an open system to maintain the consistency of its internal state through coordinated reactions directed at maintaining dynamic balance. The main property of homeostasis consists in the "desire" of a system to reproduce itself, to restore lost balance by overcoming the resistance of the external environment. In biological systems, homeostasis can be realised both through adaptive modes associated with the adjustment of the structure and functions of the organism, as well as transformation of the structural, metabolic or energetic characteristics of ecosystems. Biogeocenoses can either be in a resistant or resilient homeostasis mode. In a resistant homeostasis mode, they can maintain their structure and function relatively unchanged. Resilient homeostasis, conversely, is associated with a structural and/or functional reorganisation of a biogeocenosis during fluctuations of external influences.

In Sects. 4.1.1 and 4.1.2, it was shown that the model of the ecosystem of the photic layer remains stable in terms of biotic and abiotic characteristics, if the primary production processes are limited in terms of PAR and biogenic elements. The model of an open system, closed by a balance in material, energetic and biogenic elements, with parameters determined from observations through indicators of mineral metabolism of hydrobionts, trophic characteristics and energy equivalents of metabolism, reflects the regularities observed in nature in the distribution of living and inert components in pelagic ecosystems to a sufficient degree of adequacy. The dynamic states of the model were determined by damping fluctuations with time scales concomitant with the course of seasonal successions, while its stationary levels corresponded to the real characteristics of the distribution of biotic and abiotic components of the ecosystem. The stability of the model with respect to the concentrating characteristics of the distribution of pollutants in the environment is determined by the flux of anthropogenic impact, the biosedimentation productivity of the ecosystem, as well as the concentration and rate of the metabolism of a pollutant by hydrobionts. In the numerical experiment, each sedimentation capacity of an ecosystem and each level of concentration of a pollutant by hydrobionts corresponded to the maximum rate of anthropogenic impact at which the system is stationary.

In general, the materials can be considered as confirming Vernadsky's hypothesis that the necessary conditions of habitat are reproduced in the process of the reproduction of living matter (Vernadsky 1978), as well as being consistent with the opinion of Eugene Odum that "organisms not only themselves adapt to the physical environment, but through their own conjoint activities in ecosystems, adapt the geochemical environment to their biological needs" (Odum 1971). The results of the research have supported the hypothesis that biogeochemical interactions can maintain the stability of ecosystem states in response to various kinds of anthropogenic and climatic influences.

The study of homeostasis is important for determining the adaptive characteristics of ecosystems, their zones of tolerance, as well as the limits of ecosystem stability in relation to the effects of factors applying at different scales. In this regard, the development of theoretical laws for the realisation of homeostasis is an important task for the recently emerged scientific discipline that combines anthropogenic oceanology and biogeochemistry, aiming to solve pressing problems of nature management in implementing the sustainable development of marine aquatic areas.

4.3 Biogeochemical Homeostasis Mechanisms in Marine Ecosystems

In Sect. 4.2, it was shown that the main goal of management consists in the regulation of ecosystem stability, manifested at the level of their stationary states or periodic fluctuations within tolerance ranges, referred to in terms of homeostasis. In biological systems, homeostasis can be realised both through adaptive modes associated with the adjustment of the structure and functions of the organism, as well as sorptional, metabolic, trophic or energetic characteristics of ecosystems.

The results of the studies described in Sect. 4.1 showed that the ecosystem models of the interaction of living and inert matter with radioactive and chemical components of the marine environment were stable in terms of biotic characteristics if the primary production properties of the ecosystem were limited by PAR and biogenic elements. The dynamic states of such systems are determined by damped oscillatory processes whose periods correspond to the time scale of succession in the ecosystems of the photic layer. If the initial conditions deviate too much from the state of the ecosystems as observed in nature, a mode of relaxation oscillations arises in the system, during which time period the characteristics of the model do not reflect the state of planktonic communities in the pelagic zone. Following the completion of relaxation oscillations, the states of the simulated systems qualitatively corresponded to the real characteristics of the distribution of ecosystem components in the photic layer of waters. The stability of the systems in relation to the concentration characteristics of the distribution of pollutants in the environment was determined by their anthropogenic flux, the biosedimentation productivity of the ecosystem and the level of pollutant accumulation in hydrobionts. Each biosedimentation capacity of an ecosystem and level of concentration of a pollutant by hydrobionts corresponded to the maximum rate of anthropogenic impact at which the system was stationary. These materials testify to the existence of natural biogeochemical mechanisms for realising the homeostasis of marine ecosystems, the study of which mechanisms is considered in this section.

4.3.1   Concentrating Function of Living and Inert Matter

The interaction of living and inert matter with the radioactive and chemical components of the marine environment is determined by their absorption from water,
concentration and transfer between aquatic areas, the mineralisation and transformation of physicochemical forms, their removal into water as a result of metabolic,
trophic and sorption processes, as well as their deposition in water and geological depots. This section considers the characteristics of radioisotope and chemical
homeostasis of marine ecosystems along the parenteral and alimentary pathways of
mineral metabolism in hydrobionts. The contribution of sorption processes to homeostasis can be estimated in terms of the significance of the concentration factor (CF),
leading to a determination of the concentrating ability of living and inert matter in
solving problems concerning the sanitary and hygienic quality of waters.

Parenteral mineral and sorption exchange of hydrobionts and inert matter. In the
case of the parenteral route of mineral nutrition, the differential form of the model,
which considers one exchange pool in the hydrobiont, has the form:

$$\frac{dC_h}{dt} = p(C_w CF_S - C_h), \tag{4.19}$$

where

> C_w and C_h are the concentrations of the radionuclide in water and hydrobionts,
> respectively;
> CF_S is the stationary value of the concentration factor (at $t = \infty$);
> p is an indicator of the rate of exchange of a radionuclide by a hydrobiont.

The integral solution of this differential equation for an open system can be
expressed by the following formula:

$$C_h(t) = CF_S C_{w0} + (C_{h0} - C_{w0} CF_S)e^{-pt}, \tag{4.20}$$

where C_{w0} and C_{h0} are the concentrations of the radionuclide in the water and
hydrobiont at the initial moment of time ($t = 0$).

From relations (4.19) and (4.20), it is seen that the conditions $C_w = C_{w0}$ and $C_h(t)$
$= const$ are met when the system of mineral exchange is stationary. In case of violation of the stationarity of the state of the system due to the entry of a radionuclide
into it and an increase in C_w from C_{w0} to C_{w01} ($C_{w01} > C_{w0}$), the concentration of
the radionuclide in the hydrobiont (C_h) will increase until it reaches the level C_h
$= CF_S C_{w01}$ and until the right-hand side of relation (4.19) becomes equal to zero.
Any violation of the stationarity of the radionuclide system comprised of marine
environment/hydrobiont due to an increase in the concentration of the radionuclide
in the water will, in turn, increase the flux of its extraction from the water due to the
concentrating ability of the hydrobiont. Should the stationarity of the marine environ-
ment/hydrobiont radionuclide system change due to a decrease in the concentration

of the radionuclide in the water, the flux of the radionuclide's removal by the hydrobiont will increase. It follows from this that, in general, the system of interaction of a radionuclide in water with a hydrobiont always conduces to achieving a stationary value of CF_S. This is why the effect of radioisotope homeostasis of such systems driven by the fluxes of radionuclide extraction from water or release into the water due to its removal from the hydrobiont is always directed counter to the fluxes of external disturbance of the stationary system conditions according to the factor of radioactive contamination of the aquatic environment. It can be easily seen that, in radionuclide systems comprised of marine environment/aquatic organism, in which inanimate processes of physical and chemical sorption of radionuclides replace those carried out by living matter, homeostasis remains a significant factor in conditioning the radioisotope composition of waters.

Alimentary mineral nutrition of aquatic organisms. Section 3.3.2 presents a differential model of the kinetics of nutriment intake and metabolic excretion of a radionuclide from one exchange pool of a hydrobiont:

$$\frac{dC_h}{dt} = R(C_N q - C_h q_N) - C_h p, \tag{4.21}$$

where

C_h and C_N is the concentration of a chemical element or radionuclide in the hydrobiont and its nutriment, respectively;
q_N and q are the coefficients of assimilation of nutriment for growth and an element from nutriment, respectively;
R is the relative daily nutrient budget of the hydrobiont;
p is the indicator of the rate of exchange of an element by a hydrobiont.

From this ratio, it follows that changes in the content of ecosystem's chemical components in hydrobionts are determined by changes in their concentrations in the primary links of the trophic chains, which means that the characteristics of consumer homeostasis primarily depend on the regularities of mineral metabolism of hydrobionts along the parenteral path of their mineral nutrition. On the other hand, depending on the ratio of the values of the coefficients q_N and q, the concentration of chemical elements in each subsequent link of the trophic chains can be either higher or lower than in the previous ones. There are reports in the literature that the concentration capacity of chemicals of varying biological significance on the part of consumers decreases up the trophic chains (Romero-Romero et al. 2017). In this case, fluxes of unassimilated chemicals are formed from closed systems to open ones at each trophic level, whose transfer flux direction corresponds to the implementation of the Le Chatelier–Braun principle. However, for such toxicants as mercury and organochlorides, an increase in the levels of their content when ascending the trophic chain was observed; this is apparently associated with their concentration in the livers of marine animals (Sun et al. 2017; Van der Velden etal. 2013). Nevertheless, the general regularity of the influence of the concentrating ability of hydrobionts on the characteristics of homeostasis is preserved, since the fluxes of extracting radionuclides and their isotopic and nonisotopic carriers from water or entering the water due to excretion from the hydrobiont are always directed counter to the fluxes of external disturbance of the stationary conditions of the system by the factor of radioactive or chemical pollution of the marine environment.

Sedimentation processes. Sedimentation is one of the most significant processes of biogeochemical self-purification of water from radioactive and chemical contamination. In accordance with theoretical concepts (Polikarpov and Egorov 1986), the dependence of the deposition flux of pollutants to the surface layer of bottom sediments (F_{bs}) on v_{sed} is described by the ratio:

$$F_{bs} = v_{sed} \cdot C_{sed} \qquad (4.22)$$

or, in terms of specific fluxes (F_{sp}):

$$F_{sp} = (abP + F_{al}) \cdot CF \cdot C_w, \qquad (4.23)$$

where

P is specific primary production (kgC/m^2 per year);
a and b are parameters that take into account, respectively, the conversion of primary production (P) from carbon units into dry mass of suspended matter, as well as the part of primary production eliminated from the aquatic environment and entering the surface layer of bottom sediments in the form of sediments;
F_{al} is the specific flux of allochthonous sediments (kg/m^{2}per year);
C_w is the concentration of the pollutant in the aquatic environment;
CF is the averaged factor of concentration of pollutant by sediments.

From relation (4.23) it can be seen that the intensity of the flux (F_{sp}) is primarily determined by changes in the concentration of pollutants in the aquatic area (C_w), primary production (P), the influx of allochthonous suspensions (F_{al}) and concentrating factor (CF) of biogenic and terrigenous suspensions.

It is known (see Chap. 2) that anthropogenic impacts lead, as a rule, to an increase in the concentration of pollutants in water, as well as to a general increase in the trophicity of water associated with an increase in the flux of nutrients and terrigenous material into the aquatic area with coastal runoff. In general, anthropogenic impacts increase F_{sp}. In accordance with the natural principle of homeostasis, this flux reduces the content of pollutants in the aquatic environment; that is, its effect includes a compensation for its causes. For this reason, relation (4.23) can be seen as demonstrating the manifestation of the Le Chatelier–Braun principle under natural conditions, whose effect is directed towards a stabilisation of the ecological state of marine ecosystems under the influence of pollution factors and hypereutrophication of waters.

Figure 4.13 presents the results of studies of deposit fluxes ^{90}Sr (a), ^{137}Cs (b), $^{239+240}$Pu (c), ^{210}Po (d), $\sum$PCB$_5$ (e), $\sum$DDT (f) and mercury (g) in the thickness

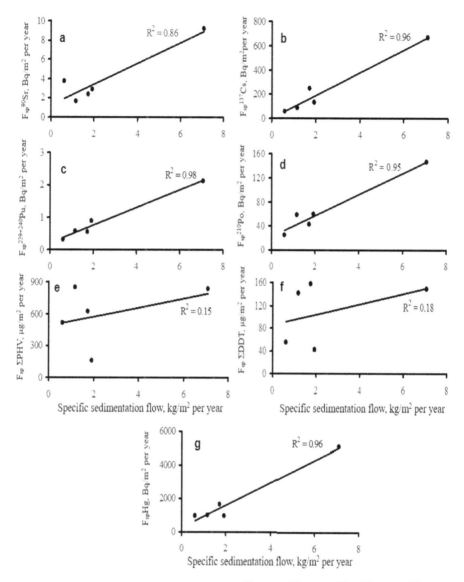

Fig. 4.13 Dependence of the deposition fluxes of ^{90}Sr (**a**), ^{137}Cs (**b**), $^{239+240}$Pu (**c**), ^{210}Po (**d**), $\sum$PCB$_5$ (**e**), $\sum$DDT (**f**) and mercury (**g**) in the stratum of bottom sediments on the corresponding specific sedimentation fluxes (kg/m^2 per year) (Egorov et al. 2018a)

of bottom sediments depending on specific sedimentation fluxes (kg/m^{2}per year) in the aquatic area of the Sevastopol bays (Egorov et al. 2018a). The figure shows that the fluxes of pollutants entering the bottom sediments (F_{sp}) increased with an increase in the sedimentation rate v_{sed}. This means that any increase in v_{sed} and C_{sed} always led to an increase in the intensity of their deposition flux into bottom sediments (F_{sp}). The presented materials thus demonstrated that the main biogeochemical mechanisms of the deposition of pollutants in bottom sediments are associated with the primary production processes initiating sedimentation with the fluxes of allochthonous suspensions in the aquatic area, as well as with the concentrating ability of sediments in relation to contaminants.

Concentration factor as an indicator of the intensity of homeostasis and the sanitary and hygienic quality of water. The review considered in the previous section showed that the specific mass (m_{sp}) and the concentrating ability of living and inert matter are characterised by the concentration factor (CF), comprising the main quantitative factors of the radioisotope and chemical homeostasis of marine ecosystems. This is primarily due to the fact that the established parameters allow a determination of the ratio of the content (or pools) of chemical substances and their radionuclides (Q_h) in the composition of living or inert substance to their pool (Q_w) in seawater. The value of the ratio $\frac{Q_h}{Q_w}$ is calculated by the formula (Polikarpov 1964; Egorov 2012):

$$\frac{Q_h}{Q_w} = \frac{100}{1 + 1/m_{sp}CF} \, (\%), \tag{4.24}$$

where

m_{sp} is specific mass;
CF is the average factor of concentration of an element or radionuclide by living and inert components of the ecosystem.

From the nomogram calculated according to formula (4.24) as illustrated in Fig. 4.14, it can be seen that in the range of measured values of the concentrating function (Table 2.8) and the specific mass of hydrobionts observed in nature, the pool of chemical elements in living and inert matter can be a significant biogeochemical factor. The problem therefore arose of how to take this factor into account when solving ecotoxicological and biogeochemical problems.

Factor used to account for the dissolved and suspended forms of pollutants when obtaining ecotoxicological and sedimentation assessments. In currently applicable normative acts, the indicator of the maximum permissible concentration of a pollutant in water (MPC) is normalised, as a rule, according to its total content in dissolved and suspended physicochemical forms (Ryabukhina et al. 2005). This indicates the acceptance of the assumption that the relevant sanitary and hygienic criteria are based on the recognition of the same ecotoxicological impact of the noted forms. As is known, suspended matter is composed of phytoplankton, zooplankton, detritus and terrigenous particles. Interacting with radioactive and chemical components of the

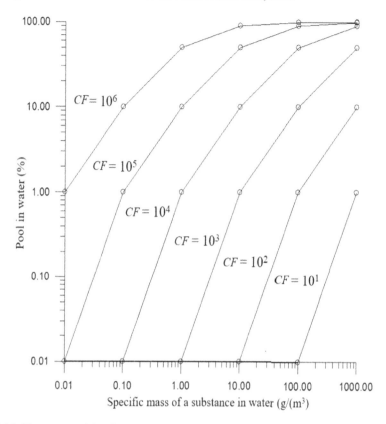

Fig. 4.14 Nomogram of the distribution of a radionuclide or its isotopic carrier between hydrobionts and water, depending on the concentration factor and the specific mass of the living or inert component of the ecosystem (constructed by V. N. Egorov based on the data of G. G. Polikarpov)

marine environment, it absorbs and exchanges contaminants, while simultaneously affecting their availability for sanitary and hygienic protection facilities. It follows that the use of the total content of dissolved and suspended forms of contaminants in water as a sanitary and hygienic criterion requires separate justifications.

In studies of the concentration of mercury in suspended matter in the Black Sea (Stetsyuk and Egorov 2018), it was shown that the factors of its concentration in suspended matter (CF_{sus}) were very high, ranging from 0.023×10^6 to 7.067×10^6. As a result, the pool of mercury calculated by the formula (4.22) on suspensions of the investigated area ranged from 1 to 98% of its total content in the aquatic environment. Examination of the dependence of changes in the concentration of dissolved and total (suspended and dissolved forms) of mercury in water (Fig. 4.15a) showed that its concentration ranked by total mercury exceeded the MAC to a greater extent than when only the dissolved form of mercury was taken into account. For this reason,

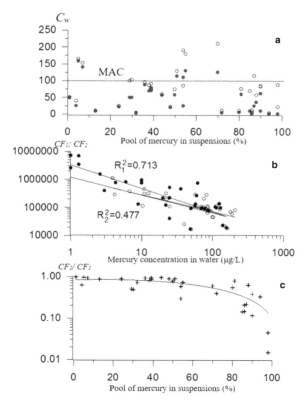

Fig. 4.15 Characteristics of mercury concentration by the Black Sea suspensions: **a** change in mercury concentration in water depending on the pool of mercury in suspensions; **b** change in the factors of concentration of mercury by suspensions when taking into account the dissolved ($CF_1 = C_{sus}/C_{sol}$) and total ($CF_2 = C_{sus}/C_{tot}$) physicochemical forms of the content of mercury in water; **c** dependence of the ratio of the concentration factors, ranked by the total (CF_2) and dissolved-only (CF_1) forms of mercury in water, on the pool of mercury on suspensions in water (Stetsyuk and Egorov 2018)

the standardisation of the sanitary and hygienic situation in terms of the total form of mercury can give overinflated estimates of the ecotoxicological hazard.

From relation (4.24) it follows that the sedimentation self-purification flux of waters depends both on the coefficients of accumulation of chemicals and on their pool on suspensions. In Fig. 4.15b it can be seen that the concentration factors $CF_2 = C_{sus}/C_{tot}$, calculated from the total (dissolved and suspended) form of mercury content in water, can be several times lower than the coefficients $CF_1 = C_{sus}/C_{sol}$, calculated taking into account only the dissolved form of mercury. This was due to the fact that with an increase in the pool of mercury on suspended matter, the value of CF_2/CF_1 decreased (Fig. 4.15c).

The material presented in Fig. 4.15a, b indicate that ranking the coefficients of mercury accumulation in suspensions can significantly reduce the real estimates of the self-purification of waters from radioactive and chemical contamination via sedimentation.

In general, the results of the studies presented in Sect. 4.3.1 showed that the concentrating ability of living matter, in parallel with the sorbing ability of inert matter, is one of the most significant factors in the biogeochemical mechanism of self-purification of waters. This made it possible to interpret the concentration factor (*CF*) not only for characterising radioecological processes (Polikarpov 1964) and solving sanitary and hygienic problems, but also as an important indicator of the intensity of biogeochemical cycles of pollutants in the marine environment.

4.3.2 Adaptive Characteristics of Aquatic Organisms

In Sect. 4.3.2, the adaptive characteristics of hydrobionts are considered on the example of the reaction of the Black Sea green alga *Ulva rigida* Ag. on the chronic and impact effects of polychlorinated biphenyls (Egorov et al. 2012), as well as on the study of the adaptive resistance of the pigment system of green algae *Entero-morpha intestinalis* (L.) in an environment where a phenol concentration was present (Egorov and Erokhin 1998).

The experimental results are shown in Fig. 4.16. According to the calculations performed in this experiment, the average concentration of chlorophyll*a* in the samples of *Ulva* thalli at the initial moment was 0.087 mg/100 g, while that of chlorophyll*b* was 0.054 mg/100 g. In the samples of *Ulva* thalli in all aquariums, an observable decrease in the concentration of chlorophylls*a* and *b* took place within 3 h (Fig. 4.16a, b). The maximum decrease was noted upon contact of *Ulva* with a PCB solution at a concentration of 100 mg/L (Fig. 4.16a); here, in relation to the control, the content of chlorophyll*a* decreased to 60%, while chlorophyll*b* decreased to 80% (Fig. 4.16b). In the subsequent period, the concentration of chlorophylls *a* and *b* in algae increased to the control level and even exceeded it within 1 h; then, after 2 h, it stabilised at a level close to the level of their content in thalli samples in control aquarium. At the same time, the concentration of pheophytin as a whole changed in antiphase (Fig. 4.16c) to the content of photopigments, which increased during the first hour and subsequently also returned to the control value in 15 h. The maximum increase in pheophytin concentration (by 60%) was observed in the aquarium having the highest PCB concentration (Fig. 4.16c).

From the literature it is known (Britton 1983) that a ratio of chlorophyll*a* to its degraded form (pheophytin) of less than 1 indicates the dying off or decay of algal chloroplasts. From the data of the autumn experiment, it can be seen (Table 4.3) that at 12 noon in the aquarium with the maximum concentration of PCBs, this ratio became 0.58, which in our case indicated a significant decay of chlorophyll*a* under the influence of a combination of factors: increased turbidity, as well as a high concentration of PCBs and acetone.

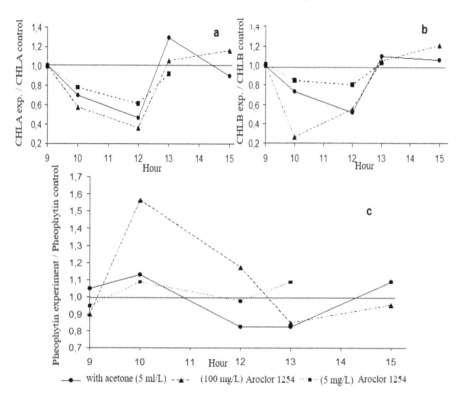

Fig. 4.16 Concentration of chlorophyll*a* (**a**), chlorophyll*b* (**b**) and pheophytin (**c**) in *Ulva rigida* (relative to control) in an experiment carried out on November 10, 2011 (Egorov et al. 2012)

Table 4.3 Average ratio between chlorophyll*a* and pheophytin in ulva (Egorov et al. 2012)

Aquarium	Time of day on 10/11/2011				
	9 h	10 h	12 h	13 h	15 h
Control	4.54	4.78	2.72	1.57	3.22
With acetone (5 ml/L)	4.68	2.35	1.48	2.15	1.90
PCB (Aroclor 1254) (5 ml/L) + acetone (5 ml/L)	5.38	3.52	1.41	1.26	−[a]
PCB (Aroclor 1254) (100 ml/L) + acetone (5 ml/L)	4.21	2.11	0.58	2.38	3.63

[a]No data

The results of experimental observations showed that when the green algae *Ulva rigida* was kept in an aquatic environment with acetone and Aroclor 1254, the mechanisms of its adaptive reactions to the effects of toxicants on a time scale of three to four hours were manifested in the decomposition of chlorophylls with the corresponding formation of pheophytin, followed by a stimulated increase in the concentration of chlorophylls on a time scale of two hours until the control level is reached. Analysis of the general kinetic regularities of the thallus reaction underlined the fact that the action of adaptation mechanisms is ultimately always directed towards reaching the control level of chlorophyll content in them. In this regard, the properties of the adaptive system of algae were observed, manifesting themselves in the inertial excess of the control levels when returning to stationary states from areas with both increased and decreased relative concentration of chlorophylls in algae. Analysis of the kinetic regularities of changes in the pigment composition of algae in general made it possible to conclude that the adaptive system of their response to the effects of toxicants in the range of concentrations considered in the experiments (100 mg/L Aroclor 1254 + 5 ml/L acetone) corresponded to the system of "exception control" with negative feedback.

Studies into the adaptive stability of the pigment system of macrophytes under the influence of chemical toxicants (Egorov and Erokhin 1998) showed that changes in the concentration of chlorophyll$a$ in marine green algae *Enteromorpha intestinalis* (L.) in a medium with a phenol concentration from 0.1 to 100.0 ml/L had an oscillatory response (Fig. 4.17).

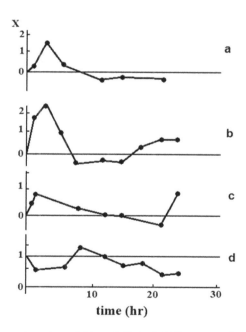

Fig. 4.17 Relative change in content of chlorophyll$a$ in thalli *Enteromorpha intestinalis* in seawater with a phenol concentration of 0.1 mg/L (**a**); 1 mg/L (**b**); 10 mg/L (**c**); 100 mg/L (**d**) depending on exposure time. $X = C_{ch}/C_{chc-1}$, where C_{ch} and C_{chc-1} are the concentrations of chlorophyll in macrophytes under the influence of phenol and in the control, respectively. The results of numerical experiments (Egorov and Erokhin 1998) are represented by solid lines

A determination of the transfer function of the system and its parameters was carried out by analysing the experimental results using the "black box" method. The model comprising differentiating and oscillatory links was found to describe the kinetic regularities of the implementation of the adaptive function of algae to the effect of phenols with a sufficient degree of adequacy. Analysis of the modeling results (Egorov and Erokhin 1998) showed that an increase in the concentration of phenol in water led to a decrease in the stability margin and an increase in the static error of the biochemical adaptation system of macrophytes. This also led to a deterioration in the ability of algae to restore the concentration of chlorophyll*a* to a normal physiological level.

4.3.3 Trophic Factor of Homeostasis

One of the factors of anthropogenic impact is water hypereutrophication, which manifests itself in almost all links of marine ecosystems. Under conditions of hypereutrophication, the primary production of phytoplankton can increase by orders of magnitude. In this connection, surplus energy supply resources arise for consumer links in the trophic chain. The dependence of the nutrient budget of marine organisms on the specific mass of food items can be described by the Ivlev equation (Ivlev 1955). The absorbed nutriment expended on basic metabolism, on somatic and generative growth becomes a component of sedimentation fluxes in the form of mineralised faeces, exoskeletons and exuvia (Petipa 1981). Under conditions of hypereutrophication, zooplankton organisms form pseudofaeces from the excess part of the diet, clumps of which, falling under the influence of gravity, accelerate the flux of sedimentary self-purification of waters (Fowler and Benayen 1979).

Experimental modelling of the processes of trophic homeostasis was carried out on the example of food intake by zooplankton organisms *Acatia clause* of unicellular algae*Prorocentrum micants* labelled with ^{60}Co. The results of observations carried out with T. E. Gulina and V. N. Popovichev showed that the daily nutrient budget of zooplankton organisms largely depended on the specific number of phytoplankton cells in the water of experimental aquariums (Fig. 4.18).

Figure 4.18, the circles show the daily rations calculated from the radioactivity of zooplankters; their estimated expenditures on various components of energy metabolism are plotted by solid lines. The figure shows that, with a small specific number of algal cells in water, the nutrient budget of *Acatia clause* was negligible. This range of specific biomass of *Prorocentrum micants* cells corresponded to the trophicity of waters, at which the energy expenditures for basal metabolism by *Acatia clause* exceeded the expenditures for locating and capturing nutriment. With an increase in the specific number of phytoplankton cells under conditions corresponding to oligotrophic waters, the daily nutrient budget was sufficient only for the basal metabolism and somatic growth of zooplankton organisms. In mesotrophic waters, energy expenditures for somatic growth were also provided, while in eutrophic waters a significant part of the energy of consumed food was transformed

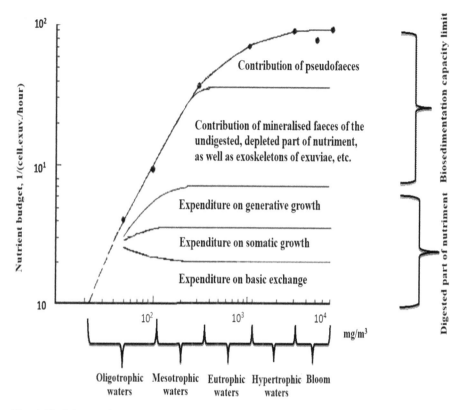

Fig. 4.18 Scheme of sedimentation function of *Acatia clause* when feeding on *Prorocentrum micants* depending on the trophicity of waters (according to T. E. Gulina, V. N. Popovichev, and V. N. Egorov)

into sedimentation fluxes involving an increase in the contribution of pseudofaeces to the circulation of organic matter in the marine environment. During hypereutrophication of waters, the limit of biosedimentation capacity from the metabolism of planktonic organisms approached with the appearance of conditions for phytoplankton bloom and subsequent die-off as a result of ecotoxicological processes.

The analysis of the results of experimental modelling from a biochemical point of view presented in Fig. 4.18 allowed their interpretation as follows. Biogenic elements that limit primary production return to the phytoplankton habitat in oligotrophic waters as a consequence of the dissipation of the energy absorbed by zooplankton and the mineralisation of organic matter occurring in it. With an increase in the trophicity of the waters, the biosedimentation flux increases, whose effect conduces to the removal of nutrients limiting the primary production from the photic layer. Thus, the effect of the trophic factor of homeostasis is that, under oligotrophic conditions, the turnover of nutrients in the aquatic environment is accelerated; with an increase in trophicity, the cycle of their turnover slows down, which in both cases increases the stability of pelagic ecosystems.

4.3.4 Population Characteristics of Biotopes

Under contemporary conditions, marine ecosystems function under the influence of a large number of anthropogenic and climatic factors. These primarily include chemical and radioactive pollution of the marine environment, hypereutrophication of waters, suffocation phenomena, deterioration of the physical, physiological and trophic conditions of existence and reproduction, as well as consumption of organic resources exceeding reproduction. The influence of such pollutants can lead to chromosome abnormalities, as well as changes in adaptive capacity, deterioration in the physiological state, metabolism, somatic growth and reproduction of organisms, as well as to an increase in mortality. To date, no methods are known for determining the complex response of ecosystems to all components of anthropogenic impact. Studies have shown that, in areas subject to anthropogenic stress, the efficiency of converting solar energy into a biological form of matter does not decrease, but that ecosystems tend to be restructured towards a decrease in the number of species having longer life cycles (Hubert 1984). In order to exclude undesirable transformations, it is necessary to understand the impact of individual factors of anthropogenic stress on the structure of marine communities, as well as their combined impact. The attempt to solve such a problem involves the assumption that the stability of the age structure of populations of marine organisms is a manifestation of the stability of ecosystems—and that violations of such stability are to be considered in terms of ecosystem failure. In this case, a quantitative measure of failures can be correlated with the survival rate of hydrobionts, with their reproductive ability, or with commercial fishing effort.

Studies using the provisions of failure theory were carried out on the example of two populations, one having a three-and the other having a five-year life cycle (Polikarpov et al. 1987). The model structure is shown in Fig. 4.19.

The model reflects the functioning of two populations competing for a specific amount of nutriment taken as an assumption under conditions of influence on the survival rate of all age groups, the reproductive capacity of older generations and the intensity of fishing effort. The quantitative measure of failure is normalisable by indicators of mortality, reproductive ability or fishing effort, as well as their relative frequency in relation to life-cycle duration.

Analysis of numerical experiments on the model showed that, in the absence of failures, the relative abundance of juvenile fish was proportional to the partial availability of nutriment (Fig. 4.20). In numerical experiments reflecting the effect of failures on the survival of hydrobionts, a high rate and intensity of failures led to the extinction of all age groups of populations. However, across a wide range of changes in the intensity and frequency of failures, their impact led to the elimination from the ecosystem of populations having longer life cycles and to the survival of populations having shorter life cycle (Fig. 4.20). This was due to the fact that, with

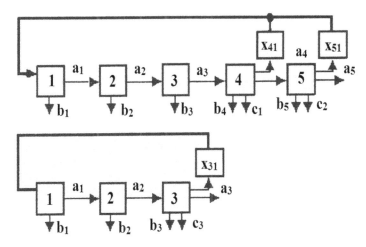

Fig. 4.19 Age structure of populations of hydrobionts with five-year (1–5) and three-year (1–3) life cycles: a_1–a_5 and a_1–a_3—age categories (years); x_{41}–x_{51} and x_{31}—reproduction rates; b_1–b_5 and b_1–b_3—mortality rates; c_1, c_2 and c_3—catch rates (Polikarpov et al. 1987)

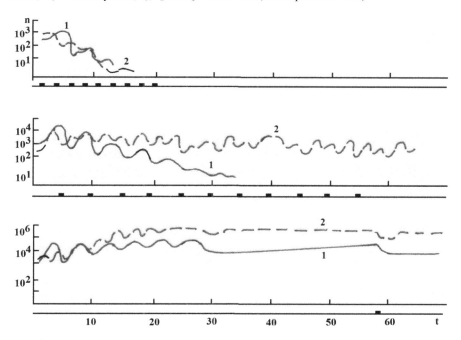

Fig. 4.20 Changes in the number of juvenile fish populations having five- (1) and three- (2) year life cycles under the influence of different failure rates on the survival of all age groups (Polikarpov et al. 1987)

a relatively high frequency of failures, exposures occurred on the scale of individual generations, leading to their increased mortality. Numerical experiments reflecting the characteristics of the impact of failures on the reproductive capacity of marine stocks and associated fisheries have yielded broadly similar results (Polikarpov et al. 1987). These materials demonstrated that the functioning of marine ecosystems under anthropogenic impacts can lead to changes observed in natural conditions, directed towards a decrease in the number of species having longer life cycles. It is generally known that hydrobionts with lower size spectra have, as a rule, higher daily nutrient budgets and greater intensity of somatic and generative growth. For this reason, they extract chemical substrates from the aquatic environment to a greater extent; consequently, the change in the size spectra of populations as a result of anthropogenic impact becomes one of the factors of radioisotope and chemical homeostasis of marine ecosystems.

4.3.5 Homeostasis of Estuarine Zones

At the geochemical barriers formed by the river-sea interface, the current slows down and the river waters are saturated with sea salts, which accompanies the precipitation of large fractions of suspended matter, the coagulation of fine clay particles, the flocculation of organic substances and metals, as well as the intensive development of production processes. It is here that the largest part of riverine suspensions is deposited and a transition from the predominance of suspended forms of chemical elements typical of river waters to the predominance of dissolved forms takes place (Lisitsyn 1982). Therefore, the study of biochemical regularities of homeostasis at the river-sea barrier is of interest in the study of the physicochemical conditions of the functioning of marine ecosystems. Within the framework of this work, the characteristics of chemical homeostasis are considered on the example of a comparative analysis of the distribution of mineral phosphorus and the exchange of its radioisotope ^{32}P by various fractions of suspended matter in the estuarine areas of the Danube in the northwestern part of the Black Sea and the Çoruh in the southeastern part (Popovichev and Egorov 2008).

The results of experiments carried out using radioactive phosphorus as a label showed that the profiles of the vertical distribution of the biotic absorption of ^{32}P in water in the northwestern part of the Black Sea correlated with the profiles of chlorophyll$a$ and fluorescence (Fig. 4.21). The experimental results also reflected processes serving to limit the primary production of phytoplankton in winter and summer in the deep-water part of the Black Sea (Fig. 4.22).

In studies of marine estuarine zones, it was determined that at the geochemical barriers formed by the Danube and Çoruh estuaries, the concentration of PO_4 also decreased with decreasing salinity (Fig. 4.23).

In the Danube area, a higher phosphorus eutrophication of waters was recorded at zero salinity; however, in waters associated with the Çoruh River having the same salinity, the concentration of PO_4 was higher. At the geochemical barrier formed by

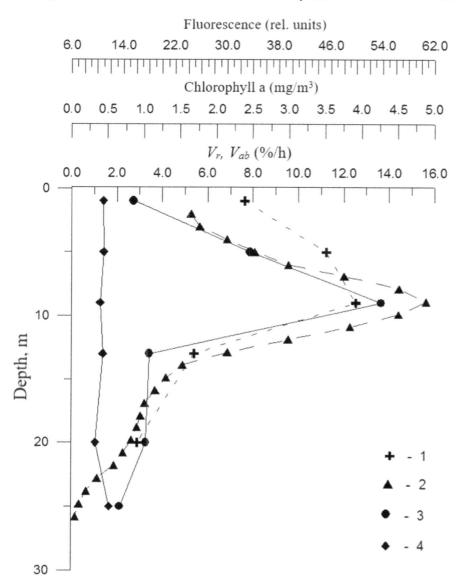

Fig. 4.21 Vertical distributions of chlorophyll$a$ (1), fluorescence (2), indicators of the total absorption rate (V_r) (3) and absorption (V_{ab}) (4) ^{32}P with a suspension fraction above 0.45 μm at the location having geographical coordinates 44° 58.73′ N, 29° 45.18′ E (26/04/1997) (Popovichev and Egorov 2000)

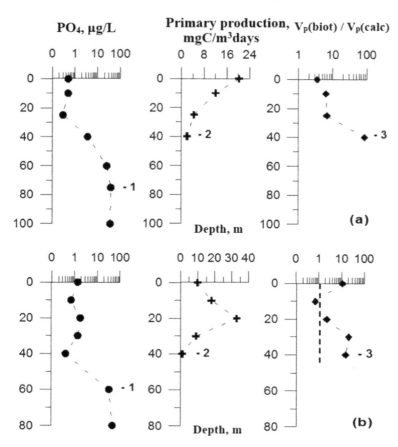

Fig. 4.22 Vertical distributions of the concentration of mineral phosphorus in water (1) and primary production (2); ratio of the rate of biotic absorption of phosphorus $[V_p(\text{biot})]$ to the rate of its absorption by producers $[V_p(\text{calc})]$ (3) in winter (a) and summer (b) periods in the centre of the western deep-water part of the Black Sea (43° 25.0′ N, 31° 10.0′ E) (Popovichev and Egorov 2000)

the Danube estuary, as the salinity of the waters increased, the contribution of large fractions of suspended matter decreased while the contribution of small fractions increased (Fig. 4.24).

A similar trend was recorded in the estuarine zone of the Çoruh River (Fig. 4.25); however, in the Prikhorokh area, the contribution of fine particles was proportionally smaller, with their specific contribution only beginning to increase significantly at salinity levels above 17‰. This was due to the mountainous flow conditions of the Çoruh River carrying significant quantities of large clay particles (Popovichev and Egorov 2008).

The research results from the Danube River area are shown in Fig. 4.26. It can be seen that in the area located at a distance of 17 km from the coastline of the hydrofront at a salinity of 2‰, the quantity of suspensions in the water sharply decreased, while

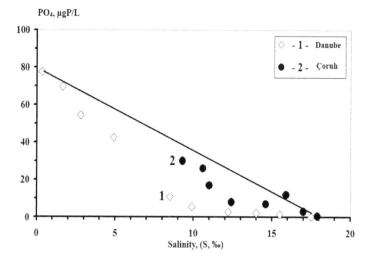

Fig. 4.23 Dependence of the concentration of phosphates (PO_4) on the salinity of surface water in the estuarine zones of the Danube and Çoruh rivers (Popovichev and Egorov 2000)

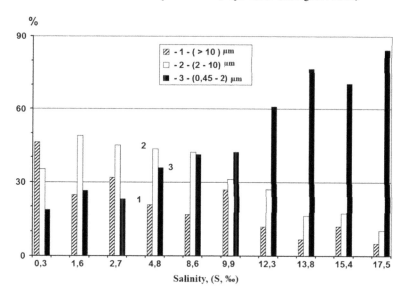

Fig. 4.24 Percentage contribution of different size fractions of suspended matter from the Danube delta region to the absorption of mineral phosphorus depending on water salinity (Popovichev and Egorov 2000)

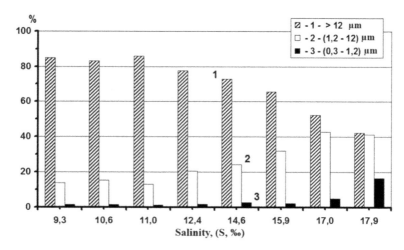

Fig. 4.25 Percentage contribution of different size fractions of suspended matter from the Çoruh estuary region to the absorption of mineral phosphorus depending on surface water salinity (Popovichev and Egorov 2000)

the rates of absorption of mineral phosphorus by suspensions began to increase. The peak of the rate of phosphorus absorption was tracked up to the seaward part of the hydrofront with its minimum value in water (at a salinity of 14‰).

In waters having higher levels of salinity, a significant increase in the rate of phosphorus absorption was observed accompanying an increase in the contribution of small living components of suspended matter (Fig. 4.27), which can be explained in terms of the greater the primary productivity of phytoplankton.

In general, it was found that at the river/sea geochemical barriers, with a decrease in water salinity, the concentration of large (clay) fractions of suspended matter decreased, while that of small fractions of suspended matter increased.

In the interval of water salinity corresponding to the hydrofront, an observed peak of the increase in the rate of mineral phosphorus was associated with both large and small fractions of suspended matter. Evidently, this was caused by a change in their chemical sorption capacity and the solubility of substrates.

Beyond the hydrofront boundaries, at a salinity above 14‰, the specific mass and indicators of the rate of biotic absorption of phosphorus increased. Under conditions of a significant decrease in the concentration of mineral phosphorus in the Black Sea water (salinity over 17‰) (Fig. 4.23), the periods of its turnover significantly decreased. This reduced the limiting role of phosphates in the production of organic matter by phytoplankton, comprising a biochemical factor of mineral homeostasis at the river/sea barriers.

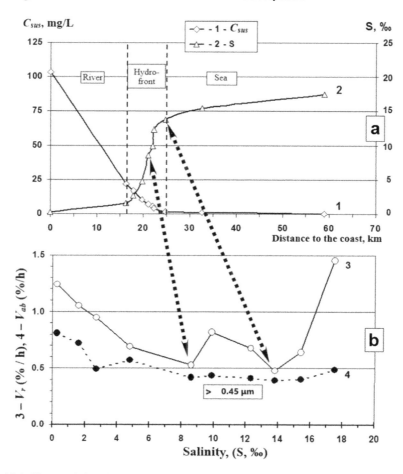

Fig. 4.26 Characteristics of the hydrofront in the area of the Danube delta (Popovichev and Egorov 2000) (along the "Sulinskoye girlo" section): **a** change in the concentration of suspensions $(1 - C_{sus})$ depending the distance to the conditional coastline; **b** change in the rate of absorption $(3 - V_r)$ and sorption

4.3.6   Homeostasis of Coastal Waters in Terms of Biogenic Elements

One of the main contemporary ecological problems affecting the Black Sea consists in water hypereutrophication (Vinogradov et al. 1992). An increase in the intensity of fluxes of nutrient inputs into the marine environment led to an increase in the primary productivity of phytoplankton and a consequent change in the species diversity of hydrobionts of coastal ecosystems (Zaitsev 1992). On the other hand, studies have shown that the concentration of nitrates in the Black Sea, which limits the primary productivity of waters, is $2\,\mu\text{g-at/L}$. At the same time, it was noted that a total content

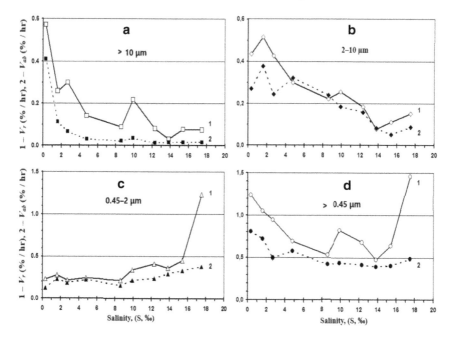

Fig. 4.27 Dependences of the rate of absorption of mineral phosphorus $(1 - V_r)$ and its sorption $(2 - V_a)$ on water salinity for different size fractions of suspended matter from the Danube delta region: **a** >10 μm; **b** 2–10 μm; **c** 0.45–2 μm; **d** >0.45 μm (Popovichev and Egorov 2000)

of nitrates and nitrites in the range of 0.06–0.97 μg-at/L did not correlate with the value of primary production (Vedernikov 1991). A four- and five-fold increase in the concentration of nitrates in the Black Sea during the first phase of eutrophication did not lead to an increase in the concentration of chlorophyll*a* in the water, which indicated the prevalence of phytoplankton mineral nutrition in the summer months, mainly due to regenerated nutrients (Yunev 2011). The results of studies carried out on phosphorus metabolism showed that a sufficiently high level of primary productivity of the Black Sea waters, up to 100 mgC/m³ per day, was provided even at extremely low phosphate concentrations in water of not more than 0.1 μg-at/L (Sorokin and Avdeev 1991). It has been suggested that the growth rate of phytoplankton in the 0–1 m layer of the Black Sea coastal waters is weakly limited by light and nutrients (Stelmakh 2017). In areas prone to eutrophication of the Black Sea, a statistically significant decrease in the average annual concentration of mineral phosphorus in the photosynthetic layer was observed, as well as an increase in its content in the underlying waters at depths of 50–100 m (Finkel'shtein and Pronenko 1991). These data indicate a manifestation of homeostasis of biogenic elements in the photic layer of water as a result of primary production processes. In an attempt to solve this problem, the influence of the primary productivity of phytoplankton as a factor of homeostasis of the content of nutrients in the coastal aquatic area of the city of

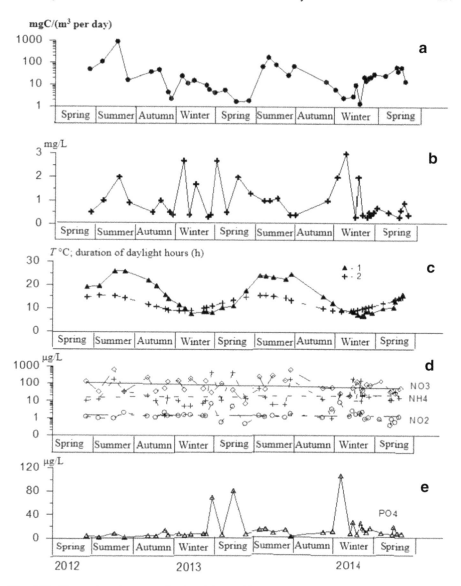

Fig. 4.28 Change in primary production (**a**), specific mass of dry suspended matter (**b**), water temperature (**1**), daylight hours (**2**) (**c**), concentration of mineral forms of nitrogen (**d**) and phosphorus (**e**) in the coastal waters of Sevastopol (Egorov et al. 2018c)

Sevastopol is considered in this section on the example of two-year observations (Egorov et al. 2018c).

Observational materials (Fig. 4.28a, b) have indicated that the primary production of phytoplankton made a certain contribution to the content of suspended organic matter (SOM) in the water, leading to its mass transfer in the photic layer.

Calculations of SOM mass transfer periods in the coastal aquatic area of Sevastopol at a ratio of 1:10 organic carbon content in phytoplankton to its wet weight (Raymont 1980; Stelmakh and Babich 2006) showed that, in the spring–summer season the period of SOM mass transfer was from 2 to 10 days, while during the autumn–winter period it was estimated at more than 100 days. In other words, the productivity of phytoplankton was a significant factor in the turnover of suspended matter in the surface waters of the coastal waters of the Black Sea during the spring-summer period (Fig. 4.29).

From the stoichiometric ratio $1P:7\ N:40C$ (Redfield 1958; Hutchinson 1969) of masses limiting the primary production of the main structural chemical elements, it was calculated that, with phytoplankton produced per 40 mg of organic carbon, 1 mg of P–PO$_4$ (phosphorus phosphates) is extracted from the aquatic environment and 7 mg N–$\sum$N (nitrogen of the total mineral complex). Taking this into account,

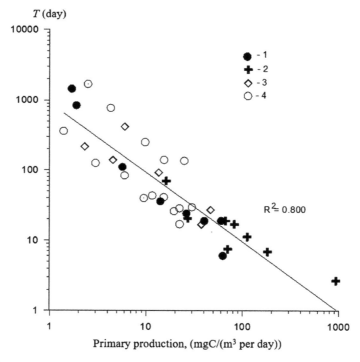

Fig. 4.29 Dependence of periods (T, days) of mass transfer in coastal waters of Sevastopol on the primary production of phytoplankton in spring (1), summer (2), autumn (3) and winter (4) seasons

it is possible to calculate the specific fluxes of phosphorus phosphate consumption $[F_{P-PO_4} = 0.025\,F\,F_{phpl}$ (mgP–PO$_4$/m^3 per day)] and mineral nitrogen compounds $[F_{N-\sum N} = 0.175\,F\,F_{phpl}$ (mgN–$\sum$N/m^3 per day)] for the production of phytoplankton. The assignment of the concentrations of P–PO$_4$ and $N-\sum N$ in water to the corresponding fluxes of their consumption by phytoplankton (P_{P-PO4} and $P_{N-\sum N}$) determined the turnover periods of biogenic elements (T_P) and (T_N) in the aquatic environment, presented as a function of both primary production (Fig. 4.30a, c) and water temperature (Fig. 4.30b, d).

It was established (Fig. 4.30a, c) that the regression dependences of changes in T_P and T_N on primary production with a high degree of statistical significance ($R^2 = 0.775$ for P–PO$_4$ and $R^2 = 0.681$ for N $- \sum$N) corresponded to the equations of a straight line on a logarithmic scale of the coordinate axes, indicating a power-law trend in the implementation of these regularities. At the same time, similar regularities of the periods of the turnover of phosphates and nitrogen compounds from water temperature appeared with less statistical significance ($R^2 = 0.395$ for P–PO$_4$ and $R^2 = 0.182$ for N–$\sum$N). In general, the presented materials (Fig. 4.30) showed that the primary production of phytoplankton is the leading factor in the conditioning of biogenic elements, primarily phosphates, in the aquatic environment during the spring–summer annual period. The essence of the process consists in the relatively small contribution of production processes to the turnover of nutrients in the aquatic environment given the absence of chemical limitation. Under the conditions of chemical limitation, the acceleration of the biotic turnover of biogenic elements leads to an increase in the flux of their remobilisation in the photic layer, which, in turn, contributes to an increase in the primary productivity of waters.

Studies in the field of mineral exchange of hydrobionts with radioactive tracers have shown that chemical elements absorbed from the aquatic environment can be removed from marine organisms in vivo as a result of metabolic processes (Barinov 1965; Polikarpov and Egorov 1986). Let us consider once again the theoretical model of the kinetics of the exchange of biogenic elements by phytoplankton obtained from relations (3.70) and (3.73):

$$\frac{dC_h}{dt} = \frac{V_{max}C_w}{K_m + C_w} - \left[p + \mu_{max}\left(1 - \frac{q_{min}}{C_h} \right) \right]C_h, \qquad (4.25)$$

where

C_h and C_w are the intracellular concentrations of nutrients in algae (μg/kg wet weight) and water (μg/L), respectively;

V_{max} is the maximum specific rate of absorption of a nutrient by algae (μg/kg wet weight per day);

K_m is the Michaelis–Menten constant (Patton 1968), numerically equal to the concentration of a nutrient in water (μg/L), at which the rate of its absorption by phytoplankton is half of the maximum;

p is the indicator of the exchange rate (lifetime release) of an element from phytoplankton cells (per day);

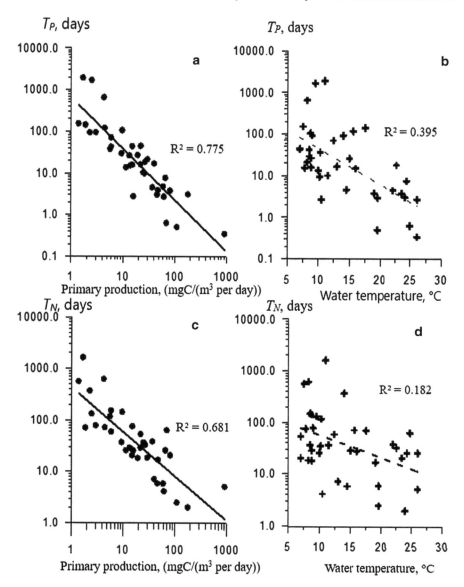

Fig. 4.30 Dependences of the periods of turnover of phosphate phosphorus (T_P, days) (**a, b**) and the total amount of mineral nitrogen compounds (T_N, days) (**c, d**) on the primary production of phytoplankton (**a, c**) and water temperature (**b, d**) (Egorov et al. 2018c)

μ_{max} is the maximum physiologically possible specific growth rate of phytoplankton (per day);

q_{min}—limiting concentration of the nutrient in phytoplankton (μg/kg wet weight).

The term $\mu_{max} \cdot (1 - q_{min}/C_h) = P_{spec}$ entered into Eq. (4.25) characterises the chemical limitation of the specific rate of cell division of algae (Droop 1974). From this expression, it can be seen that the growth rate is limited by the ratio q_{min}/C_h. When the intracellular concentration of a nutrient decreases to a value of $C_h \approx q_{min}$ production processes cease. From Eq. (4.25) it follows that production is limited from below by that element for which the value of q_{min}/C_h is close to unity. In this case, all other elements can be accumulated by phytoplankton cells to levels significantly exceeding the values of q_{min}.

According to relation (4.25), at a stationary state of the system comprised of the biogenic element in the marine environment and unicellular algae, when $\frac{dC_h}{dt} = 0$ and $P_{sp} = const$, the coefficient of its accumulation by phytoplankton is:

$$CF = \frac{V_{max}}{(K_m + C_w)(p + P_{sp})} \tag{4.26}$$

From Eq. (4.26) it follows that, in the general case, the concentration factor (CF) depends on the ratio of the values of the Michaelis–Menten constant (K_m) and the concentration of the nutrient in water (C_w). At $C_w \gg K_m$, the value of C_w in relation to K_m in expression (4.26) can be disregarded to permit the assumption that $CF \approx CF_{max}$, where CF_{max} is the maximum possible relative level of biogenic element concentration by phytoplankton. At $C_w \gg K_m$, the value of K_m in relation to C_w can be disregarded to permit the assumption that the coefficient of the accumulation of a nutrient by phytoplankton decreases in inverse proportion to the value of C_w with an increase in its concentration in water.

To date, a significant number of experimental observations have been accumulated in experiments with radioactive phosphorus (^{32}P) on the parameterisation of Eq. (4.26) (Popovichev and Egorov 1992, 2000, 2003, 2008). In a summary from 2008 (Popovichev and Egorov 2008) it was reported that the Michaelis–Menten constant depends on salinity, PO_4 and chlorophyll$a$ concentrationin water; estimates of K_m obtained for different seasons of the year ranged from 1 to 22 μgP/L. In complex experimental studies with suspended matter and ^{32}P for Black Sea phytoplankton, the following was obtained (Egorov et al. 1992): $K_m = 4.44$ μgP/L; $V_{max} = 6.45$ μgP/mg per day; $p = 0.1$–0.4 (per day). When modelling quasi-stationary states of marine aquatic areas in different seasons of the year at a variation in the concentration of mineral phosphorus (P–PO_4) in water from 1.0 to 80 μgP/L, the specific production rate (P_{sp} fluctuated within 0.1–1.0 (per day). Substitution of these data into Eq. (4.26) showed that, for the selected maximum and minimum values of the parameters, the coefficient of phosphorus accumulation (CF_P) by phytoplankton could be estimated as 0.05×10^6–16.1×10^6.

The optimal levels of the content of nutrients limiting production processes in phytoplankton of the coastal sea area near the city of Sevastopol were compared

(equal to 80 g of carbon, 2 g of phosphorus, and 14 g of nitrogen per 1000 g of organic matter) and the range of variation of their concentrations in water: 1.1–105.1 µg/L for phosphates (PO_4) [or 0.4–33.9 µgP/L for phosphorus phosphates ($P - PO_4$)] and 23.2–770.0 µg/L for the nitrogen complex were determined ($\sum N$): nitrites + nitrates + ammonium [or 13.6–341.6 µgN/L for nitrogen from this complex ($N-\sum N$)]. The analysis showed that, over the observation period, the coefficients of phosphorus accumulation (CF_P) in algae could vary within a range of 0.06×10^6–5.0×10^6, while nitrogen compounds (CF_N) varied from 0.04×10^6–1.0×10^6.

According to calculations using formula (4.16), with a wet phytoplankton mass of 10 g/m^3, which on average corresponds to the eutrophic water level (Kitaev 1984), the phosphorus pool in algae ranged from 37 to 98%, while nitrogen compounds varied from 29 to 91% of content in the marine environment, with the upper estimates corresponding to the spring–summer values of the primary productivity of phytoplankton. For this reason, the daily extraction of nutrients can reach 91–98% of their pool in the photic layer with partial remobilisation into the aquatic environment as a result of metabolic processes and the mineralisation of dying phytoplankton cells.

According to contemporary concepts (Stelmakh 2013), up to 70% of the primary production of the photic layer is consumed by microzooplankton, which is a trophic source of energy and mineral nutrition for the consumer and decomposer units of the ecosystem. Since, according to the results of experiments with diatoms (Solomonova and Akimov 2014) showing that the lifespan of phytoplankton cells can also occur on an hourly time scale, the return of nutrients as a result of remobilisation processes can maintain the stability of their biotic turnover in the marine environment on hourly as well as daily time scales.

It was determined (Fig. 4.31) that at high concentrations of phosphorus (Fig. 4.31a) and nitrogen (Fig. 4.31b) in water, the intensity of their biotic turnover was low, while at low concentrations it increased.

Theoretically, in relation to the element limiting production processes, biogeo-chemical processes taking place in the ecosystem proceed as under oligotrophic conditions, while in relation to non-limiting processes, they proceed as under eutrophic conditions. From expressions (4.25) and (4.26), it follows that phyto-plankton accumulates biogenic elements at $C_w \gg K_m$ up to a value of $C_h \gg q_{min}$ during the eutrophication of the marine environment. This explains why the content of biogenic elements in the sedimentation flux eliminated from the photic layer is maximised, leading to a relative increase in the intensity of water de-eutrophication. Under oligotrophic conditions, when $C_w \gg K_m$, the intracellular concentration of nutrients in phytoplankton tends towards the minimum ($C_h \approx K_m$), while, in rela-tion to the element limiting production, the concentrating ability of phytoplankton tends towards the maximum ($CF \approx CF_{max}$). Thus, the maximum possible supply of biogenic elements to the consumer and reduction units of the ecosystem is achieved, which, due to dying off and mineralisation, ensure the return of nutrients to the photic layer under conditions according to which their content in the sedimentation flux is minimal.

Primary production in the coastal waters of the city of Sevastopol was observed to change by three orders of magnitude over the course of a year, reaching the level

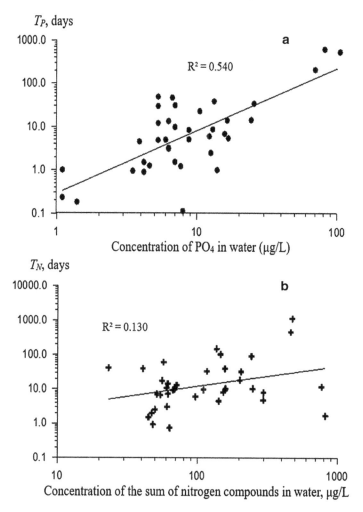

Fig. 4.31 Change in the turnover periods of mineral phosphorus (**a**) and nitrogen (**b**) depending on their concentration in the coastal waters of Sevastopol (Egorov et al. 2018c)

of water hypertrophy in the spring-summer period. Mineral phosphorus was the key factor in the chemical limitation of the primary production of phytoplankton taking place under the influence of the duration of daylight hours and water temperature. In the studied aquatic area near the city of Sevastopol, the periods of mass turnover of suspended organic matter and turnover of nutrients in the water of the photic layer as a result of the primary production of phytoplankton ranged from 0.1 to 1700 days. The high turnover rate of nutrients in the spring-summer period coincided with their concentration by phytoplankton with concentration factors (CF) 10^4–10^6. When the primary production of phytoplankton changes, the mass exchange of suspended organic matter and the turnover of biogenic elements within the photic layer are accelerated or slowed down in such a way that the combined effect of production and elimination processes always conduces to a weakening of the chemical limiting factor. These data allowed us to conclude that the role of phytoplankton in the homeostasis of biogenic elements is determined by the implementation under natural conditions of the Le Chatelier–Baun principle of negative feedback, which increases the stability of the ecosystem both under conditions of the limitation of production processes by biogenic elements and during hypereutrophication of waters.

4.3.7   Biogeochemical Mechanisms of Compensatory Homeostasis

As already noted, the functioning of marine ecosystems can be described by a scheme of forward and backward connections, implemented according to the Le Chatelier–Braun principle (Odum 1986). Biogeocenoses can either be in a resistant or a resilient homeostasis mode. In a resistant homeostasis mode, they can maintain their structure and function relatively unchanged. Resilient homeostasis, conversely, is associated with a structural and/or functional reorganisation of a biogeocenosis during fluctuations of external influences. In marine ecosystems, homeostasis is provided by biogeochemical mechanisms of the reproductive unity of living matter and the conditions of its habitation (Vernadsky 1965), whose sphere of action is limited by a variety of biotic and abiotic factors.

Over the past several decades, the Black Sea has undergone significant anthropogenic impact. According to recent estimates, more than 80 tonnes of mercury, 4500 tonnes of lead, 12,000 tonnes of zinc and 50,000 tonnes of detergents, as well as a large quantity of petroleum and chlorinated hydrocarbons, enter the Black Sea annually from the catchment area (Zaitsev 2006). As a result, for example, the concentration of DDT in the Danube Delta became 35 times higher than in the open sea (Vinogradov and Simonov 1989). Following the accident at the Chernobyl nuclear power plant, radioactive contamination of the Black Sea aquatic area with atmospheric fallouts in 1986 as measured by ^{90}Sr amounted to 100–300 TBq; with the runoff of the Dnieper and Danube rivers in 1986–2000—(160 ± 28) TBq; for ^{137}Cs in 1986 with atmospheric fallouts: 1740–2400 TBq; with the runoff of the Dnieper and Danube rivers for 1986–2000—(23 ± 5) TBq (Polikarpov and Egorov 2008). The annual release of organic matter into the northwestern part of the Black Sea increased from 2350 in the 1950s to 10,488 tonnes to the 1980s (Garkavaya et al. 1997). By 1997, the annual flow of organic matter into the sea area already exceeded 1 million tonnes (Mee 1997). As a result, the specific allochthonous intake of organic substances into the Black Sea for 50–60 years reached 40 g/m^2 per year, while their concentration in water increased by 2–3 times to 8–12 mg/L (Vinogradov et al. 1992). Oil pollution of the sea, as estimated in 1994 and 1996, (Mee 1997) reached 110,840 tonnes per year. Between the 1950s and 1980s, the supply of phosphates to the northwestern part of the Black Sea increased from 13,940 to 55,000 tonnes per year, while nitrates increased from 154,000 to 340,000 tonnes per year (Zaitsev 1998).

The high rate of anthropogenic nutrient loading significantly increased the primary productivity of the Black Sea waters. Along with the anthropogenic impacts described above, the Black Sea also came under the influence of fluctuations in climatic processes. 20–25-year cycles of nutrient input and content in the surface water column of the Black Sea were discovered to be related to periodic changes in natural hydrometeorological factors (Gulin 2000). Water trophicity increased from an oligotrophic level involving average primary production of phytoplankton up to 100 mgC/m^2 per day (Vinogradov et al. 1992) to mesotrophic and eutrophic levels with primary productivity of 190–240 mgC/m^2 per day (Finenko et al. 2011; Yunev et al. 2005). In processes of primary productivity, smaller forms of phytoplankton came to predominate (Bodeanu 1992). In open waters of the Black Sea, phytoplankton biomass was several times higher than those values typical for deep-aquatic areas (Finenko et al. 2011). In 1948–1984, it averaged 6.1 g/m^2, while between 1985 and 1994, the period of greatest anthropogenic impact, it increased to 24.9 g/m^2 (Parkhomenko and Krivenko 2011). Even higher levels of eutrophication occurred in the area around the port of Constanţa (Romania). Between 1971 and 1975, the specific biomass of phytoplankton here was measured at 720 mg/m^3; between 1978 and 1989, it averaged 2240 mg/m^3; while in the period 1983–1988, it reached 4780 mg/m^3 (Bodeanu 1992). In general, in the aquatic area of the northwestern part of the Black Sea, the phytoplankton content increased from 670 mg/m^3 in the 1950s to 1030 mg/m^3 in the 1960s, 18,690 mg/m^3 in the 1970s and up to >30,000 mg/m^3 in the 1980s (Zaitsev 1998). Eutrophication also affected the bacterial heterotrophic link. In 1989, bacterial production in the Black Sea had increased by 5–8 times as compared with 1964, while the biomass was also 5 times higher (Vinogradov et al. 1992). In 1946–1947, the level of *E. coli* in the coastal seawater and sand of the beaches of Odessa was measured at between 10 and 200 cells/L, while by the 1960s it had risen to up to 90,000 cells/L, and at the end of the 1980s from 250,000 to 2,400,000 cells in 1 L of water (Zaitsev 1998).

Simultaneously with anthropogenic pollution, the Black Sea was exposed to the influence of an aggressive invasive species in the form of the *Mnemiopsis leidyi* comb jelly. In 1989, the biomass of Mnemiopsis reached 1 million tonnes (4.6–12.0 kg wet weight per m^2). By 2004, this species was responsible for the consumption of more than 16% of the zooplankton of the coastal shelf waters (Vinogradov et al. 1992; Finenko et al. 2011).

According to recent estimates (Zaitsev and Polikarpov 2002), parts of the Black Sea ecosystem have undergone various kinds of reorganisation. Since the early 1970s, extensive areas of mass blooming of unicellular algae—so-called red tides—began to appear in the northwestern part of the Black Sea. The subsequent dying off of algae led to mass deaths of marine organisms associated with a deficiency of oxygen in the water (Zaitsev 1992). There was a decrease in plankton biodiversity, which negatively affected all trophic levels. A sharp fall in the number of large native species in the plankton community led to the increasing predominance of smaller species (Zagorodnyaya and Moryakova 2011). By 1992, the biomass of a number of species and groups of mesozooplankton had decreased by 3–5 times (Vinogradov et al. 1992). The negative reaction of planktonic larvae to the contaminated substrate is also illustrated by the reduction in rock colonies of mussels (Revkov 2011). The total stocks of macrophytes comprising species from *Cystoseira*, *Phyllophora* and *Zostera*genera decreased significantly (Milchakova et al. 2011). It was established that, under the conditions of anthropogenic impact, the restructuring of the Black Sea phytocoenoses was conducive to increasing the number of macrophyte species whose morphological characteristics are expressed in a higher specific surface area (Minicheva 1996). Another adaptive feature consisted in the change in the indicator of the maximum absorption rate of phosphate by the brown alga *Cystoseira crinita* under conditions of hypereutrophication. It was noted that, with a significant excess of the concentration of phosphates in water, the value of the Michaelis–Menten constant also increased (Popovichev and Egorov 2008). Over the past decades, the number of organisms occupying the upper levels of the trophic chains—namely, commercial fish species—has sharply decreased. While, during the 1960s, the Black Sea boasted 12 major commercial fish species, by 2008–2009, the spawning importance of the area had significantly decreased. At the present time, the basis of fishing in the Black Sea consists two small short-cycle species—sprat and anchovy. The factors that had a decisive influence on the deformation of the reserves and the structure of the region's commercial resources were chronic pollution, overfishing, physical destruction of biotopes, the introduction of aggressive allochthonous aquatic organisms and a decrease in the natural water balance of water bodies as a result of regulation of river flow (Boltachev and Eremeev 2011).

Thus, the complex response of the Black Sea to anthropogenic impact boiled down to a reorganisation of ecological structures and functions whose functionality consists in increasing the primary productivity of waters, along with a reduction in phytoplankton cells (Bodeanu 1992), a reduced contribution of macrophytes to bioproductivity and increased significance of the bacterial heterotrophic link the in mineralisation of organic matter. In addition, this process has been accompanied by a reduction in the number and size characteristics of consumer trophic links, especially zooplankton and fish. In addition, so-called critical zones (Zaitsev and Polikarpov 2002), within which the concentration of pollutants in the components of ecosystems permanently exceed natural levels, have arisen in those areas experiencing significant anthropogenic pressure, especially in the northwestern part of the Black Sea.

From a systems perspective, the stability of the Black Sea ecosystems up until the end of the 1970s can be seen as having been maintained by resistant homeostasis (Odum 1971), while under contemporary conditions it began to be characterised by compensatory homeostasis associated with the ability to change the structure and function of biogeocenoses. A consideration of the characteristics of compensatory homeostasis required a study of the influence of interactions of bioinert and inert matter with radioactive and chemical components of the marine environment under the conditions of a restructuring of the Black Sea biocenoses.

A striking feature of the Black Sea consists in the differentiation of its water column into two types of ecosystems. The ecosystem of the oxic zone is located at depths from the surface down to 120–180 m. Below, to the maximum depths of the Black Sea, the recently identified (Polikarpov 2012) ecosystem of the anoxic hydrogen sulphide zone occupies up to 87% of its volume.

According to expertspecialists (Sorokin 1982; Vinogradov et al. 1992), the main processes of transformation of matter and energy in the Black Sea occur in the ecosystem of its oxic zone. Studies carried out in recent decades have demonstrated the energetic significance of chemosynthetic processes taking place in the ecosystem of the sea's anoxic zone (Ivanov et al. 1991; Pimenov 2006; Gulin and Artemov 2007; Rusanov 2007; Lein and Ivanov 2009; Egorov et al. 2011).

From a biochemical point of view, the ecosystem of the oxygen zone receives mineral nutrition and organic matter from the underlying layers, from the drainage basin of the Black Sea and from the atmosphere, producing living matter within the photic layer and exchanging it with the ecosystem of the anoxic hydrogen sulphide zone and with the seabed. At certain time scales, bottom sediments for the ecosystem of the oxic layer comprise a geological depot, while the pelagial zone consists of a water depot. The ecosystem of the hydrogen sulphide zone produces chemosynthetic products (Rusanov 2007; Polikarpov 2012), exchanging matter with the oxic zone as well as with the seabed, which also comprises a geological depot for it. Given these circumstances, we can conclude that the primary anthropogenic impact currently occurring in the Black Sea is reflected in the ecosystem of the oxic zone. At the same time, it is necessary to take into account the environmental threat of anthropogenic impact on the anoxic zone of the sea due to the high development rates of deep-water mineral and hydrocarbon resources that occur there (Pererva et al. 1997; Shnyukov and Ziborov 2004; Shnyukov et al. 2001).

The involvement of pollutants in biogeochemical cycles depends on a variety of abiotic and biotic factors (Polikarpov and Egorov 1986). For different types of pollution and physicochemical conditions, the contribution of individual factors to the formation of biogeochemical cycles will obviously differ. In what follows, our consideration will be limited to those factors related to the characteristics of the mineral and radioisotope exchange of living, bioinert and inert matter. The factor of sedimentation conditioning of the marine environment is also taken into account. The relative importance of this is due to the fact that, according to the results of radio-trace measurements (Stokozov 2003), the deepening of the gradient layer in the Black Sea under the influence of hydrological processes occurs at an average speed of about 10–35 m/year or 0.03–0.09 m/day; sedimentation rate of suspended matter of size fractions from 2 to 60 μm is 0.12–1.10 m/day; dead phytoplankton—1–510 m/day; zooplankton—36–875 m/day; faecal pellets—36–376 m/day (Alekin and Lyakhin 1984).

Radioactive contamination. The primary source of radioactive contamination of the Black Sea waters over the past decades occurred as a result of the accident at the Chernobyl nuclear power plant on 26th April 1986. The primary influx of long-lived fission products ^{137}Cs and ^{90}Sr, which amounted to 2–6% of the release from the damaged reactor, occurred at the beginning of May 1986 due to atmospheric fallouts occurring with a southern flow of air mass transfer. Subsequently, ^{137}Cs and ^{90}Sr were also carried into the Black Sea waters with the flow of rivers, mainly the Dnieper and

Danube, as well as via the Dnieper irrigation system (Polikarpov and Egorov 2008). Observations in the large-scale monitoring system showed that, following the Chernobyl accident, the concentration of long-lived fission radionuclides in the surface layers decreased with their simultaneous migration into the deep layers of water. Although the radioactive contamination of biological, bioinert and inert components of the Black Sea ecosystems increased in proportion to the accumulation rates of radionuclides by hydrobionts, the radiation-ecological hazard in the region did not exceed the permissible sanitary levels according to dose and cytogenetic criteria (Polikarpov et al. 2007). Nevertheless, an increase in the content of ^{137}Cs and ^{90}Sr in the geological depot of the bottom sediment strata led to the emergence of potentially critical zones in the estuaries of rivers in the northwestern part of the Black Sea (Stokozov et al. 2008); in this connection, it is possible that any environmental threat will only manifest itself on a longer time scale. The study of the concentrating ability of hydrobionts showed that the stationary values of the concentration factors were largely independent of the content of the corresponding radionuclides and their isotopic carriers in the aquatic environment; while, within individual taxonomic groups, they were determined by the surface mass ratios of marine organisms. Evidently, the increase in the trophicity of waters associated with the response of the Black Sea ecosystems to anthropogenic impact, with a concomitant decrease in the size spectra of hydrobionts, led to an increase in the rate of extraction of post-Chernobyl radionuclides from the aquatic environment and to an acceleration of the biogeochemical cycles of their migration and elimination in geological depots. The main trend in the change in the concentration of radionuclides in the Black Sea ecosystems is characterised by an exponential decrease in radioactive contamination of all its components to pre-Chernobyl levels on a time scale of 15–25 years. The dwell time of Chernobyl ^{137}Cs in the 0–50 m layer of the Black Sea was estimated at 25–35 years, while that of ^{90}Sr was estimated at 40–60 years. According to these estimates, the influence of biogeochemical processes reduced the effective "lifetime" (150 years) of 97% of the atoms of these radioisotopes to 25–40% (Polikarpov and Egorov 2008).

Chemical pollution. From the point of view of the stability of physical and chemical forms and toxicity, chemical pollution can be divided into conservative and non-conservative types. Since non-conservative pollution in the marine environment can undergo lysis and bacterial mineralisation, the hypereutrophication of waters accompanying anthropogenic impact and an increase in biological oxygen consumption (BOD_5) unequivocally indicate an increase in the intensity of the flux of biogenic mineralisation of non-conservative pollution. Conservative pollution, conversely, does not undergo alteration in its chemical properties and toxicity with participation in biogeochemical cycles. In contrast to the concentrating ability of radionuclides found in the marine environment in practically weightless amounts, the concentrating ability of chemical contaminants by living, bio-inert and inert matter depends on their concentration in water. The characteristics of the absorption of conservative chemical pollutants from the aquatic environment and their elimination into geological depots have different effects in terms of the sorption and alimentary paths of mineral nutrition of hydrobionts, as reflected in the previous sections. Studies have shown that an increase in the accumulation of pollutants along the food chain can be observed for polychlorinated biphenyls (PCBs) (Nifon and Coussins 2007) and mercury (Phillips et al. 1980; Evans et al. 2000). It is self-evident that marine organisms primarily assimilate from food the necessary range of substrates for their energetic and mineral nutrition. From this point of view, the increase in the concentration of the noted pollutants along the trophic chain described in the literature can be explained in terms of the transformation of physicochemical forms of PCBs and the methylation of mercury, as well as the concentration of these contaminants in the livers of marine animals. For many other chemical substances of varying biological significance, an increase in their concentration along the trophic chains was not observed. It was determined that the average concentration of Cd in phytoplankton is $2.0\ \mu g/g$, while in fish it is $0.2\ \mu g/g$ dry weight; the coefficient of Fe accumulation by algae is 20,000 units; whereas by fish it is 1000; the concentration factor of Sr by algae—100; by fish—5; the concentration factor of Pb by phytoplankton ranges from 10,000 to 3,000,000 units; while by zooplankton the coefficient is between 200 and 10,000 units; and by fish it is from 5 to 1000 units (Polikarpov and Egorov 1986). It was also noted (Magnusson et al. 2006), that the bioaccumulation levels of PCBs did not correlate with the nutrient budgets of hydrobionts but could instead be caused by differences in biotransformation or age, as well as in the size of the analysed marine animal specimens. According to L. V. Malakhova's data (2006), the total amount of organochlorine compounds concentrated in mussels was 429 ng/g, while the concentration in their biodepositions was 496 ng/g. The enrichment of mussel biodepositions with carotenoids was also noted (Pospelova and Nekhoroshev 2003). From the presented materials it follows that, with decreasing levels of food concentration of contaminants by consumers along the food chain, the faecal flux of elimination of pollutants in geological depots becomes more important (Fowler and Small 1972). With an increase in the trophicity of waters, the intensity of energy supply of consumers, as a rule, exceeds the expenditure on their basic metabolism, which leads to an increase in the faecal flux representing undigested food. Thus, during the alimentary pathway of absorption of pollutants by consumers, the relative intensity of biotic involvement of pollutants increases with a decrease in the size spectra and with an increase in the nutrient budgets of hydrobionts.

Biogenic elements. It is evident that the distribution of biogenic elements between hydrobionts and water has changed under the conditions of contemporary transformations occurring in the ecosystems of the Black Sea. It has been shown that one of these changes consists in a fivefold decrease in the concentration of phosphates in the 0–50 m water layer accompanied by an 11-fold decrease in the concentration of silicon (Vinogradov et al. 1992). Between the 1950s and 1980s, the supply of phosphates to the northwestern part of the Black Sea increased from 13,940 to 55,000 tonnes per year, while nitrates increased from 154,000 to 340,000 tonnes per year (Zaitsev 1998).

Allochthonous organic matter. Organic matter of allochthonous origin, which enters the seas and oceans with river and underground runoff, includes aeolian, abrasion and cosmogenic material as well as anthropogenic pollution in the form of dissolved organic matter (DOM), colloids and suspensions, which, in turn, include both living and nonliving components. Allochthonous sources account for 2.5–5.0% of the organic carbon produced in the ocean (Alekin and Lyakhin 1984). According to M. E. Vinogradov (Vinogradov et al. 1992), bacterial microflora, fungi and protozoa develop on allochthonous organic matter and solid suspensions, which, in turn, sharply accelerates recycling of nutrients, increasing the primary production, but not the "harvest", of organic matter. On the basis of a large quantity of experimental material and natural observations, E. F. Shulgina and colleagues (Shulgina et al. 1978) showed that, when sufficient oxygen is present in the first phase of transformation, non-conservative pollution forms comprising carbon-containing compounds can mineralise to CO_2 and H_2O. The second phase of oxidation that follows involves transformation to nitrates and nitrites. Since biochemical oxidation is provided by bacteria, we can consider the standard value of biological oxygen demand in 5 days (BOD_5) as an indicator of the concentration of pollution, since the number of heterotrophic bacteria is proportional to the degree of water pollution. The authors showed that, given sufficient oxygen content in the water, the BOD_5 constant rate of change, which lies in the range of 0.10–0.12 (per day), does not depend on the initial content of allochthonous organic matter, but is determined only by temperature. This indicates that the process of biochemical oxidation of organic substances in the Black Sea occurs on a scale of 10–12 days in accordance with the law of monomolecular reactions, or the first order of metabolic reactions. A similar order of reactions corresponds to the regeneration of mineral phosphorus and nitrogen compounds. According to existing estimates (Shulgina et al. 1978), the contribution of biological and chemical processes to the total change in hydrochemical parameters reaches 70%. As a result of these processes, the decay of unstable organic matter, the utilisation of decay products and chemical transformations lead to the limited restoration of the natural characteristics of marine waters—that is, to their self-purification. The monomolecularity of the reaction rate indices indicates that the rate of mineralisation is greater the higher the initial concentration of allochthonous organic matter in water. Hence it follows that the increase in the flux of allochthonous organic matter that occurred in recent years has led to an increase in the intensity of its biogeochemical circulation in the Black Sea.

Red tide and die-off. As already noted, algal blooms (so-called red tides) and associated hypoxia have increasingly been observed in the Black Sea in recent years (Zaitsev 1992). The primary production of phytoplankton in the red tide zone reached 1.5 gC/m^3 per day, with a corresponding biomass of 30 g/m^3 (Vinogradov et al. 1992). In August 1973, approximately 500,000 tonnes of benthic animals died in a hypoxic zone occurring in an area of 3500 km^2. In subsequent years, the area in which hypoxia and die-offs occurred increased to 30–40 thousand km^2. In the period from 1993 to 1990, the loss of benthic animals on the northwestern shelf reached 60 million tonnes (Zaitsev 1998). Taking into account that the north-western shelf occupies an area of 68,390 km^2 (Goncharov et al. 1965), it can be concluded that the die-offs affected up to 58% of its area.

From a biochemical point of view, red tides and hypoxia can be considered as factors of the self-purification of the marine environment, leading to the deposition of organic matter and pollutants in geological depots. With a biomass of 30 g/m^3 and a coefficient of its concentration by phytoplankton $CF = 1000$, the substrate pool will be 3%; at $CF = 10,000$—29%; at $CF = 30,000$—47%. From this it follows that, upon the death of phytoplankton cells (as noted, for phosphorus K = 34,000), water de-eutrophication of 35–40% will occur. The coefficients of accumulation by phytoplankton of chemical substances of various biological significance generally lie within the range of 10^2–10^6 units (Polikarpov and Egorov 1986), which means that in case of die-offssalvo self-cleaning of the marine environment can occur. With an average concentration of Zn in marine animals lying in the range of 38–1700 µg/g dry mass (Polikarpov and Egorov 1986), it is relatively straight forward to calculate that the die-off of 500,000 tonnes of organic matter (or 50,000 tonnes of dry mass) in an area of 3500 km^2 will lead to the elimination of between 1.9 and 85 tonnes of zinc from the marine environment—or 5.4–24.3 kg/km^2 of the die-off area. If we designate the annual flux of Zn input (12,000 t) to the entire surface of the Black Sea as equal to 547,015 km^2 (Goncharov et al. 1965), we can obtain an estimate of its specific annual input as equal to 30 kg/km^2. A comparison of these calculations shows that die-offs can be an important biogeochemical factor in the adaptation of the Black Sea ecosystems to anthropogenic impact.

The conducted analysis has shown that, in general, under an anthropogenic impact exceeding the conditioning capacity of the marine environment, the Black Sea ecosystems have passed from resistant to resilient (or compensatory) homeostasis. This general tendency has applied particularly to the critical zones. In all the considered cases, the structural and functional reorganisations conduced to an increase the productivity of primary production links along with the role of allochthonous organic matter in energy transformation, along with changes in the biodiversity of ecosystem components involving a reduction in the contribution of consumer links of higher orders and the size spectra of marine organisms. Such structural transformations increased the intensity of self-purification of waters to reduce the impact of the processes to which they can be seen as responding. When applied to complex systems, this control mechanism is referred to in terms of negative feedback—or the Le Chatelier-Braun principle (Odum 1971). In our opinion, it is possible to distinguish two phases of resilient homeostasis. The first is accompanied by a structural and functional reorganisation of ecosystems aimed at accelerating the biogeochemical cycles of transformation of matter and energy along with an associated increase in the rate of self-purification of waters. The second phase is characterised by a periodic degradation or collapse of ecosystem components, leading to a salvo self-purification of the marine environment.

In general, the response of the Black Sea ecosystems to negative factors can be illustrated by the structural diagram shown in Fig. 4.32.

Figure 4.32 shows the stability modes of the Black Sea ecosystems, depending on the impact of anthropogenic influence, climatic changes and the introduction of aggressive species of hydrobionts. It is estimated that the period before the 1970s was

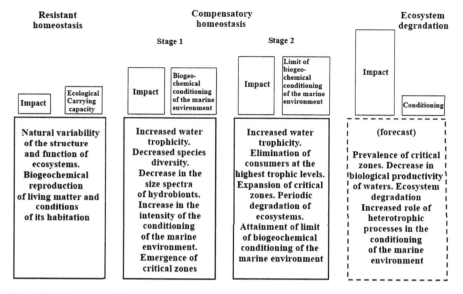

Fig. 4.32 Structure of the conceptual model of the biogeochemical response of the Black Sea ecosystems to anthropogenic impact, climatic changes and invasive species of hydrobionts (Egorov 2012)

characterised by resistant homeostasis, while in subsequent years its mode of homeostasis was transformed to resilience. Figure 4.32 shows that, with resistant homeostasis, the cumulative negative impact was lower than with the compensatory or resilient form. Subsequently, in the first phase of resilient homeostasis, the increased negative impact was counterbalanced due to the restructuring of ecosystems and the related increase in the intensity of biogeochemical processes of conditioning of the marine environment. In the second phase of resilient homeostasis, compensation for further growth of negative impact was provided not only by transformative reorganisations in ecosystems, but also by their periodic degradation. It can be safely predicted that, upon reaching the limit of the ability to condition the marine environment due to the restructuring of ecosystems, their degradation will become irreversible, which, in turn, will reduce the intensity of biogeochemical cycles of water conditioning, since they will be provided mainly only by heterotrophic processes.

4.4   Ecological Carrying Capacity and the Assimilative Capacity of the Marine Environment in Relation to Pollutants

In general, the functionality of the system comprised of anthropogenic pollution/biogeochemical self-purification of the marine environment functions can be described as follows. The concentration of pollutants in the aquatic environment is established on the basis of a balance in terms of the fluxes of their input, absorption and turnover as a result of biogeochemical interactions, as well as the carrying capacity of water masses. A significant part of the contaminants entering the aquatic area can be removed from the aquatic environment through their concentration by living and inert matter in the course of sorption, metabolic and trophic interactions, leading to the fluxes of their elimination into adjacent aquatic areas and bottom sediments.

From the standpoint of contemporary theories of mineral and radioisotope exchange, as well as sorption interactions, the metabolic and sorption reactions of living and inert matter can change from first-order to zero-order in the event of an increase in the concentration of a pollutant in an aqueous medium. Here, the first order of metabolic or sorption reactions corresponds to the biosedimentation or biotransformation purification capacity of the aquatic environment proportional to a change in the concentration of a pollutant in it. With the zero order of metabolic reactions, a growth in the concentration of a pollutant in the environment results in an increase in the period of its biotic cycle along with a decrease in the corresponding ability to self-purify water to reach a limiting level. When the pollution flux exceeds this limit, the system of biogeochemical system of water self-purification loses its stability and is no longer able to condition the marine environment for pollutants. Hence, it follows that an important factor in solving the problem of the carrying capacity of marine areas consists in limitingpollutant fluxes according to biogeochemical criteria that reflect the concentration and sorption function of living and inert matter, the intensity of mass transfer with adjacent areas, as well as the production and sedimentation characteristics of marine ecosystems (Egorov 2001).

In the previous sections, it was shown that, as a result of biogeochemical interactions in marine ecosystems, two modes of conditioning the radioisotope and chemical composition of waters can be observed. The first mode corresponds to the observance of the conditions $C_w \ll K_m$ and $C_w \ll 1/k$, under which the biochemical conditions of its habitat are reproduced in the process of reproduction of living matter. In this case, the relationships between the concentration of chemicals in water (C_w) and the concentrating ability of living and inert matter, characterised by the concentration factors, are linear. In the second mode, at $C_w \gg K_m$ and $C_w \gg 1/k$, the conditioning capacity of ecosystems decreases with the result that the fluxes of biogeochemical self-purification of waters can reach limiting values.

In conditioning modes, three interrelated problems are defined: (1) To assess the production capacity and the maximum permissible concentration of pollutants in biotic components of ecosystems under various conditions of energy metabolism and mineral nutrition at levels of pollutants in the aquatic environment reaching the MAC. (2) To establish the maximum (in terms of biogeochemical and ecotoxicological criteria) fluxes leading to the elimination of pollutants from aquatic areas as a result of the impact of hydrodynamic processes, biotic and abiotic interactions. (3) To assess the maximum permissible assimilation capacity of water volumes according to the factors of their receptive capacity and the fluxes of pollutant elimination into geological depots as a result of biogeochemical processes. Hence, the ambiguity of the concepts of "self-cleaning ability", "ecological carrying capacity" and "assimilation capacity" of the marine environment in relation to radioactive and chemical pollutionnoted in the literature (Goldberg 1981; Zaika 1981; Polikarpov and Egorov 1986; Izrael and Tsyban 1983; Izrael and Tsyban 2009; Jernelov 1983).

On the one hand, the assimilation capacity of the marine environment refers to the amount of a pollutant that can be diluted in the water of aquatic areas such that the concentration of the pollutant in the critical biotic components of ecosystems does not exceed the maximum permissible values. In relation to radioactive contaminants, the term "radiocapacity" is used, which also takes into account the assimilation fluxes into water masses due to radioactive decay. On the other hand, self-cleaning ability implies a differential criterion, that is, the limiting flux of pollution biotically eliminated from aquatic areas to aqueous or geological depots. Here, considering the results of the functioning of ecosystems as a source of fluxes of elimination of pollutants from aquatic areas, it is more appropriate to use the concept of "ecological capacity". If the object of research is the assessment of the limiting fluxes of contaminants entering water or geological depots as a result of gravitational, abiotic and biotic processes, then the term "assimilation capacity or capacity" will more fully reflect the biogeochemical mechanisms of self-purification of the environment.

In the process of solving the problems of ecological and assimilation capacity, it became obvious that non-conservative pollution can be eliminated from the aquatic environment as a result of mineralisation or lysis. At the same time, the concentration of conservative contaminants in the study areas was seen to decrease as a result of radioactive decay, biogeochemical transformation of their physicochemical forms, as well as their transfer to adjacent waters or geological depots. For this reason, the problem of determining the ecological or assimilation capacity for conservative pollutants should always be solved in conjunction with an assessment of the carrying capacity of water volumes and the maximum permissible inflow fluxes into geological depots formed by bottom sediments. It should be noted that, since the processes of biogeochemical conditioning of the chemical and radioisotope composition of marine ecosystems are interrelated, the formation of estimates of their radiocapacity, ecological carrying capacity or assimilation capacity is typically indicative of the considered leading factor of self-purification of the marine environment. Within the framework of this section, examples of assessments of the radiocapacity, ecological carrying capacity and assimilation capacity of marine ecosystems are considered separately.

4.4.1   Metabolic Factor of Ecological Carrying Capacity

Studies of the characteristics of self-cleaning of the marine environment in general showed (Polikarpov and Egorov 1986) that in the range of $C_w \ll K_m$ and $C_w \ll 1/k$, the change in the concentration of pollutants depends linearly on the pollution fluxes of aquatic areas or bottom sediments. With an increase in the flux of pollutants to the levels $C_w \gg K_m$ and $C_w \gg 1/k$, the self-cleaning fluxes reach their limiting values.

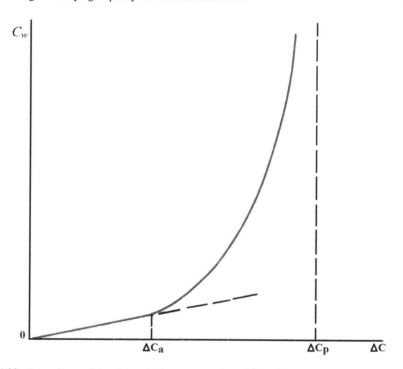

Fig. 4.33 Dependence of the change in the concentration of the pollutant in the water of the photic layer C_w on the intensity of the marine pollution flux (ΔC): ΔC_p is the limiting pollution flux; ΔC_a is the boundary of the ΔC interval, in which the relationship between ΔC and C_w is linear (Polikarpov and Egorov 1986)

Under these conditions, the stationarity of the pollution/self-purification systems of aquatic areas according to the parameters $\Delta C_{poll} - C_w$, where ΔC_{poll} is the flux of pollutant entering the aquatic area, collapses. Graphically, such a dependence, constructed from the results of numerical modelling, has the form shown in Fig. 4.33.

From Fig. 4.33 it can be seen that the relationship between ΔC and C_w is linear at low rates of pollutant input into the marine environment. Within this range of changes in ΔC values, the influence of biotic and abiotic processes is a significant factor in the self-purification of waters. With an increase in ΔC, starting from a certain level ΔC_a, the concentration of the contaminant in the water increases sharply; when approaching the limiting flux $\Delta C_a \rightarrow \Delta C_p$, the stability of the biogeochemical conditioning system for the chemical composition of waters collapses.

From the standpoint of contemporary theories of mineral and radioisotope exchange, as well as sorption interactions, the metabolic and sorption reactions of living and inert matter can change from first-order to zero-order in the event of an increase in the concentration of a pollutant in an aqueous medium (Polikarpov and Egorov 1986). The structure of the transition of an ecosystem's functionality from the first order of metabolic reactions to zero-order is shown in Fig. 4.34.

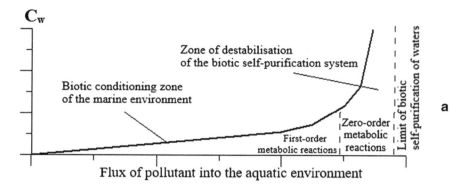

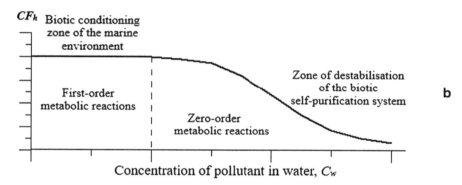

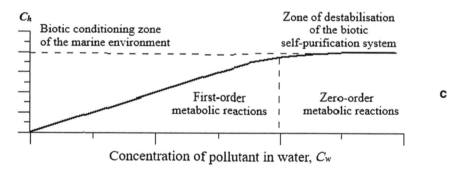

Fig. 4.34 Change in stationary levels of pollutant concentration in water of the photic layer C_w (**a**), concentration factors CF_h (**b**) and concentration of pollutant in hydrobionts C_h (**c**) as a result of the functioning of the ecosystem of the photic layer under the influence of a flux of anthropogenic water pollution (Polikarpov et al. 2012)

On the left side of Fig. 4.34, it is shown that the first order of metabolic or sorption reactions corresponds to the zone of biotic conditioning of the chemical or radioisotope composition of waters, in which the concentrating ability of hydrobionts (CF_h) does not depend on the concentration of the pollutant in the water (C_w), while the concentration of pollutants in organisms (C_h) is proportional to C_w. With the zero order of metabolic reactions (shown in the right part of Fig. 4.34), the concentration of the pollutant in the medium (C_h) reaches its maximum value (C_{wmax}); the concentrating ability of marine organisms decreases with an increase in C_w; the relative capacity for self-purification of waters tends to the limiting value. When the pollution flux exceeds this limit, the system of biogeochemical system of water self-purification loses its stability and is no longer able to condition the marine environment for pollutants. From Fig. 4.34 it can be seen that one of the most important factors in solving the problem of the ecological or assimilation capacity of the marine environment consists in a regulation of the fluxes of pollutants entering the aquatic area such that the concentration of pollutants in the water does not exceed the maximum permissible concentrations, standardised both by ecotoxicological and biogeochemical criteria.

4.4.2 Assimilation Capacity

The factor of assimilation capacity is considered in this section on the example of assessing the deposition of mercury in bottom sediments of Sevastopol Bay.

The results of research and analytical evaluations are presented in Figs. 4.35 and 4.36. Comparison of the data showed (Kostova et al. 2001) that the concentration of mercury in the bottom sediments of the Sevastopol Bay was 2–3 times higher than in the bottom sediments of the western halistatic region of the Black Sea.

These data testified to the anthropogenic nature of the mercury pollution of the bay. According to observations carried out from 1988 to 1999 (Fig. 4.35a), the highest mercury content exceeding its maximum permissible concentration in water (100 ng/L) was recorded in the early 1990s. Subsequently, due to a sharp decline in economic activity in the region, a tendency towards a decrease in the content of mercury in the composition of waters and bottom sediments of Sevastopol Bay was noted (Fig. 4.35b) (Kostova et al. 2001).

At the same time, the relationship between the concentrations of mercury in water and in bottom sediments was revealed to be non-linear (Fig. 4.35c). The dependence of the ratio of the concentrations of mercury in bottom sediments and in the aquatic environment was described with a sufficient degree of adequacy by the Langmuir equation at the values of the parameters $C_{max} = 2740$ ng/g and $k = 0.128$L/ng. The obtained results indicated that the sorption saturation of bottom sediments with mercury took place in the aquatic area of Sevastopol Bay during a period of intensive economic activity (Fig. 4.35d). Calculations showed that at a mercury concentration in 1999 in the surface layer of sediments was equal to 369 ng/g, while, at the rate of sedimentation in the area of the Pavlovsky Cape, Sevastopol Bay, estimated at

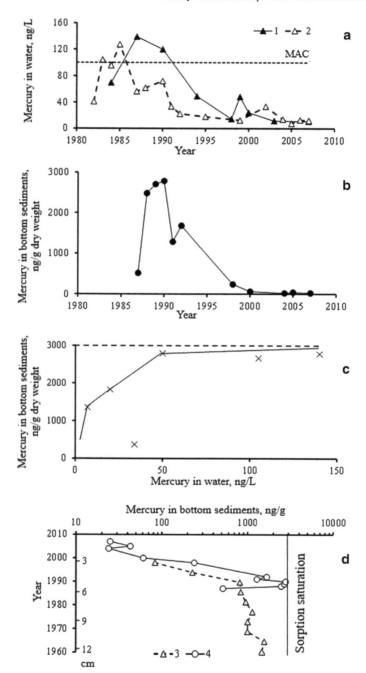

◀**Fig. 4.35** Characteristics of the distribution of mercury in the aquatic area of Sevastopol Bay: **a** in the surface water of the Northern Bay (△—average for the bay; ▲—only in the Pavlovsky Cape area); **b** in the surface layer of bottom sediments in the Pavlovsky Cape area; **c** changes in the average mercury concentrations in the surface layer of bottom sediments and in the water of the Northern Bay; **d** profiles of the distribution of mercury in the thickness of bottom sediments in the area of Cape Pavlovsky, where △ is determined by radio trace dating of bottom sediments (Zherko et al. 2001) and ○—constructed according to distribution data (**b**), taking into account the annual depth of 2.4 mm upper layer of bottom sediments

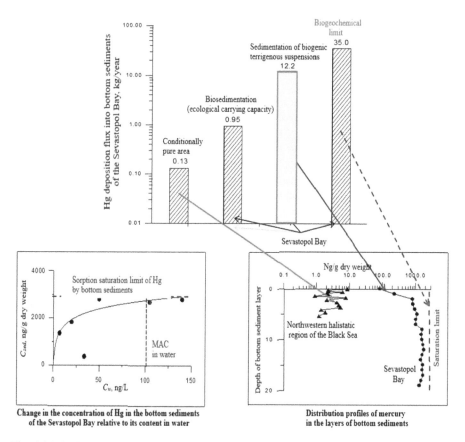

Fig. 4.36 Estimates of mercury deposition fluxes in the bottom sediments of Sevastopol Bay (Egorov et al. 2018a)

0.24 cm/year, or 607 g/m^2 per year (Zherko et al. 2001), the flux of mercury deposition into the bottom sediments of the bay in this year was 224 μg/m^2 per year. At C_{max} = 2740 ng/g, the limiting flux of mercury deposition is estimated at 1660 μg/m^2 per year. By normalising these fluxes, taking into account the area of the Sevastopol Bay and the intensity of production and sedimentation processes (Egorov et al. 2018a), it was determined that 0.95 kg of Hg is deposited per year, while biosedimentation due to biogenic and terrigenous suspensions in the thickness of the bottom sediments of the bay was 12.2 kg Hg per year. The biogeochemical limit characterising the assimilation capacity of the bay associated with the sorption saturation of bottom sediments was 35 kg of Hg per year (Fig. 4.36).

The considered materials indicate that the leading factor in the self-purification of the waters of Sevastopol Bay from mercury consisted in the sedimentation of terrigenous suspensions. Between 1988 and 1999, the limit of sedimentation saturation of bottom sediments with respect to mercury was reached.

4.4.3 Radiocapacity

The problem of assessing radiocapacity is considered in this section using the example of the oxygen and hydrogen sulphide zones of the Black Sea in relation to ^{90}Sr and ^{137}Cs (Gulin et al. 2017).

The problem of the radiocapacity of the marine environment arose with the onset of the nuclear era. Its relevance has increased due to the possibility of technogenic radionuclides entering various components of ecosystems as a result of the impact of physical and biogeochemical processes. Depending on the biological significance of these components, the accumulated radioactivity in them can exhibit a wide range of properties ranging from sanitary and hygienic hazards to the environmentally safe flux of pollutants into geological depots.

As noted in the previous sections, ecological capacity as a factor of water self-purification can manifest itself only if there are mechanisms for the elimination of pollutants from the considered aquatic areas. Objectively, there are several possible means of purifying water from conservative pollutants, among which are included radionuclides. One of these consists in natural radioactive decay, which is characteristic of each radioisotope and does not depend on any physical and biogeochemical processes. In aquatic areas for which radioactive contamination is ecologically dangerous for critical groups of the population, the use of sorbent-based methods of water purification using can be recommended. An alternative approach is associated with the technogenic deposition of radionuclides in the bottom layers of the deep-water oceanic regions. Initially, it was assumed that technogenic deposition, referred to as dumping, would allow the elimination of waste materials generated from the use of nuclear technologies into bottom sediments for a period of time until the complete decay of radionuclides. However, later it became clear that the implementation of dumping is only possible when taking into account a number of factors having both technical and biogeochemical aspects.

One of the key natural processes of self-purification of aquatic areas from conservative pollution is their migration and transfer as a result of hydrodynamic or biochemical processes into adjacent, less significant or less accessible aquatic areas for possible exploitation. In these areas, considered in terms of water depots, radioactive decay occurs under less environmentally hazardous conditions. The maximum permissible fluxes of radionuclide intake into water depots, referred to in terms of radiocapacity, are limited by sanitary and hygienic criteria expressed in maximum allowable concentrations (MACs), which indicate the rate of their radioactive decay as well as the hydrodynamic characteristics of water exchange.

The task of assessing the radioactive capacity of the Black Sea waters was first raised in 1958 in connection with the intention of various foreign and international organisations—in particular the US Congress and the UN—to dump radioactive waste into the deep layers of the Black Sea based on the notion that the vertical circulation of its waters was insignificant (Vodyanitsky 1958). However, calculations carried out at that time, which showed that the period of water exchange in the deep-sea basin of the Black Sea ranges from 70 (Vodyanitsky 1948; Bogdanova 1959) up to 150–200 years (Skopintsev 1975), implied that long-lived radionuclides from deep layers can reach the surface waters of the oxidising zone of the Black Sea relatively quickly.

Radioecological studies carried out in these years by Academician G. G. Polikarpov (Polikarpov 1961, 1964) demonstrated the significant radionuclide-concentrating capability of hydrobionts. This would serve as the basis for an inter-governmental decision to exclude the dumping of radioactive waste into the Black Sea and later be used in the preparation of the London Convention on the Prevention of Marine Pollution by Dumping of Wastes and Other Matter (Polikarpov 2006; Polikarpov and Egorov 2008).

The 1986 nuclear disaster at the Chernobyl nuclear power plant led to the influx of a significant quantity of long-lived radionuclides into the Black Sea via atmospheric precipitation. In subsequent years, these radionuclides migrated throughout the entire sea area, gradually penetrating into the depth of the water column and into bottom sediments (Fig. 2.8). At the same time, the problem arose of assessing and forecasting the environmental hazard and the time scale of the radioecological response of the Black Sea to the radioactive contamination of its waters (Polikarpov and Egorov 2008).

Monitoring carried out since 1986 made it possible to assess the threat posed by the impact of the Chernobyl ^{90}Sr and ^{137}Cs radionuclides on living organisms of the Black Sea ecosystems (Mirzoeva et al. 2008). In addition, the study of the migration of these radionuclides in the marine environment made it possible to use them as tracers of large-scale vertical circulation of the Black Sea waters (Egorov et al. 2001; Gulin et al. 2012).

The objective of this study was to compare the radiocapacity of the waters of the oxic and hydrogen sulphide zones of the Black Sea based on the results of studying the dynamics of the vertical distribution of ^{90}Sr and ^{137}Cs in the water column for the entire period following the Chernobyl accident. With an area of 423 thousand km^2 and a maximum depth of 2212 m, the volume of the Black Sea comprises 547,015 km^3 of water, of which 475,890 km^3 (87%) comprises its hydrogen sulphide zone (Goncharov et al. 1965; Zaitsev 1998). The dependence of the distribution of the relative volume and areas of the sea at different depths is shown in Fig. 4.37. The figure shows that, on average, about 90% of the water volume of the hydrogen

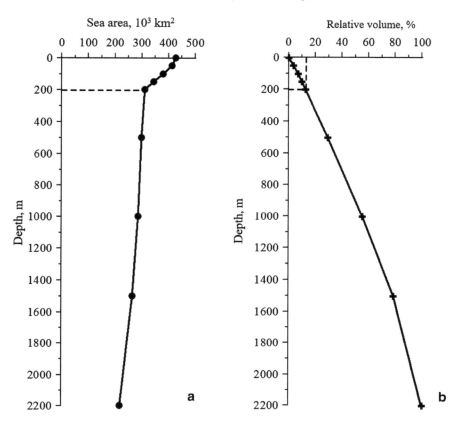

Fig. 4.37 Distribution of surface area (**a**) and specific volume (**b**) of the Black Sea by depth (Zaitsev 1998; Goncharov et al. 1965)

sulphide zone of the Black Sea corresponds to depths of more than 200 m, while the area of the oxidative mass of surface waters comprises about 312,000 km^2.

The monitoring results showed that, by 1987, the bulk of the readily seawater-soluble Chernobyl radionuclides ^{90}Sr and ^{137}Cs (Polikarpov and Egorov 2008) had been distributed throughout the upper quasi-uniform layer of the Black Sea, while the maximum concentration gradients of these radionuclides were confined to the seasonal thermocline layer (Fig. 2.8).

The graphs represented in Fig. 2.8 show the observed decrease in the concentration of ^{90}Sr and ^{137}Cs in the upper quasi homogeneous layer (UQHL) and deepening of the maximum gradients of their vertical distribution in the water column in the post-Chernobyl period. These data show that the Chernobyl radionuclides quickly penetrated into the deep-water masses of the Black Sea as a result of the vertical circulation of waters. The dynamics of the deepening of the gradient layers is shown in Fig. 4.38. It can be seen that this did not depend on the concentration of ^{90}Sr and ^{137}Cs and was generally the same for both radionuclides. This fact reflects the

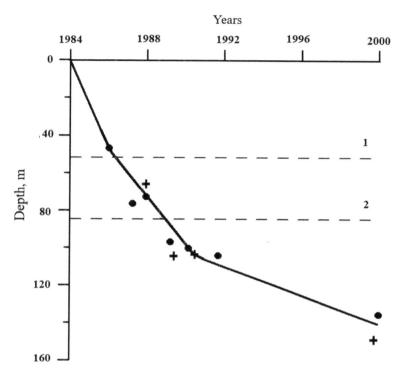

Fig. 4.38 Dynamics of changes in the depth of the maximum vertical distribution gradients ^{90}Sr (+) and ^{137}Cs (●) in the centre of the western cyclonic gyre of the Black Sea; 1—depth of maximum salinity gradient from April–August; 2—depth of maximum salinity gradient in January–February (Egorov et al. 2008)

influence of large-scale vertical circulation of water masses in the deep-water basin of the Black Sea. In this case, the maximum intensity of vertical mixing of waters related to the water layer 0–50 m; a decrease was observed in the layer of the seasonal pycnocline, while a further decrease was experienced at the lower boundary of the oxic zone of the Black Sea. Similar data were obtained for the zone of the Black Sea Rim Current (Stokozov 2003).

Based on the results of these studies, the period of vertical circulation of the Black Sea waters in the 0–50 m layer can be estimated at 5 years, the period of complete water renewal in the lower part of the pycnocline—at 15–25 years, and the period of the total volume of the sea—at 140 years (Stokozov et al. 2008). Analysis of long-term measurements of the vertical distribution of ^{90}Sr and ^{137}Cs in the water of various waters of the Black Sea (Stokozov 2003; Stokozov et al. 2008) showed that, near the lower boundary of the oxygen zone, the rate of deepening of gradient layers in the zones of cyclonic gyres was 10–12 m per year, while in the zone of the Main Black Sea Current, the rate was 30–35 m per year. An assessment of the rate

Table 4.4 Indicators of large-scale water circulation in the oxic and hydrogen sulphide zones of the Black Sea (Gulin et al. 2017)

Oxidising zone

Depth interval (m)	Surface area at the lower border of the zone (km^2)	Volume (km^3)	Water exchange with underlying layers		Water exchange period T_{ex} across the lower boundary (years)
			m/year	T_{ex} (km^3/year)	
0–200	423,000	71,125	10–35	3120–10,920 5050.1[a]	6.5–23.0 14.1[a]

Hydrogen sulphide zone

Depth interval (m)	Surface area at the lower border of the zone (km^2)	Volume (km^3)	Water exchange with underlying layers		Water exchange period T_{ex} of the hydrogen sulphide zone (years)
			m/year	T_{ex} (km^3/year)	
200–2212	312,000	475,890	10–35	3120–10,920 5051[a]	43.0–152.0 94.2[a]

[a]Estimate obtained from the results of modelling (Egorov et al. 1993)
T_{ex}—water exchange period

of water exchange in the oxic and hydrogen sulphide zones of the Black Sea, taking into account these data, is shown in Table 4.4.

The materials presented in Table 4.4 indicate that, as a result of the complex impact of hydrodynamic processes, the flow of water across the border of the oxygen and hydrogen sulphide zones of the sea, with large-scale averaging, can be 3120–10,920 km^3 per year. In this case, the period of water exchange in the oxidising zone can be estimated at 6.5–23 years, while in the hydrogen sulphide zone, the period of 43–152 years generally corresponds to the previously published integral estimates of the rate of vertical circulation of the Black Sea waters (Vodyanitsky 1948; Bogdanova 1959; Skopintsev 1975).

As can be seen from the above data, the estimates of the intensity of water exchange in the oxygen and hydrogen sulphide zones of the Black Sea, carried out on the basis of the results of monitoring the vertical migration of post-Chernobyl radionuclides, are burdened with rather large errors. These errors arise due to the impossibility of covering all the characteristic waters of the Black Sea on a timescale of decades when investigating the profiles of the vertical distribution of ^{90}Sr and ^{137}Cs in water.

The periods of water exchange in the oxidation and reduction zones were refined using the model of the large-scale radioisotope-, salt- and water-balance of the Black Sea (Egorov et al. 1993). To parameterise this model, we used data on changes in the profiles of the vertical distribution of fission radionuclides ^{90}Sr and ^{137}Cs in water during the post-Chernobyl period. The components of the water balance –including river runoff, evaporation, precipitation and water exchange through the Black Sea straits—were determined based on the results of long-term hydrometeorological observations (Voitsekhovitch et al. 1998), while the averaged profiles of the vertical distribution of water salinity were estimated from numerous data from hydrological surveys. The purpose of the modelling was to use numerical experiments to establish the stationary profiles of the vertical distribution of water salinity using the estimates obtained from the study of the vertical distribution of fragmentation radionuclides in water as indicators of the rate of vertical water exchange.

Verification of the results of calculations and modeling (Egorov et al. 1993) showed that, with an average value of water exchange between the oxygen and hydrogen sulphide zones of the Black Sea through the 200 m horizon equal to 5051 km^3 per year, the period of the water cycle in the oxic zone was about 14 years, while the corresponding period in the hydrogen sulphide zone was 94 years. At the same time, the data of numerical experiments on the model fell into the confidence zone of changes in the profiles of the vertical distribution of Chernobyl radionuclides ^{90}Sr and ^{137}Cs in water on a 20-year time scale; the forecast estimates obtained in this way in 1991 completely coincided with the results of subsequent observations (Egorov et al. 1999). For this reason, in relation to the problem of radioactive contamination of the marine environment with ^{90}Sr and ^{137}Cs radionuclides, the water mass of the oxic zone of the Black Sea (volume—71,125 km^3) can be considered as a zone of deposition of these radionuclides on a time scale of the order of several decades, while the water mass of the hydrogen sulphide zone (475,890 km^3) can be seen as a zone of their deposition on a time scale measured in centuries.

The study of the radioecological response of the Black Sea to the Chernobyl accident (Polikarpov and Egorov 2008) showed that the main factors in the formation of radioactive contamination fields comprised atmospheric fallout on the sea surface, the influx of radionuclides into the oxic zone with river runoff, their removal through straits, migration to the deep-water zone and removal into bottom sediment strata. Atmospheric fallout of ^{90}Sr in the Black Sea in 1986 comprised 100–300 TBq, while that of ^{137}Cs was 1700–2400 TBq. For the period 1986–2000 (160 $\pm$ 28) TBq of ^{90}Sr and (22.6 $\pm$ 5.4) TBq of ^{137}Cs entered the Black Sea with the runoff of the two largest rivers—the Dnieper and the Danube (Egorov et al. 2008).

Thus, the total input of ^{90}Sr into the Black Sea from 1986 to 2000 can be estimated at 260–460 TBq, while that of ^{137}Cs is determined to be 1723–2423 TBq. By 2000, the sedimentation deposition of ^{90}Sr in the strata of the Black Sea bottom sediments amounted to only 0.4 TBq or 0.09–0.15% of the total input of this radionuclide into the sea for the specified period, while for ^{137}Cs, the corresponding figure was 40 TBq or 1.6–2.3% (Gulin et al. 2012; Egorov et al. 2008). From these estimates it follows that the deposition of ^{90}Sr and ^{137}Cs in the strata of bottom sediments was insignificant compared to the content of Chernobyl radionuclides in the oxic and hydrogen sulphide zones of the Black Sea, implying that the water masses of these zones—especially the hydrogen sulphide zones—can be considered as the main depots of Chernobyl ^{90}Sr and ^{137}Cs.

Obviously, the degree of environmental hazard of water mass contamination by any radionuclide is determined primarily by its concentration (C_w) in comparison with its maximum allowable concentration (C_{MAC}). The value of C_w is equal to the ratio of the radionuclide pool in this water mass (P) to its volume (V). Hence it follows that the maximum allowable pool (P_{max}) of a radionuclide with its homogeneous distribution in the water mass characterises its highest radio capacity (R_{max}):

$$R_{max} = C_{MAC} \cdot V, \tag{4.27}$$

while the value

$$R_{rel} = \frac{P}{P_{max}} = \frac{C_w}{C_{MAC}} \tag{4.28}$$

can serve as an indicator of the relative depletion of radio capacity.

Since the levels and time scales of the deposition of radionuclides in water depots depend on the time constant of the radioactive decay of the radioisotope (λ) and on the time period of water circulation in the considered water mass (T_{ex}), the assimilation capacity of the water depot (R_{ac}) can be calculated by the formula:

$$R_{ac} = \lambda P + \frac{P}{T_{ex}}, \tag{4.29}$$

while the maximum assimilation capacity of the water mass can be determined from the ratio:

$$R_{ac-max} = \lambda P_{max} + \frac{P_{max}}{T_{ex}}. \tag{4.30}$$

Calculations using formulas (4.27)–(4.30) allowed a determination of the relative and maximum radiocapacities of water masses, along with an assessment of their assimilation capacity and changing trends in the radioactive contamination of waters. The calculations can be used to assess the environmental hazard of radioactive contamination of the marine environment, with the caveat that assessments made according to formulas (4.27)–(4.30) may not be an absolute criterion, but rather reflect the sanitary regulatory limits for individual countries.

The results of calculations of the radiocapacity of the oxygen and hydrogen sulphide zones of the Black Sea, performed in accordance with the sanitary and hygienic criterion of maximum permissible concentration, equal for ^{90}Sr and ^{137}Cs to 2 Bq/L, are given in Table 4.5.

The results obtained on the basis of the data given in Table 4.4 also took into account that the half-life of ^{90}Sr is 29.1 years ($\lambda = 0.0238$) per year), while the half-life of ^{137}Cs is 30.2 years ($\lambda = 0.02295$) per year). These results give an idea of the maximum radiocapacity of the Black Sea, including its oxic and hydrogen sulphide zones.

Referring to the data presented in Table 4.5, it is necessary to pay attention to the fact that the assimilation capacity of the oxygen and hydrogen sulphide zones in total exceeds the radioactive decay of the radionuclides entering the Black Sea following the Chernobyl accident, although, with the exception of the sedimentation factor not considered here, it is the only factor for the elimination of radionuclides from the aquatic environment. This is due to the fact that the assimilation capacity

Table 4.5 Parameters of the maximum radiocapacity of the Black Sea water masses (Gulin et al. 2017)

Parameters	Oxic zone 0–200 m, V = 71,125 km^3	Hydrogen sulphide zone 200–2200 m, V = 475,890 km^3	Black Sea 0–2200 m, V = 547,015 km^3
According to ^{90}Sr			
Radiocapacity R_{max}, PBq	142.250	951.780	1094.030
Radioactive decay R_{dec}, PBq^{-1}	3.387	22.662	26.049
Assimilation ability R_{ac-max}, PBq/year	13.369	32.738	26.049
According to ^{137}Cs			
Radiocapacity R_{max}, PBq	142.250	951.780	1094.030
Radioactive decay R_{dec}, PBq^{-1}	3.264	21.843	25.108
Assimilation ability R_{ac-max}, PBq/year	13.350	31.951	25.108

of individual zones is determined not only by the intensity of the radioactive decay of the radionuclide in them, but also by the water exchange of these zones with adjacent water layers. This factor is important in the event that only separate aquatic areas for the elimination of radionuclides are considered as water depots. Thus, the radiocapacity of these depots is determined both by radioactive decay and the intake of radionuclides from them into adjacent water masses.

A general idea of the magnitude of the Black Sea radiocapacity in relation to long-lived technogenic radioisotopes ^{90}Sr and ^{137}Cs can be obtained from Table 4.6.

Table 4.6 also provides estimates of the Black Sea radiocapacity in relation to the most significant levels of radioactive contamination of the environment over the entire period of the nuclear era. These data indicate that the release of ^{90}Sr and ^{137}Cs into the environment during the nuclear explosion in Hiroshima was no more than 0.1% of the maximum radiocapacity of the Black Sea waters determined according to sanitary standards (2 Bq/L) for these radionuclides. The release of ^{90}Sr and ^{137}Cs, which is equivalent to the influx during atmospheric tests of nuclear weapons, would have an order of magnitude equal to its maximum radiocapacity. The injection of ^{90}Sr into the environment following the Chernobyl accident was in the order of tenths of a percent, while its level of contamination was in the order of hundredths of a percent of the radiocapacity of the Black Sea.

Table 4.6 Comparison of various levels of environmental pollution ^{90}Sr and ^{137}Cs with the radiocapacity of the Black Sea waters (Gulin et al. 2017)

Radionuclide	Sources of release of radionuclides into the environment							
	Input (I) from the nuclear explosion at Hiroshima (Gudiksen et al. 1989)		Input (I) from nuclear tests in the atmosphere (Gudiksen et al. 1989)		Input (I) from the Chernobyl emergency reactor (Gudiksen et al. 1991)		Input (I) of Chernobyl radionuclides into the Black Sea in 1986–2000 (Egorov et al. 2008)	
	I (PBq)	I/R_{max} (%)	I (PBq)	I/R_{max} (%)	I (PBq)	I/R_{max} (%)	I (PBq)	I/R_{max} (%)
^{90}Sr	0.085	7.8×10^{-2}	650–1300	59.4–118.8	1.3–1.8	0.12–0.16	0.26–0.46	0.02–0.04
^{137}Cs	0.100	9.1×10^{-2}	1300–1500	118.8–137.1	85–100	7.77–9.14	1.723–2.423	0.16–0.22

4.5 Problems of Marine Ecosystem Management

The attempt to solve problems of marine nature management is aimed at optimising fishing effort and the extraction of other organic, chemical, hydrocarbon and recreational marine resources. The main tasks of nature management are associated with the development of regulatory legislative and administrative acts that ensure the appropriate use of marine resources, taking into account the characteristics of their exhaustion and renewability, as well as the preservation of a safe environment for mankind along with the species diversity of marine organisms and the necessary conditions of their habitat. Contemporary state and departmental acts in the field of marine environment protection are based on both anthropocentric and ecocentric approaches to the protection of the biosphere (Aleksakhin and Fesenko 2004). With the anthropocentric approach, interactions between humans and nature are governed according to the rules established by human beings themselves, which do not extend to the management of the existence of natural biological communities. The main principle of the implementation of the anthropocentric principle of nature management can be summed up in the statement "our task is to take everything from nature" (I. V. Michurin). The result of this approach is already known, consisting in a marked decrease in species diversity in nature. The ecocentric approach, conversely, proceeds from the objective existence of a unified system, within the framework of which people and all other living organisms interact with each other, as well as with their environment. It is this principle of integrity that underlies the contemporary understanding of the relationship between man and nature (Kinne 1997).

4.5.1 The Principle of Sustainable Development of Aquatic Areas

The deterioration of the ecological state of the environment has brought to the fore the need to develop scientifically grounded methods for regulating anthropogenic impact. To this end, the UN has officially adopted the concept of so-called sustainable development. For marine areas, the implementation of the principle of sustainable development consists in ensuring a balance between the use and reproduction of their resources. At the same time, when regulating the state of the marine environment according to the factors of radioactive and chemical pollution of waters, the implementation of the concept has its own particular characteristics.

4.5.2 Implementation of the Principle of Sustainable Development of Aquatic Areas in Terms of Marine Pollution Factors

According to contemporary understandings, marine ecosystems are evolutionarily formed natural objects in which interactions between biotic and abiotic environmental factors occur within a unified spatio-temporal scale. The energy source underpinning their existence consists of solar radiation, while the material source is chemical substances and their compounds of various biological significance. From a thermodynamic point of view, the functioning of any marine ecosystem is associated with the assimilation and dissipation of part of the solar energy, transformed into physiologically active radiation. In accordance with views informed by biogeochemistry, the impact of biotic and abiotic mechanisms of ecosystem functionality is conduces to the consumption, transformation of physicochemical forms and the return to geological depots of chemical elements having various biological significance. From an ecological perspective, these processes pertain to the reproduction of organic, mineral and recreational resources.

For material—including organic and mineral—resources, the implementation of the principle of sustainable development is associated with the adoption of international agreements governing the extraction of natural marine resources, including the licensing of fishing operations, with the creation of the Red Data Books of rare or endangered species of organisms, as well as with the development of standards for maximum allowable concentrations (MAC) of pollutants in the marine environment both for the human population and the biotic components of ecosystems. Recreational resources are primarily determined by the quality of the marine environment as regulated by sanitary and hygienic criteria. According to the wording, MAC is a criterion used to determine the measure of acceptable risk, whether for humans or for critical groups of marine animals (Ryabukhina et al. 2005). The practice of environmental regulation according to MAC criteria takes into account the following features:

1. According to the literature, there are over 2028 animal species currently living in the Black Sea (Zaitsev 2006). The MAC list (Water quality criteria summary 1991) covers more than 190 types of pollutants. However, it is known that over 1000 new contaminants enter the marine environment every year. Under such conditions, monitoring the quality of marine waters according to MPC criteria constitutes not only a scientific, but also an economic problem.

2. MAC criteria apply the dimension of units of contaminant concentration in living components of ecosystems or in the marine environment. For this reason, their role is limited to diagnostic criteria for indicating the degree of ecotoxicological hazard of aquatic areas.

3. The concentration of pollutants in water (C_w) is always established as a result of the balance of pollution fluxes (F_{poll}) and biogeochemical self-purification (F_{elim}) aquatic areas, taking into account their absorptive capacity, the dynamic characteristics of water exchange, regularities of mineral and

radioisotope metabolism by hydrobionts, as well as migration fluxes and transfer of contaminants to adjacent waters and geological depots.

For this reason, estimates of fluxes of maximum permissible pollution (F_{poll}) and self-purification of waters (F_{elim}) can be used for the purposes of regulation and management of the quality of the marine environment. It is clear that the value of F_{poll} characterises the flux of water quality consumption, while F_{elim} can be seen to represent the flux of its reproduction. In this regard, the task of implementing the principle of sustainable development of marine areas in terms of pollution factors impacting the environment is reduced to the determination of such stationary values of the fluxes F_{poll} and F_{elim}, at which the concentration of the contaminant in the water (C_w) does not exceed the MAC.

In the previous sections, it was shown that the stationary values of F_{poll} and F_{elim} in marine ecosystems are established as a result of the functioning of the entire complex of interactions of living and inert matter with radioactive and chemical components of the marine environment. These interactions can be ranked not only by the mineral balance of ecosystems, but also by the factor of the maximum permissible concentration of the contaminant in water, as well as by the value C_w, limited by sorption, metabolic and sedimentation criteria or stages of the homeostasis of biogeocenoses. At the same time, it is important to know not only the limiting fluxes of F_{poll} and F_{elim}, but also the intervals of changes in their values, which can be a measure of the depletion and reproduction of water quality resources. This is especially important when implementing the concept of sustainable development of recreational zones. It can be easily seen that, in relation to the problems of implementing sustainable development in terms of pollution factors of the marine environment, the ratio between the real concentration of the contaminant in water (C_w) and the MAC can be a measure of the consumption of the water quality resource. The largest value of this difference will correspond to the minimum resource for water quality consumption, while the critical value ($C_w \rightarrow$ MAC) will reflect the depletion of this resource. Similarly, the maximum difference between the real and limiting fluxes of biogeochemical self-purification of waters indicates the largest resource for the reproduction of water quality as a result of the influence of biogeochemical processes. The equality of the real and limiting fluxes of self-purification of waters will indicate the exhaustion of the resource of reproduction of water quality.

4.5.3 Biogeochemical Criteria

In Sect. 4.1, it is shown on the basis of the results of numerical experiments that models of ecosystems, closed by the balance in fluxes of matter, energy, mineral elements and contaminants, can pass into stable states following the completion of nonstationary processes. One of the main tasks of modelling such ecosystems is to study the fluxes of self-purification of the marine environment in relation to eutrophying elements and chemical pollution, including nuclear, as a result of the impact of natural ecological and physicochemical processes, as well as biogeochemical cycles, under anthropogenic impact. Under conditions of stationarity, the left side of the differential equations describing the state of ecosystems taking the form (3.101)–(3.102) can be equated to zero. For this reason, they can be transformed into algebraic equations, whose solution permits the derivation of functional connections between the fluxes of anthropogenic impact and the biogeochemical reaction of ecosystems to these fluxes.

The study of Eqs. (3.101)–(3.102) showed that almost all of their parameters have physical and biological meanings and can therefore be unambiguously determined from the results of experimental and natural observations. Accordingly, the solution of the algebraic equations makes it possible to obtain the rather simple formulas illustrated below, which are suitable for use as criteria for biogeochemical regulation of anthropogenic water pollution fluxes.

Turnover of chemicals or radionuclides in hydrobionts. When parenterally and alimentarily assimilated by hydrobionts, chemical elements and their radionuclides are returned to the marine environment as a result of mineralisation and lysis of chemicals and their compounds taking place alongside desorption and metabolic processes. Considered in relation to individual mineral elements, the intensity of these processes is characterised by a period of circulation in hydrobionts. If the metabolic processes are stationary, the cycle period of the radionuclide or its isotopic carrier in the hydrobiont can be determined from the relation (Kuenzler 1965):

$$T_{ex} = \frac{C_h}{v_{ex}}, \tag{4.31}$$

where

C_h is the stationary value of the concentration of an element in a hydrobiont;
v_{ex}—element exchange rate.

Let us consider the application of the compartment theory, as implemented by Eq. (3.101), used to determine the cycle period of inorganic radioactive and chemical substances in hydrobionts under conditions of changing concentration of these elements in the aquatic environment. In the simplest case, from relation (3.101) it follows that, with parenteral intake of an element into one exchange pool of a hydrobiont, without taking into account production processes and with $C_w \ll K_m$, the balance equality of changes in its concentration in the ecosystem component has the form:

$$\frac{dC_h}{dt} = p(C_w C F_S - C_h). \tag{4.32}$$

Since from expression (4.32) it can be seen that the rate of release of an element under stationary conditions is equal to the rate of its elimination, any of these flux rate components can be used as an estimate of v_{ex}, for example, $v_{ex} = C_h \cdot p$. Substituting the value v_{ex} into the formula (4.31), we obtain:

$$T_{ex} = \frac{1}{p}.$$ (4.33)

Thus, the period of turnover of a radioactive or stable element through the exchange pool of a hydrobiont is equal to the reciprocal of the rate of its removal from the exchange pool. In this connection, it becomes important to know whether the mineral elements remain in the exchange pools for an equal amount of time. If the residence time of each atom or molecule of an inorganic substance in the exchange pool were the same, a linear decrease in the radioactivity of the hydrobiont with time would be observed when investigating the elimination of the radionuclide of this substance. Since the elimination from the pool proceeds according to an exponential dependence of the form (3.2), at $n = 1$ for a time interval equal to the half-decay period of the exponent τ_{ex} in expression (3.3), the concentration of the radionuclide in the exchange pool of the hydrobiont is only reduced by half. From (3.2) at $n = 1$ it can be seen that the maximum residence time of the mineral substance in the pool (T_{max}) can only be determined approximately. Generally, for T_{max}, the time is taken as $T_{max} = 5 \cdot \tau_{ex}$, during which at least 97% of the pool volume is exchanged. Since $p = 0.693/\tau_{ex}$, that is, $T_{max} = 3.5 \cdot T_{ex}$, the maximum dwell time of an element in the exchange pool is 3.5 times greater than the value T_{ex}. From this it follows that there are always molecules of matter in the exchange pool, whose circulation takes place in a period less than T_{ex}. For this reason, the parameter T_{ex}, determined by the formula (4.33), should be considered as the average time of the turnover of a mineral element or its radionuclide through the exchange pool. Consequently, the model of the form (4.32) reflects such a kinetic mechanism of exchange processes, as a result of which the flux of an element through the pool consists of atoms or molecules exchanged by the pool at different rates. If there are several exchange pools, the entire flux of matter exchanged by the hydrobiont should be considered as the sum of fluxes, each of which is characterised by its own parameter T. If there are two exchange pools in the hydrobiont, we obtain $C_h = C_w \sum B_i$; $v_{ex} = C_w \sum B_i p_i$. Substituting these values into (4.31), we determine the value of the average circulation of inorganic matter in the hydrobiont:

$$T_{ex} = \frac{B_1 + B_2}{B_1 p_1 + B_2 p_2}.$$ (4.34)

From (3.101) it follows that, if the pollutant can enter the exchange pool in the hydrobiont, both with food and from the aquatic environment, then, taking into account the growth processes at $n = 1$, we get:

$$T_{ex} = \frac{1}{p + P_h/B_h}.$$ (4.35)

From Eq. (4.35), it can be seen that the average period of the cycle does not depend on the route of entry or the concentration of the element in the hydrobiont and in the medium, but rather is determined only by the intensity of metabolic and production processes in the hydrobiont. In relation to radionuclides, taking into account radioactive decay, we can write:

$$T_{ex} = \frac{1}{p + P_h/B_h + \beta}. \tag{4.36}$$

According to the literature data, the period of the turnover of mineral elements is estimated by the formula (4.33) (Kuenzler 1965; Lowman et al. 1971), which is equivalent to accepting the hypothesis of the presence of one exchangeable pool of each element in the hydrobiont. Our studies have shown that the flux of inorganic matter through a hydrobiont consists of atoms or molecules passing through both one and two exchange pools at different average rates; the maximum residence time of an element in the aquatic pool can be 3.5 times longer than the average. The cycle period of a radionuclide in a hydrobiont is determined by the intensity of metabolic and production processes and does not depend on the route of its entry or the concentration of the isotopic carrier in the hydrobiont and water.

Biotic circulation of chemicals or radionuclides in the marine environment. Considered in relation to conservative chemicals and their compounds, a quantitative characteristic of the intensity of biochemical processes in the ocean consists in the average time of their biotic cycle in the marine environment (Small and Fowler 1973), which can also be seen as characterising the time of their biogenic mobilisation (T_b) (Bogdanov et al. 1983). The value of T_b corresponds to the period of passage of all of the radionuclide or mineral element dissolved in water through the biogenic cycle of absorption/elimination by living matter of the marine environment. This period is calculated using a formula (Kuenzler 1965) analogous to (4.31), in the numerator of which, in place of C_h, the value C_w is substituted.

Let us consider how expression (4.31) is transformed in the case of accepting the hypothesis of the applicability of the generalised model (3.101) and (3.102) for describing the kinetic regularities of the mineral metabolism of hydrobionts. In the simplest case, in the absence of anthropogenic pollution and physical transfer, when hydrobionts do not produce organic matter but have only one parenterally filled exchangeable pool of a mineral element, it follows from expression (3.102) that the rate of change in the concentration of a chemical or radioactive substance in the medium as a result interaction with hydrobionts in the range $C_w \ll K_m$ can be described by the equation:

$$\frac{dC_w}{dt} = m_{sp}(C_h p - C_w C F_S p). \tag{4.37}$$

Substituting the value $v_{ex} = m_{sp} C_w C F_S p$ from the expression (4.37) in the formula (4.31), we obtain:

$$T_b = \frac{1}{m_{sp} C F_S p}. \tag{4.38}$$

Relation (4.38) shows that the higher the specific biomass, concentration factor and rate of exchange of the element by hydrobionts, the more intense the biocirculation. Taking into account the dependence of the change in the value of CF_S during the mineral exchange of hydrobionts in accordance with the Michaelis–Menten equation, we obtain:

$$T_b = \frac{K_m + C_w}{m_{sp} V_{\max}}. \qquad (4.39)$$

Hence it follows: if, during anthropogenic pollution of seas and oceans, the concentration of a radionuclide or an isotopic carrier in the water reaches a level commensurate with the value of K_m, then the period of the element's biotic cycle will increase. As applied to primary production problems, it follows that the relative capacity for the biotic assimilation of an excess of nutrient input decreases and the environmental hazard of water hypertrophication consequently increases. If we consider the relationship between C_w and T_b in relation to radionuclides, we can conclude that the influence of the biotic factor on the transformation of the radionuclide decreases with an increase in the concentration of the isotopic carrier in water.

Transport of chemicals (radionuclides) along the trophic chain. The need to develop methods for predicting the migration of pollution along trophic chains, including their intake by humans in the form of seafood, is one of the main problems stimulating the creation of a theoretical basis for describing the kinetics of interaction of hydrobionts with radioactive and chemical substances of the marine environment. Numerous studies have shown that the content of mineral elements in aquatic organisms comprising potential pollutants of the marine environment generally decreases with an increase in the trophic level. For example, it was found that, while the concentration of selenium in shrimps was 4.01 µg/g dry weight, in fish from the same waters, the corresponding figure was 1.49–1.74 µg/g (Maher 1985). The content of microelements Cd, Hg, Cu, Pb, Zn, Se, Cr, Ag, Sb, Cs, Co and Fe in organisms of the southern part of the Baltic was shown to decrease in the order phytoplankton → zooplankton → zoobenthos → fish (Brzezińska et al. 1984). The amount of trace elements (Fe, Mn, Zn, Cu, Co, Ni, Pb) contained in the Black Sea zooplankton measurably higher than in corresponding fish species (Rozhanskaya 1983). Similar data were obtained in studies of the content of heavy metals in marine organisms (Clerck et al. 1984), as well as in the study of the transfer of fission radionuclides along trophic chains (Osterberg and Pirsi 1971). An opposite trend was particularly noted only in the case of mercury content present in aquatic organisms. In shrimp from the coastal waters of Belgium, the concentration of this metal was lower than in corresponding fish populations (Clerck et al. 1984). In measurements carried out in the North Atlantic, the mercury content of phytoplankton was 0.005 µg/kg wet weight; in zooplankton—0.01 µg/kg; in fish fry—0.01 µg/kg; in planktophagous fish—0.01 µg/kg; in squid—0.24 µg/kg (Morozov and Petukhov 1979).

We will use compartment theory to obtain a ratio applicable for assessing the migration of radionuclides and isotopic carriers along trophic chains. If a radioactive or chemical pollutant is transmitted along the alimentary trophic chain, the coefficient of its accumulation by each subsequent link (CF_h) from the concentration factor of the previous link (CF_N) can be determined from Eq. (3.75). In this case, from the stationarity condition, we obtain:

$$CF_h = CF_N \frac{Rq}{Rq_N + p}.$$ (4.40)

From equality (4.40), it can be seen that an increase in the levels of accumulation of a pollutant along the trophic chain during the course of production processes is possible at $Rq > (Rq_N + p$; that is, only if its absorption from food exceeds the degree of its assimilation for growth, as well as removal of the pollutant from the hydrobiont as a result of metabolism. In the absence of growth of hydrobionts (at $q_N = 0$), an excess of the accumulation of the pollutant in each subsequent link is possible at $Rq > p$, when its assimilation from food is greater than excretion as a result of metabolic processes. Interpreting the above empirical data in accordance with the theoretical relationship (4.40) leads to the conclusion that, in cases where the accumulation factors of pollutants in higher trophic levels decrease, it can be predicted that the assimilation of nutriment for growth will exceed the degree of assimilation of these pollutants from food.

Comparison of parenteral and alimentary routes of absorption of inorganic substances by hydrobionts. Until now, there has been no consensus on the role of food and the aquatic environment in the absorption of minerals and radionuclides by hydrobionts. In studies carried out with ^{65}Zn and *Artemia salina* it was found that this radionuclide was absorbed from food around seven times more efficiently than from water (Rice 1965). The predominance of the alimentary pathway was also noted in studies of the intake of ^{65}Zn and ^{59}Fe along the food chain of *Fucus serratus* → *Littorina obtusata* (Young 1975). Similar conclusions were obtained when studying the uptake of ^{60}Co by shrimps (Weers 1975a). On the other hand, the data obtained by Renfro et al. (1975) on the absorption of ^{65}Zn from the aquatic environment and from food for three months by crabs and shrimps showed that parenteral accumulations were no less significant than nutritional. The same authors found that the amphipods *Gammarus locusta* held for 27 days in an environment with ^{65}Zn along with Artemia brine shrimp labelled with the same radionuclide accumulated ^{65}Zn to a value of $CF = 1000$. Animals kept without food had aconcentration factor of 2400, while amphipods feeding on Artemia labeled ^{65}Zn accumulated the radionuclide to a value of $CF = 200$ units. The principal contemporary approach to the study of the role of food and water in mineral nutrition consists the study of the kinetic regularities of the absorption of a radioactive label of inorganic substances by hydrobionts (Polikarpov 1966; Weers 1975b; Young1975) in some cases involving the use of mathematical models to interpret metabolic processes (Kowal1971; Pentreath1973).

Let us consider the possibilities of using the generalised Eq. (3.101) to assess the role of food and the aquatic environment in the absorption of radionuclides and isotopic carriers by consumers. Our studies (Egorov and Ivanov 1981) demonstrated the possibility of describing parenteral absorption exchange kinetics of ^{65}Zn by copepods *Euchirella bella* with double-compartment models having the parameters $B_1 = 150$, $p_1 = 3.46$ per day and $B_2 = 1360$, $p_2 = 0.102$ days^{-1}. The kinetics of alimentary absorption of ^{65}Zn by the same species of copepods when feeding on unicellular *Peridinium trochoideum* algae in parallel experiments (Fig. 3.48) was satisfactorily described by a model having parameters $B_1 = 6 \times 10^{-3}$, $p_1 = 4.15$ per day and $B_2 = 52 \times 10^{-3}$, $p_1 = 0.102$ days^{-1} (Piontkovsky et al. 1983). In these experiments, the nutriment radioactivity of 20 cpm/mg was constant over the course of the observation period. In both experiments, the copepods did not increase in size over the entire observation period. A comparison of parameters of models of alimentary and parenteral absorption pathways of ^{65}Zn by *Euchirella bella* showed that the ratio of zinc exchange pools in animals was approximately the same. The exchange rate constants of these pools were also close in value. This indicated that the rate of parenteral and alimentary metabolism was the same for zinc in *Euchirella bella*. The differences in the absolute values of the exchange pools in this case could be explained by the difference in specific concentrations of ^{65}Zn in nutriment and water.

The identity of the metabolic constants p_1 and p_2, which reflect the exchange kinetics of ^{65}Zn supplied to animals with nutriment and from the aquatic environment, was used to derive a combined model (Ivanov et al. 1986):

$$\frac{dC_1}{dt} = C_w B_1 p_1 + \alpha Rq C_N + C_1 p_1;$$
$$\frac{dC_2}{dt} = C_w B_2 p_2 + (1 - \alpha) Rq C_N - C_2 p_2;$$
$$C_h = C_1 + C_2, \qquad (4.41)$$

in which the parameters B_1 and B_2 corresponded to the parameters for parenteral zinc absorption by a hydrobiont. From expressions (4.41) it can be seen that the intake rate of ^{65}Zn from water is $v_w = C_w \cdot (B_1 p_1 + B_2 p_2)$, while the rate from nutriment is $v_N = Rq \cdot C_N$. Hence, the percentage contribution of the alimentary pathway in the process of absorption of an element by a hydrobiont can be described by the expression:

$$\frac{v_N}{v_w + v_N} = \frac{Rq C_N}{C_w (B_1 p_1 + B_2 p_2) + Rq C_N}. \qquad (4.42)$$

It is clear that formula (4.42) is valid both for radionuclides and their stable analogues. Meanwhile, the use of the radioactive tracer method only reflects the true ratio of the routes of entry of isotopic carriers when the specific radioactivity of food reaches the stationary level $C_N = CF_N \cdot C_w$, where CF_N is the stationary factor of concentration of the element in nutriment. Substituting the value C_N in (4.42), we obtain the equation:

$$\frac{v_N}{v_w + v_N} = \frac{RqCF_N}{B_1 p_1 + B_2 p_2 + RqCF_N}, \tag{4.43}$$

whose parameters do not depend on the concentration of the element in water. Calculations using the formula (4.43) showed that at $CF_N = 15{,}000$ for phytoplankton (Lowman et al. 1971), $q = 1$ and the minimum ($R = 0.1$) and maximum ($R = 0.5$) nutrient budgets of copepods, the contribution of the alimentary pathway to the absorption of zinc by *Euchirella bella* lay within the range of 68–92%. These experiments showed that, in the absence of growth, the alimentary pathway was prevalent in the absorption of zinc by marine animals.

Under natural conditions, the most characterising aspect of marine animals is the state of their production of organic matter. Let us consider the solution to the problem of determining the role of nutriment and that of the aquatic environment itself in the mineral nutrition of hydrobionts using a model of the form (3.101). With regard to the conditions in a closed system when an inorganic substance enters parenterally (at $C_w \ll K_m$) and alimentarily into one exchange pool of a hydrobiont, Eq. (3.101) is transformed into the expression:

$$\frac{dC_h}{dt} = C_w C F_S p + R C_N q - C_h (p + P_h / B_h), \tag{4.44}$$

where CF_S is the stationary factor of concentration of the mineral element, which would be established in relation to the hydrobiont if this element was absorbed from the environment only parenterally.

When a marine organism gains mass due to the assimilation of nutriment and the absorption of an element directly from the aquatic environment, Eq. (4.44) can be written as:

$$\frac{dC_h}{dt} = C_w C F_S p + R(C_N q - C_h q_N) - C_h p. \tag{4.45}$$

Hence, the ratio used to compare the rates of parenteral and alimentary intake of elements is written as follows:

$$\frac{v_N}{v_w + v_N} = \frac{R}{C F_{Sh} p}(C F_N q - C F_{Sh} q_N), \tag{4.46}$$

where CF_{Sh} is the stationary factor of concentration of an element when it is absorbed by an hydrobiont directly from the aquatic environment and with nutriment.

Relationship (4.31) shows that the efficiency of the alimentary pathway is largely determined by the factor of concentration and the degree of assimilation of the element from nutriment. Depending on the sign of the difference $(C F_N q - C F_{Sh} q_N)$, the role of the alimentary pathway can be positive or negative. The latter means that the role of nutriment within a particular time frame can be reduced to a decrease in the concentration of inorganic matter in the hydrobiont.

We will use Eq. (3.5)–(3.15) to estimate the dependence of CF_{Sh} on the values of CF_S and CF_N. From the stationarity condition, taking into account the fact that $C_N = CF_N \cdot C_w$, we obtain:

$$CF_{Sh} = CF_S \frac{p + \frac{CF_N}{CF_S}q}{p + Rq_N}. \tag{4.47}$$

From the formula (4.47) it can be seen that if the hydrobiont does not produce organic matter ($q_N = 0$), in the case of alimentary and parenteral absorption of an element, the coefficient of its accumulation by the hydrobiont CF_{Sh} is always higher than CF_S. The value of CF_{Sh} is also higher than CF_S if $CF_N > CF_S$. In the limit, if $CF_N \gg CF_S$, the value of CF_{Sh} tends to the value CF_N. Thus, if $CF_N < CF_S$, then $CF_{Sh} < CF_S$. In our opinion, the analysis of Eqs. (4.45) and (4.47) helps to explain the sources of inconsistency in the results of studies by a number of authors, which were obtained without taking into account the nutrient budgets, the degree of assimilation of nutriment for growth, the assimilation of mineral elements from nutriment, as well as the metabolic activity of hydrobionts in relation to these elements.

The study carried out on a mathematical model showed that, on the whole, in the case of parenteral and alimentary routes of absorption of mineral elements and their radionuclides by hydrobionts, the role of the alimentary pathway is determined by the ratio of the concentrations of the element in the consumer and nutriment, by the nutrient budget, as well as the degree of assimilation of nutriment and the assimilation of the element from nutriment. In hydrobionts producing organic matter, the alimentary pathway can both increase and decrease the concentration of mineral elements, in contrast to the effect of only the parenteral mechanism of their absorption.

Biotic transformation of physical and chemical forms of elements. In Sect. 3.2.9, using the example of an empirical study and a mathematical description of the concentration kinetics of mono- and pentavalent ^{131}I by ulva, it was shown that the kinetics of exchange of different physicochemical forms of pollutants by hydrobionts can be considered as a result of their entry into the exchange pools of hydrobionts with the indicator of the speed of metabolic reaction corresponding to each physicochemical form. The rates of elimination of different physicochemical forms of inorganic substances from the exchange pools in hydrobiont are the same. Thus, the ratio of the physicochemical forms of elements in the marine environment changes as a result of metabolic processes occurring in hydrobionts.

For a stationary system with one exchange pool of an element in a hydrobiont, described by the balance equalities of the generalised model (3.101), taking into account relations (3.62), the rate of biotic transformation of physicochemical forms of elements (V_{bt}) during the parenteral route of mineral nutrition can be determined by the formula:

$$v_{bt} = d_m \cdot m_{sp} \cdot V_{\max} \cdot C_w / (K_m + C_w), \tag{4.48}$$

where

C_w is the concentration of a chemical element (or radionuclide) in water;
K_m is the Michaelis–Menten constant;
V_{max} is the maximum rate of absorption of an element by a hydrobiont;
m_{sp} is the specific biomass of organisms in the marine environment;
d_m is the coefficient of biogenic transformation of the physicochemical form of the element.

With the alimentary pathway of mineral nutrition, the intensity of biotic transformation of the physicochemical forms of elements (ΔC_{bt}) is calculated by the formula:

$$\Delta C_{bt} = d_m \cdot m_{sp} \cdot C F_N \cdot C_w \cdot p/(R q_N + p). \tag{4.49}$$

From Eqs. (4.48) and (4.49), it follows that the intensity of transformation of physicochemical forms depends on the ratio of K_m and C_w. At $K_m \gg C_w$, the value of ΔC_{bt} increases in proportion to C_w. At $K_m \ll C_w$, the biotic transformation rate reaches its limit:

$$\Delta C_{bt}(\text{max}) = d_m \cdot m_{sp} \cdot V_{max}. \tag{4.50}$$

Biosedimentary self-purification of waters. From the equalities of the generalised model (3.101) applied to one generalised exchange pool in hydrobionts and to stationary conditions, it follows: if the non-alimentary route prevails in the absorption of the pollutant, then the flux of sedimentation elimination of chemicals or radionuclides (ΔC_{bs}, V_{sed}) from the photic layer can be calculated by the formula:

$$V_{sed} = d_{sed} \cdot P \cdot C_w \cdot C F_h = d_{sed} \cdot C_w \cdot P \cdot V_{max}/(r \cdot (K_m + C_w)), \tag{4.51}$$

where

C_w and CF_h are the concentration in water and the accumulation factor, respectively, of the element (or radionuclide) by marine organisms;
d_{sed} is part of the primary production (P) of the ecosystem, which is eliminated by sedimentation from the territorial limits of the ecosystem under consideration or deposited in the seabed layers;
r is the indicator of the average exchange rate of an element (radionuclide) by a hydrobiont.

When the alimentary pathway of mineral nutrition of hydrobionts predominates, the rate of sedimentation from the photic layer can be estimated from the ratio:

$$\Delta C_{bs} = \frac{d_c R C F_N C_w P q_N}{(K_m + C_w) p}. \tag{4.52}$$

where

d_c is the portion of primary production eliminated from the aquatic area;
q_N is part of the assimilated nutrient budget;
CF_N is the nutrient absorption coefficient of a radioactive or chemical substrate.

From expression (4.51) it follows that at $C_w \gg K_m$ the rate of sedimentary self-cleaning of the photic layer reaches the limiting value:

$$V_{sed}(\max) = d_{sed} \cdot P \cdot V_{\max}/r. \tag{4.53}$$

If the sedimentation flux is deposited in seabed layers, then the concentration factor CF_{sed} of chemicals or radionuclides by bottom sediments is calculated by the formula:

$$CF_{sed} = \frac{C_{sed}}{C_w} = \frac{C_{\max}}{C_w + 1/k_L}, \tag{4.54}$$

where

C_{sed} is the concentration of the element in bottom sediments;
$C_{\max}$ is the maximum concentration of the element in bottom sediments;
k_L is the constant of the Langmuir equation.

The thickness of the layer of deposition of radionuclides in bottom sediments (h) can be estimated from the ratio:

$$h = 5 \cdot T_{05} \cdot b \cdot V_d, \tag{4.55}$$

where

V_d is the rate of sedimentation;
b is the coefficient that takes into account the compaction of bottom sediments;
T_{05} is the half-life of the radioisotope.

The flux of deposition of pollutants into bottom sediments:

$$F_{dep} = V_d C_{sed} = V_d K_{sed} C_w = \frac{V_d C_{\max} C_w}{C_w + 1/k_L}. \tag{4.56}$$

Limiting flux of pollutants entering bottom sediments (at $C_w \gg 1/k_L$):

$$F_{\max} = V_d \cdot C_{\max}. \tag{4.57}$$

The radiocapacity of bottom sediments is determined by the fluxes of radioisotopes entering the bottom sediments, as well as by their radioactive decay (Q). If these fluxes are equal, the stationary level of the radionuclide content in bottom sediments can be calculated by the formula:

$$Q = \int\limits_{0-}^{h} C_{sed}(h)dh, \tag{4.58}$$

where $C_{sed}(h)$ is the distribution function of the radionuclide concentration in the stratum of bottom sediments.

The ratios (4.31)–(4.58) include parameters reflecting the metabolic characteristics of hydrobionts, the concentrating and sorption functions of living and inert matter in relation to various chemicals, including marine pollutants, the production and sedimentation capacity of ecosystems, as well as the depositing capacity of bottom sediments. Since all these criteria reflect the influence of biological and geological processes on the turnover of chemical substances and their isotopic carriers in hydrobionts and the aquatic environment, as well as the characteristics of its self-purification, they can be termed as biogeochemical. Since these criteria conform to the dimensionality of fluxes, they may be used for the purposes of environmental regulation. Formulas (4.31)–(4.58) show that the concentration of pollutants in the aquatic environment, marine organisms and bottom sediments depends on the stationary levels of the pollution/self-purification systems of aquatic areas. When limiting the maximum permissible concentrations of pollutants in the components of marine ecosystems, the ratios (4.31)–(4.58), by substituting the values $C_w = C_w(MAC)$, $C_h = C_h(MAC)$ and $C_{sed} = C_{sed}(MAC)$, allow us to estimate the limiting fluxes of pollution of aquatic areas by biogeochemical criteria, that is, those fluxes of pollutants that can be assimilated by aquatic areas without prejudice to their biological and other water resources.

4.6 Environmental Regulation of Aquatic Pollution

4.6.1 Sevastopol Bay

Sevastopol Bay represents an aquatic area of increased ecological risk due to the limitation of its water exchange with the outer roadsteads as a result of the construction of protective breakwaters, as well as intensive anthropogenic activity on the coast, wastewater discharges (Ovsyany et al. 2001) and the influx of pollutants from the Chernaya River (Ovsyany et al. 2007). The need to control the ecological state of the bay requires the implementation of the principle of sustainable development based on the regulation of the balance between the use and reproduction of the quality of its water resources.

The aim of the study was to develop criteria for standardising the anthropogenic impact on the aquatic area of Sevastopol Bay in terms of the fluxes of deposition into the strata of post-Chernobyl bottom sediments (^{90}Sr, ^{137}Cs, ^{239,240}Pu) and natural (^{210}Po) radionuclides, as well as chlorinated hydrocarbons, under contemporary conditions of natural biogeochemical cycles that condition the quality of its

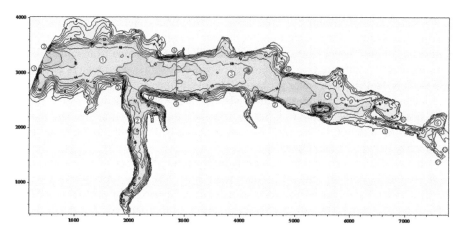

Fig. 4.39 Bathymetric map of the Sevastopol Bay (Stokozov 2010) (1–5—numbers of sounding boxes; ●—locations of wastewater outlets)

waters. The work was carried out in the aquatic area of Sevastopol Bay—as well as, in comparative terms, in the outer roadstead of Sevastopol. The zoning of the bay aquatic area according to hydrodynamic, hydrochemical and morphological characteristics as presented in Fig. 4.39 was performed in accordance with the criteria described in works (Ivanov et al. 2006; Stokozov 2010). For the measurement of ^{90}Sr, ^{137}Cs, ^{239,240}Pu, ^{210}Po, Hg and organochloride compounds, conventional methods were used (Marei and Zykova 1980; Polikarpov and Egorov 2008; Unifitsirovannye metody monitoringa... 1986).

For dating the bottom sediments, the peaks ^{137}Cs and ^{90}Sr from of the maximum fallout of bomb-testing and post-Chernobyl radionuclides were identified in them in 1962 and 1986, respectively. The sedimentation rate was determined from the distance between these peaks and the surface layer of core samples, corrected for the density of bottom sediments (Gulin et al. 2008; Zherko et al. 2001; Mirzoeva et al. 2013). The considered radioactive and chemical substances comprise conservative pollutants of the marine environment. When assessing the quality of water according to hygienic and biogeochemical criteria, the main mechanisms of self-purification of the marine environment from conservative pollution are associated with a decrease in their concentrations in water due to migration through aquatic areas and with sedimentation elimination into the composition of bottom sediments.

It is known that the influence of hydrodynamic processes is determined by the transfer of dissolved conservative pollutants under the influence of currents, advection, diffusion, as well as tidal and seiche phenomena with the resulting effect of decreasing gradients in their concentration distribution fields in water in terms of aquatic areas and depth. Hydrodynamicmigration of pollutants, which are contained in the composition of particles with a density different from the density of water, is reduced to the occurrence of zones of their thickening or rarefaction under the influence of eddies occurring at different scales. The main contribution of these processes to the conditioning of the marine environment is determined by the transport of dissolved and suspended pollutants into water masses according to various timescales of their water exchange. The transfer of contaminants from aquatic areas with increased anthropogenic load into the open sea or into deep waters is equivalent to their deposition in water geological depots, whose pollution localisation time scales can be measured in tens or hundreds of years (Polikarpov and Egorov 2008). For chemically conservative radioactive contamination, water masses can also be considered as a depot, whose pool decreases annually according to the quantity of radioactive decay of radionuclides.

A detailed study of the influence of hydrodynamic processes of a synoptic scale on the redistribution of pollutants in the aquatic area ofSevastopol Bay (Ivanov et al. 2006) showed that winds prevailing over it are transformed mainly into northeastern (36%) and southern (20%) airstreams due to the coastal orography. The northeastern winds create latitudinal transport, leading to the self-cleaning of Severnaya Bay (boxes 1, 3, 4 and 5 in Fig. 4.39) and embaying Yuzhnaya Bay (box 2), while the southern winds contribute to the self-cleaning of the latter.

The results of calculations of horizontal water exchange with winds of latitudinal directions, which were based on the morphometric characteristics of the bay (Stokozov 2010) and on the velocities of the eastward currents indicated in (Ivanov et al. 2006), are shown in Table 4.7.

From the data presented in Table 4.7, it can be seen that, without taking into account the profiles of changes in the speed of currents in depth, the latitudinal exchange of waters of Sevastopol Bay due to drift currents occurs on a daily time scale. In addition to processes occurring over a small timescale, all sources of water balance formation also affect the exchange of waters of Sevastopol Bay. The elements

Table 4.7 Period of latitudinal exchange of the waters of Sevastopol Bay due to drift currents (Egorov et al. 2018a)

Number of box	Morphometric characteristics of sounding box (Stokozov 2010)				Flow rate from east to west (cm/s)	Estimation of the period of horizontal water exchange (h)
	Length by latitude (m)	Average depth (m)	Surface area (m^2)	Volume (m^3)		
1	2600	12.0	2,831,750	33,825,650	40–60	1.2–1.8
2	2500	12.8	806,900	10,235,990	−[b]	−
3	2100	13.1	1,748,870	22,802,850	10–15	3.9–5.8
4	2200	8.7	1,487,970	12,935,640	8–10	6.1–7.6
5	600	4.7	322,130	1,512,410	4–6	2.8–4.2
Entire bay	7500[a]	10.3	7,197,620	81,312,540	−	14.0–19.4 ≤1 day

[a]Excluding box number 2
[b]Not evaluated

Table 4.8 Elements of the water balance of Sevastopol Bay (Egorov et al. 2018a)

	Million m³ per year	Source
Water inflow sources (+)		
1. Average annual flow of the Chernaya river	56.8	(Ivanov et al. 2006)
2. Sewer outlets	4.56[a]	
3. Slope runoff	1.50[a]	
4. Precipitation	2.04	(Zaitsev 2006)
Aggregate (+)	64.90	
Outflows (−)		
1. Evaporation	5.69[b]	(Zaitsev 2006)
2. Hydrological efflux (estimate)	56.65	
Total (−)	62.34	
Bay water exchange period due to external sources and outflows	1.3 years	

[a] According to the State Inspectorate for the Protection of the Black and Azov Seas
[b] Recalculated on the basis of estimates for the entire Black Sea

of the water balance of the bay due to sources and outflows are presented in Table 4.8.

In the calculations, it was assumed (Table 4.8) that the removal of water from the bay was equal to the discrepancy between sources and outflows. From the data presented in this table, it can be seen that, with the average annual averaging, the period of replacement of the waters of Sevastopol Bay is 1.3 years. The materials considered in Tables 4.7 and 4.8 in general showed that the wind-driven exchange of water in Sevastopol Bay can occur on a daily basis, while the replacement of water occurs on an annual time scale.

We assessed the degree of influence of hydrodynamic processes on the transfer of pollutants by the example of studying the distribution fields of the sum of the concentration of such conservative pollutants as p,p'-DDT and its metabolites p,p'-DDD and p,p'-DDE (hereinafter—ΣDDT), in the water and in the thickness of bottom sediments of Sevastopol Bay. The results of these determinations are presented in Fig. 4.40.

From Fig. 4.40a, it can be seen that in the spring season the concentration of ΣDDT in the water decreased from the zone at the mouth of the Chernaya River to the exit from the bay. The results of these observations testified that the main source of ΣDDT water pollution was the runoff of the Chernaya River. Calculations showed that in 2008 the total content of ΣDDT in the water of Sevastopol Bay was 277.6 g, while the equivalent figure for ΣPCB$_6$ was 355.2 g. From the estimates of water circulation (Tables 4.7 and 4.8) it follows that such an amount of organochlorine compounds could be carried out from the bay towards the outer roadstead due to the positive component of the water balance on an annual—and, with winds of latitudinal directions, daily—timescale. At the same time, a consideration of materials presented in Fig. 4.40b showed that the main amount of ΣDDT in Sevastopol

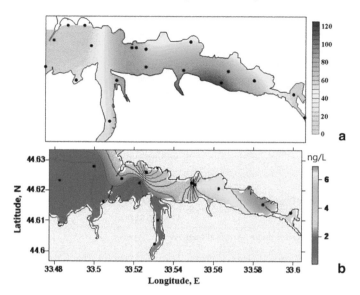

Fig. 4.40 Distribution of ΣDDT (**a**) and ΣPCB₅ (**b**) (ng/g dry mass) in the surface layer of bottom sediments of Sevastopol Bay in 2006–2011. ●—location of sampling. The average concentration of ΣDDT is (64 ± 10) ng/g; ΣPCB_5—(402 ± 45) ng/g (Egorov et al. 2018a)

Bay was deposited in the surface layer of bottom sediments in the part of its aquatic area where wastewater was discharged (Fig. 4.39). These data allow us to conclude that the effect of hydrodynamic processes did not lead to equalisation of ΣDDT concentrations throughout the entire aquatic area of the bay.

Distribution fields ^{90}Sr, ^{137}Cs, $^{239.240}$Pu, ^{210}Po and Hg in the surface layer of bottom sediments of Sevastopol Bay are shown in Fig. 4.41. It can be seen that the maximum concentrations of pollutants in bottom sediments were largely confined to the locations of wastewater discharge (boxes 1, 2, 3 and 4 in Fig. 4.39), which indicates the prevailing role of the influence of sedimentation processes on the self-purification of waters in these areas.

The results of determining the sedimentation rate (v_{sed}) in Sevastopol Bay using radio-tracer technologies (Gulin et al. 2008; Zherko et al. 2001; Mirzoeva et al. 2013) are shown in Table 4.9.

The results of these estimates showed that the sedimentation rate in different waters of the bay varied from 2.3 mm/year, or 664 g/(m² year), to 9.3 mm/year, or 7094 g/(m² year). It should be noted that the vertical distribution profiles of ^{137}Cs in Yuzhnaya Bay could not be methodologically used for geochronological purposes, due to the seabed structure having been significantly disturbed by mooring anchors over its entire area. For this reason, the assessment of sedimentation in Yuzhnaya Bay was made based on the results of averaging data for sounding boxes 1, 3, and 4 of Sevastopol Bay. Taking this into account, the sedimentation in the entire aquatic area of the bay is estimated at 12,850.9 tonnes/year.

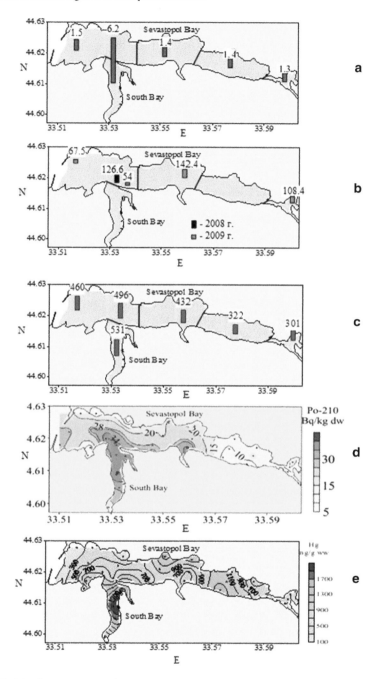

Fig. 4.41 Distribution in the surface layer (0–5 cm) of bottom sediments of Sevastopol Bay: **a** ^{90}Sr (Bq /kg dry weight) in 2008–2010; **b** ^{137}Cs (Bq/ kg dry weight) in 2008–2009; **c** ^{239,240}Pu (mBq/kg dry weight) in 2009–2011; **d** ^{210}Po (Bq/ kg dry weight) in 2003; **e** total mercury (ng/g wet weight) in 2001 (Egorov et al. 2018a)

Table 4.9 Sedimentation rate in various aquatic areas of the Sevastopol Bay (Egorov et al. 2018a)

Bay area	Coordinates, N E	Depth (m)	H (%)	SR (mm/yea)r	MAR (g/m² per year)
Inkerman	44.6127; 33.5980	4	66.1	9.3	7094
Gollandiya	44.6225; 33.5605	15	71.8	3.3	1727
Pavlovsky Cape	44.6182; 33.5335	15	60.1	2.4	607
Konstantinovsky Ravelin	44.6248; 33.5150	13	57.1	4.6	3253
Outer Sevastopol roadstead	44.6242; 33.4901	22	50.2	2.3	664

H—relative humidity of the upper layer of bottom sediments
SR is the sedimentation rate
MAR is the mass accumulation rate

The specific fluxes of the deposition of radionuclides and chemical pollutants into the strata of bottom sediments in various aquatic areas of Sevastopol Bay were calculated using the materials presented in Figs. 4.40 and 4.41 consisting of observations and averaged estimates of the rate of sedimentation. The distribution of specific fluxes according to boxes is shown in Fig. 4.42.

From Fig. 4.42, it can be seen that the largest specific fluxes of deposition of pollutants belonged to box 2, that is, to the aquatic area of the most polluted bay, Yuzhnaya. Another leading factor of water self-purification consisted in the most intense solid runoff in the estuarine zone (box 5) of the Chernaya River.

Balance calculations for the self-purification of Sevastopol Bay waters from pollution are summarised in Table 4.10. These calculations consist of estimates of sedimentation deposition of ^{90}Sr, ^{137}Cs, $^{239.240}$Pu, ^{210}Po, Hg, ΣPCB$_6$ and ΣDDT in the thickness of bottom sediments at the current level of biochemical pollution ecosystems of the bay. In expressing the characteristics of self-purification through the dimensions of fluxes, it is obvious that they can be used as quantitative criteria for standardising the maximum permissible influx of pollutants into the aquatic area.

4.6.2   Sea of Azov

This section analyses the data on the content of Pb, Zn, Cu, Cd, and Hg in water and in the surface layer of bottom sediments, as well as on the rate of sedimentation processes occurring in Taganrog Bay, the central part of the Sea of Azov and the Kerch Strait (Matishov et al. 2017). Estimates of their maximum permissible fluxes into the aquatic area of the sea are made and the time scales of the relevant processes of sedimentation self-purification of waters are determined.

The Sea of Azov comprises a water basin to the west of Southern Russia, disposing mineral and biological resources important for the country's economy, as well as

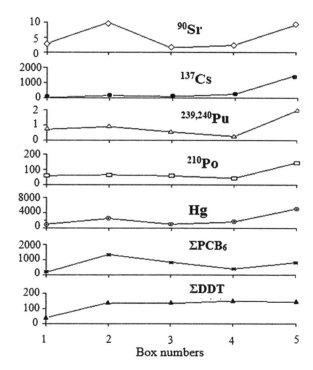

Fig. 4.42 Specific fluxes of deposition of ^{90}Sr, ^{137}Cs, ^{239,240}Pu, ^{210}Po (Bq/(m^2 year)), Hg, ΣPCB$_6$ and ΣDDT (μg/(m^2 year)) (Egorov et al. 2018a)

transport communications. Due to its status as a fishery reservoir and its high recreational potential, it is of great interest as an object of research and environmental monitoring.

The present work used data provided by FSI "Azovmorinformcenter" on the concentration of Pb, Zn, Cu, Cd and Hg in water and bottom sediments relating to the period 2010–2014. Samples for analysis were taken with a PE-1220 sampling system according to GOST R 51592-2000 in the surface layer at 32 points (Fig. 4.43).

Water samples were taken annually in spring, summer and autumn seasons. Chemical analysis of water samples for lead content was carried out in accordance with the PND F 14.1:2:4140–98 method with a lower sensitivity limit (LSL) of 0.0002; for cadmium—PND F 14.1:2:4.140–98 at an LSL of 0.00001; for copper—PND F 14.1:2:4.140–98 at an LSL of 0.0001; zinc—M-MVI-539–03 at an LSL of 0.001. All the indicated heavy metals were measured with the AAS KVANT-Z-ETA instrument.

Bottom sediment samples for analysis of Pb, Zn, Cu, Cd and Hg were taken with a DCh-0.034 grab in accordance with GOST 17.1.5.01-80 in the surface layer (Fig. 4.43). Bottom sediment samples were taken annually, mainly in the summer months. Chemical analysis of bottom sediment samples for the content of lead, cadmium, copper and zinc was carried out in accordance with the M-MVI-80-2008

Table 4.10 Fluxes of sedimentation deposition of pollutants in bottom sediments of Sevastopol Bay (Egorov et al. 2018a)

Box no.	90Sr		137Cs		239,240Pu		210Po		Hg		∑PCB6		∑DDT	
	C_{sed}^a, Bq/kg dry mass	N_{bot}, MBq/year	C_{sed}^a, Bq/kg dry mass	N_{bot}, MBq/year	C_{sed}^a, Bq/kg dry mass	N_{bot}, MBq/year	C_{sed}^a, Bq/kg dry mass	N_{bot}, MBq/year	C_{sed}^a, µg/kg wet weight	N_{bot}, kg/year	C_{sed}^a, µg/kg dry weight	N_{bot}, kg/year	C_{sed}^a, µg/kg dry weight	N_{bot}, kg/year
1	1.5	8.20	67.5	368.89	0.75	4.1	30.5	166.73	493	2.694	83	0.453	22	0.120
2	6.2	8.04	98.7	128.01	0.55	0.71	40.3	52.27	1546	2.070	844	1.095	90	0.117
3	1.4	2.86	75.0	153.07	0.45	0.92	50.2	102.55	837	1.708	725	1.480	120	0.245
4	1.4	3.60	142.4	365.97	0.32	0.82	24.7	63.51	948	2.436	360	0.952	91	0.234
5	1.3	2.97	94.1	215.02	0.28	0.64	20.7	47.30	720	1.645	118	0.270	21	0.048
Entire bay	C_{av} = 2.4	$\sum P$ = 25.67	C_{av} = 95.5	$\sum P$ = 1230.96	C_{av} = 0.47	$\sum P$ = 7.19	C_{av} = 33.1	$\sum P$ = 432.36	C_{av} = 909	$\sum P$ = 10.553	C_{av} = 426	$\sum P$ = 4.25	C_{av} = 69	$\sum P$ = 0.764

[a] Average concentration in bottom sediments
[b] Flux of pollutants into the bottom sediments

Fig. 4.43 Boxes and scheme for sampling water and bottom sediments in 2010–2014 (Matishov et al. 2017)

method using an LSL of 0.0005 mg/g for lead, copper and zinc and an LSL for cadmium of 0.00005 mg/g. All indicated heavy metals were measured using the AAS KVANT-Z-ETA instrument. Determination of mercury in bottom sediments was carried out according to the PND F 16.1:2.23-2000 method using an RA-915 + mercury analyser with an LSL of 5×10^{-6} mg/g. In order to establish interannual trends, the literature data on the content of heavy metals in water and bottom sediments of the Sea of Azov between 1986 and 2009 were also used (Matishov et al. 2006).

In the Sea of Azov, three boxes were distinguished [Taganrog Bay, open aquatic area of the Sea of Azov (the sea itself) and the Kerch Strait including the pre-strait zone (Fig. 4.43)], each differing in their morphometric and hydrological characteristics. The box parameters are presented in Table 4.11.

The sorption of heavy metals by bottom sediments depends on the characteristics of their composition. In the Azov Sea, sedimentation is distributed as follows (Fig. 4.44). The predominant sediments are of a clayey-silty type (0.01 mm fraction coomprises more than 70%). These sediments are distributed mainly on the Panov accumulative plain. In addition to the central and southern parts of the shelf, grey and dark grey silts accumulate locally in the depressions of estuaries and bays, as well as in elongated troughs situated between banks. A characteristic narrow area of silts lines the bottom of the axial trough of Taganrog Bay at a depth of 5–10 m. All

Table 4.11 Box parameters (GOST R 51592-2000 2008; PND F 14.1:2:4.140-98 2013) (Matishov et al. 2017)

Box	Area (km^2)	Volume (km^3)	Average depth (m)	Average speed of sedimentation (g/(m^2 year))
I—Taganrog Bay	5600	25	4.9	700
II—Azov Sea proper	33,400	231	7	300
III—Kerch Strait	675	12	18	500

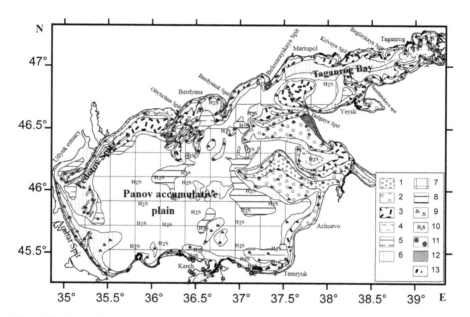

Fig. 4.44 Map of bottom sediments of the Sea of Azov (Klenkin et al. 2007): 1—medium and fine-grained sand with shells (fraction 1–0.1 mm → 70%); 2—silty-silty sand (fraction 1–0.1 mm → 50–70%); 3—mixed type of sediment (silty-silty-sandy); 4—aleurite (fraction 0.1–0.01 mm → 70%); 5—silty silt (fraction 0.1–0.01 mm → 50–70%); 6—silts with impurities (fraction less than 0.01 → 50–70%); 7—clayey silt (fraction less than 0.01 → 70%); 8—clayey silt (fraction less than 0.01 → 85%); 9—shell and shell detritus; 10—hydrogen sulphide contamination; 11—mud volcanoes; 12—rock bench; 13—pebbles, gravel (Matishov et al. 2017)

silty sediments are high in organic content. The sedimentogenesis of the Sea of Azov is characterised by a mixed type of bottom sediments. Their distinctive feature is a mixture in close proportions (from 25 to 40%) of silt and aleurite fractions, including detritus. Areas of mixed sediments tend to the coastal shelf, to the foot of all significant banks of the open sea, as well as to the centre of the seabed depression in large bays. The sand zone (fraction 1.0–0.1 mm, comprising more than 50%) extends on the Azov shelf in a narrow plume in the coastal area at a depth of up to 2–6 m, as well as on the underwater coastal slopes of spits. Shell-sand deposits form submarine

Table 4.12 Admissible concentrations of heavy metals in water and bottom sediments (cited by Matishov et al. 2017)

Characteristics	Metal				
	Pb	Zn	Cu	Cd	Hg
Hazard class	3	3	3	2	1
Maximum allowable concentration in seawater (MAC_w), μg/L	10.0	50.0	5.0	10.0	0.1
Maximum permissible concentration in bottom sediments according to the Dutch Lists (MAC_{bs}), μg/g dry weight	85.0	140.0	35.0	0.8	0.3

banks at a depth of 1–9 m, as well as narrow, gently sloping sandy bars and ridges. In many locations of the banks, the sediments are comprised of shelly limestone with sandy-silty filler (M-MVI-539-03 2003).

In the Russian Federation, the maximum permissible pollution of sea waters is currently regulated by sanitary and hygienic criteria (GOST 17.1.5.01-80 2002).

However, maximum permissible concentrations of heavy metals in bottom sediments in Russia have not been approved at the federal level. In international practice, the Dutch Lists (M-MVI-80-2008 2008) serve as an example of sedimentation quality regulation. In Russia, the Dutch Lists are referred to, for example, in the annual publication "The Quality of Sea Waters according to Geochemical Indicators".

The permissible concentration levels for the considered heavy metals in seawater according to sanitary and hygienic standards and in bottom sediments according to the Dutch Lists are presented in Table 4.12.

Analysis of the summary materials on the content of heavy metals in water and bottom sediments showed the following. Figure 4.45a shows the results of long-term observations of changes in the average lead concentrations in the water of the Taganrog Bay (●), in the open part of the Sea of Azov (O) and (in the period 2010–2014) in the Kerch Strait (+). The figure shows that from 1991 to 2006–2007, the lead content in the water of these regions was an order of magnitude less than MAC_w. Since 2007, there has been a tendency towards an increase in the water pollution of the open part of the Sea of Azov with lead; in 2012–2013 the concentration of lead in the water of the Taganrog Bay was close to MAC_w. In the same period, water pollution by lead in the Kerch Strait exceeded MAC_w or was close to it. However, as shown in Fig. 4.45b, the results of determinations of lead concentrations in the surface layer of bottom sediments carried out between 1991 and 2014 show that their contamination with lead did not exceed the permissible levels of MAC_{bs} for any years in this period. At the same time, lower levels of lead concentration in water corresponded to higher levels of its content in the surface layer of bottom sediments. From 1991 to 2010, (Fig. 4.45b) there was an almost synchronous change in the concentration of lead in the sediments of Taganrog Bay and the open part of the Sea of Azov, while from 2011–2014, the lead content was higher in the bottom sediments of the open part of the Azov Sea.

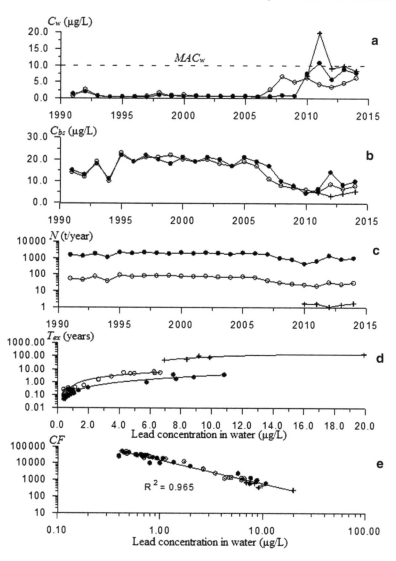

Fig. 4.45 Geochemical characteristics of the distribution of lead in the open part of the Sea of Azov (●), Taganrog Bay (○) and the Kerch Strait (+): **a** concentration of lead in water (µg/L); **b** concentration of lead in the surface layer of bottom sediments (µg/g dry weight); **c** flux of lead deposition in bottom sediments (tonnes per year); **d** period of sedimentation turnover of lead in water (years); **e** change in the accumulation factor of lead in bottom sediments depending on its concentration in water (Matishov et al. 2017)

Observations presented in publications (PND F 16.1:2.23-2000 2005) showed that the content of heavy metals in bottom sediments is determined both by the intensity of sedimentation processes and the concentration characteristics of suspended matter in relation to heavy metals. Taking this into account, we used the following formula to estimate the fluxes (F) of the annual deposition of heavy metals in bottom sediments:

$$F = C_{bs} \cdot S \cdot v_{sed}, \tag{4.59}$$

where

C_{bs} is the concentration of metal in the surface layer of bottom sediments ($\mu g/g$);
S is the size of the considered aquatic area (km^2);
v_{sed} is the specific rate of sedimentation ($g/(m^2 \ year)$).

The results of calculations according to the formula (4.59) of the fluxes of lead input into the bottom sediments of Taganrog Bay, the Azov Sea proper and the Kerch Strait are presented in Fig. 4.45c. Elimination of lead from the waters of the open part of the Sea of Azov was in the range of 450–2200 tonnes per year; its deposition in bottom sediments of Taganrog Bay was 20–90 tonnes per year; the bottom sediments of the Kerch Strait absorbed less than 2 tonnes per year.

It is known from the literature (Bufetova 2015) that some remobilization of pollutants into the water column can be observed under stormy conditions due to the agitation of bottom sediments. Nevertheless, studies of their content in bottom sediment columns have shown that the main part of pollutants is sufficiently well deposited in soils (PND F 16.1:2.23-2000 2005); therefore, the sedimentation self-cleaning of water can be adequately characterised by estimates of the deposition fluxes of pollutants in bottom sediments. The turnover period of heavy metals in the aquatic environment (T, years) was calculated using the formula:

$$T = C_w \cdot S \cdot h_{av}/F \quad \text{or} \quad T = C_w \cdot V/F, \tag{4.60}$$

where S, V, h_{av} and C_w are, respectively, area (km^2), volume (km^3), average depth (m) and concentration of heavy metal ($\mu g/L$) in the water of the analysed aquatic area.

The results of calculations showed that the period of sedimentation self-purification of Taganrog Bay from lead ranges from hundredths of a year to one year, while the period for the open part of the Sea of Azov ranges on a scale of 0.1–5 years and that of the Kerch Strait on a scale of over 10 years (Fig. 4.45d). Figure 4.45d shows that with an increase in the concentration of lead in seawater, the periods of its turnover increased. This indicated that the sorption saturation of bottom sediments with lead is a factor that reduces self-purification capacity. The study of the trend of change in the factor of concentration of lead by bottom sediments ($CF = C_{bs}/C_w$) showed that, with a high degree of statistical reliability (coefficient of determination $R^2 = 0.965$), this dependence falls on a straight line on the graph with logarithmic scales along the ordinate axes (Fig. 4.45e). The materials presented in the figure testify to the increased intensity of sedimentation self-purification of waters at low concentrations of lead in water being provided by the high (at $CF > n \times 10^4$ units) concentrating capacity of bottom sediments. With an increase in the degree of water pollution with lead to 10 µg/L, the value of CF decreased by more than two orders of magnitude, resulting in a decrease in the contribution of sedimentation processes to water self-purification.

Geochemical characteristics of zinc distribution in the Sea of Azov are presented in Fig. 4.46. The figure shows the trends of increasing Zn concentration in water with a peak occurring in 2010–2014 (Fig. 4.46a) along with the dependence of the corresponding decrease in its content in bottom sediments during this time period (Fig. 4.46b). The concentration of zinc in water only exceeded MAC_w in 2014, while in bottom sediments it did not reach tolerance levels throughout the entire observation period. The results of evaluating the fluxes of zinc deposition in bottom sediments using formula (4.59) demonstrated (Fig. 4.46c) that the flux of sedimentation self-purification of waters from this heavy metal in the sea proper was 1800–9800 tonnes/year, in Taganrog Bay it was 110–435 tonnes per year, while in the Kerch Strait, it was less than 2 tonnes per year. From Fig. 4.46d it can be seen that the period of mercury turnover in the open part of the sea was 0.20–2.00 years, that in Taganrog Bay, it was 0.08–1.00 years, while in the Kerch Strait, it exceeded 10 years. The graphic materials presented in Fig. 4.46e demonstrate that the dependence of the change in the coefficients of zinc accumulation by bottom sediments at various concentrations in water with a sufficient degree of adequacy ($R^2 = 0.824$) is described by the equation of a straight line on a logarithmic scale along the ordinate axes.

Analysis of the geochemical characteristics of the distribution of copper (Fig. 4.47) showed that in the period 1991–1995, its concentration in the open water of the Azov Sea and Taganrog Bay decreased (Fig. 4.47a); then, in both boxes, as well as in the Kerch Strait, a trend of increasing copper pollution of water was observed. The concentration of copper in the water of the Sea of Azov exceeded MAC_w in different years. The distribution of copper in the surface layers of bottom sediments corresponded to the opposite trend: the higher its concentration in water, the lower its content in bottom sediments (Fig. 4.47b). This indicated the manifestation of the effect of saturation of bottom sediments with copper as its content in water increased. At the same time, the concentration of copper in the period from 1995 to 2005 reached the critical limits of the content of this metal in the surface layer of bottom sediments. Calculations using formula (4.59) showed that the flux of copper deposition into bottom sediments in the open sea varied from 750 to 3800 tonnes per year, that in Taganrog Bay, it was 46–153 tonnes/year, while in the Kerch Strait, it did not exceed 10 tonnes/year (Fig. 4.47c). It was estimated that the sedimentation turnover of copper in the open part of the Sea of Azov and in Taganrog Bay took place on less than an annual time scale, while in the Kerch Strait the exchange period was two orders of magnitude higher (Fig. 4.47d). The dependence of the coefficients of copper accumulation by bottom sediments on its content in the aquatic environment (Fig. 4.47d) is described, as for lead (Fig. 4.45d) and zinc (Fig. 4.46d), by the equation of a straight line in logarithmic scales along the ordinate axes, but with less statistical significance ($R^2 = 0.794$).

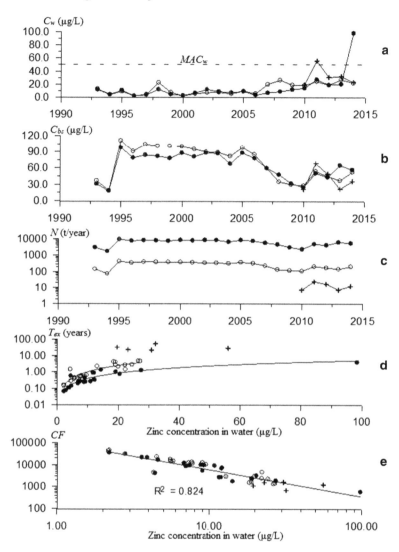

Fig. 4.46 Geochemical characteristics of the distribution of zinc in the open part of the Sea of Azov (●), Taganrog Bay (○) and the Kerch Strait (+): **a** concentration of zinc in water (μg/L); **b** concentration of zinc in the surface layer of bottom sediments (μg/g dry weight); **c** flux of zinc deposition in bottom sediments (tonnes/year); **d** period of sedimentation turnover of zinc in water (years); **e** change in the coefficient of zinc accumulation in bottom sediments depending on its concentration in water (Matishov et al. 2017)

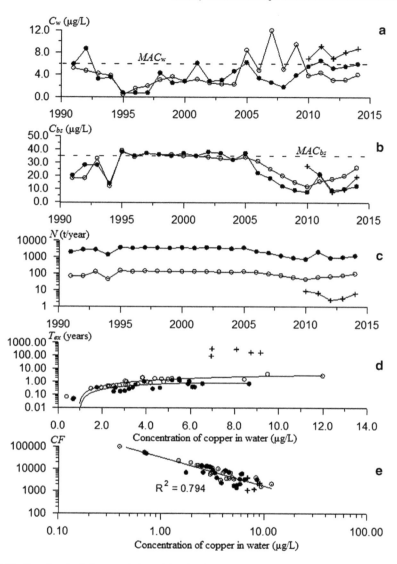

Fig. 4.47 Geochemical characteristics of the distribution of copper in the open part of the Sea of Azov (●), Taganrog Bay (○) and the Kerch Strait (+): **a** concentration of copper in water (μg/L); **b** concentration of copper in the surface layer of bottom sediments (μg/g dry weight); **c** flux of copper deposition in bottom sediments (tonnes/year); **d** period of sedimentation turnover of copper in water (years); **e** change in the coefficient of copper accumulation in bottom sediments depending on its concentration in water (Matishov et al. 2017)

A consideration of materials on the distribution of cadmium showed that from 1993 to 2009 there was a trend of a slow decrease in its concentration in the open water of the Sea of Azov and in Taganrog Bay, followed by a peak in 2010–2014. (Fig. 4.48a). In 2010–2014, the concentration of cadmium did not exceed MAC_w in any parts of the sea under consideration. Up until 2010, the distribution of cadmium in the surface layer of bottom sediments (Fig. 4.48b) decreased; after that, a peak of its increase was noted in the open waters of the Azov Sea and in the Kerch Strait. The cadmium content in bottom sediments remained within the maximum allowed concentrations over the entire observation period. In the open waters of the Sea of Azov, sedimentation self-purification was within the range of 9–60 tonnes/year; in Taganrog Bay, it was between 0.5 and 2.4 tonnes/year, while in the Kerch Strait, it was close to 100 kg/year (Fig. 4.48c). The periods of cadmium turnover in the Azov Sea at different concentrations in water ranged from 0.5 to 30 years (Fig. 4.48d). The dependence of the coefficients of cadmium accumulation on changes in its concentration in water was described by the equation of a straight line with a logarithmic scale along the ordinate axes (Fig. 4.48e) with a statistical availability of data having a coefficient of determination $R^2 = 0.805$.

The data on the distribution of mercury in the Sea of Azov were more variable than the data on other studied heavy metals. In 1991–2005, its content in water exceeded MAC_w. The peaks of maximum mercury pollution of the waters of the sea proper and Taganrog Bay were noted in 1994 and 1998. (Fig. 4.49a). The concentration of mercury in the surface layer of bottom sediments was at a maximum in 1991 and 1999. (Fig. 4.49b). In 1999, the concentration of mercury in the bottom sediments of Taganrog Bay exceeded the allowable concentration, amounting to 0.3 μg/g. However, the concentration of mercury in the water and bottom sediments of the Kerch Strait during the observation period did not exceed the allowable concentrations. The flux of sedimentation self-purification of waters in the open part of the sea was 3–18 tonnes/year, while that of Taganrog Bay was 0.12–1.18 tonnes/year (Fig. 4.49c). The study showed that, with an increase in the concentration of mercury in the open water of the Sea of Azov (Fig. 4.49d), the periods of self-purification of waters from this heavy metal increased from 1 to 12 days, indicating an increase in the degree of saturation of mercury in the bottom sediments of this region. In Taganrog Bay (Fig. 4.49d), a tendency towards a decrease in the period of sedimentation turnover of mercury in water was recorded. Apparently, this indicates that the bottom sediments of Taganrog Bay had a higher sorption capacity for mercury. The dependence of the coefficients of accumulation of bottom sediments by mercury was described by the equation of a straight line on the graph with a logarithmic scale along the y-axes (Fig. 4.49e). However, the higher variability of these data ($R^2 = 0.353$) was due to the difference in the sorption capacity of bottom sediments in different parts of the Sea of Azov.

The consideration presented in Figs. 4.43, 4.44, 4.45, 4.46, 4.47, 4.48 and 4.49 of the data as a whole made it possible to identify a period of stable sanitary and hygienic conditions for a number of heavy metals in the open part of the Sea of Azov and in Taganrog Bay in the last decade of the twentieth century and in the first decade of the twenty-first century, as well as the tendency for their concentration

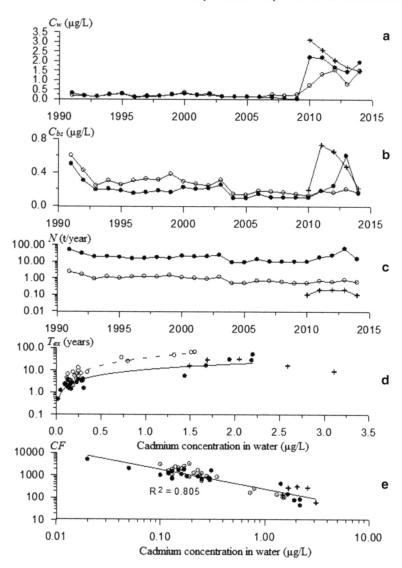

Fig. 4.48 Geochemical characteristics of the distribution of cadmium in the open part of the Sea of Azov (●), Taganrog Bay (○) and the Kerch Strait (+): **a** concentration of cadmium in water (µg/L); **b** concentration of cadmium in the surface layer of bottom sediments (µg/g dry weight); **c** flux of cadmium deposition in bottom sediments (tonnes/year); **d** period of sedimentation turnover of cadmium in water (years); **e** change in the coefficient of cadmium accumulation in bottom sediments depending on its concentration in water (Matishov et al. 2017)

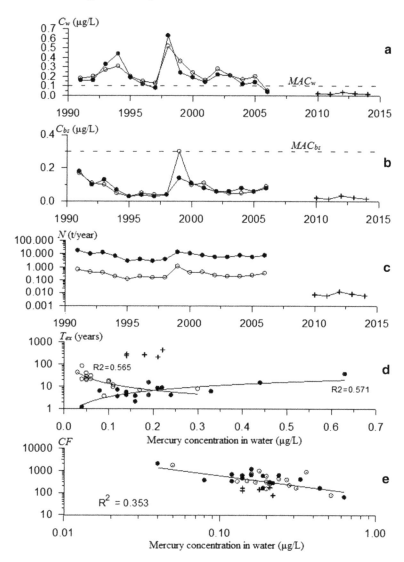

Fig. 4.49 Geochemical characteristics of the distribution of mercury in the open part of the Sea of Azov (●), Taganrog Bay (○) and the Kerch Strait (+): **a** concentration of mercury in water (μg/L); **b** concentration of mercury in the surface layer of bottom sediments (μg/g dry weight); **c** flux of mercury deposition in bottom sediments (tonnes/year); **d** period of sedimentation turnover of mercury in water (years); **e** change in the coefficient of mercury accumulation in bottom sediments depending on its concentration in water (Matishov et al. 2017)

Table 4.13 Estimates of the maximum allowable fluxes of heavy metals into the Sea of Azov (Matishov et al. 2017)

Metal	Freundlich equation parameters		MAC_w (μg/L)	Calculated values of CF	Maximum allowable fluxes (tonnes/year)			
	A	B			Box I	Box II	Box III	Total
Pb	15,000.0	−1.40	10.0	597	234	598	20.15	852.15
Zn	998,900.0	−1.23	50.0	814	1595	4078	137.36	5807.76
Cu	33,200.0	−1.29	5.0	4165	816	2086	70.28	2902.28
Cd	245.8	−0.85	10.0	40	15.7	40.1	1.35	57.15
Hg	71.1	−0.90	0.1	575	2.25	5.8	0.194	8.244

to increase in 2010–2014. Against this background, significant fluctuations in the content of mercury and tendencies for an increase in the concentration of copper in seawater were noted. The excessive concentration of heavy metals in the sea water relative to MAC_w indicates an unfavourable environmental situation in the region. Analysis of the rate of sedimentation and the content of Pb, Zn, Cu, Cd and Hg in bottom sediments showed that sedimentation processes occur on seasonal and annual time scales; moreover, the fluxes of deposition of heavy metals into seabed layers are significant factors of sedimentation self-purification of waters. For this reason, the estimates of the deposition fluxes of heavy metals in bottom sediments can be used for the purposes of environmental regulation.

The practical task of such regulation obviously consists in the establishment of regularities between the concentrations of heavy metals in bottom sediments, which depend on their concentration in water and on the intensity of aquatic self-purification sedimentation fluxes. When solving this problem, it is possible to determine the deposition fluxes of pollutants into bottom sediments, at which their concentration in water does not exceed MAC_w, from the condition of stationarity of the system comprised of heavy metal in water/heavy metal in bottom sedimentsю The results of calculating the maximum permissible fluxes of heavy metals in different parts of the Sea of Azov, subject to the observance of sanitary and hygienic standards for water pollution, are given in Table 4.13.

4.6.3 Critical Areas in the Black Sea

From the 1970s onwards, so-called critical zones began to form in the Black Sea (Zaitsev 1998; Zaitsev and Polikarpov 2002), within which the concentration of pollutants in the components of the Black Sea ecosystems exceeded natural levels or were observed in relatively clean areas. The confinement of critical zones to aquatic areas experiencing significant negative pressure testified to the fact that the ratio between the inflow and elimination fluxes of pollutants in these areas determined a higher stationary level of their content in the marine environment. In this regard, the problem of optimal use of recreational and commercial resources necessitated a study of biogeochemical mechanisms of the arising and self-purification of critical zones of the Black Sea. This section discusses the biogeochemical mechanisms of conditioning the radioisotope and chemical composition of waters in relation to ^{90}Sr, ^{137}Cs, ^{239,240}Pu, ^{210}Po, Hg and organochloride compounds.

The list of registered critical zones in the Black Sea, ranked by pollution factors and anthropogenic impact, is given in Table 4.14. The localisation of critical zones in the Black Sea is shown in Fig. 4.50.

Comparison of our own research results (Fig. 4.50a) with literature data (Fig. 4.50b) showed that critical zones are mainly confined to locations with highly developed anthropogenic activity on the coasts and shelf, as well as to aquatic areas with intensive shipping and recreational use of natural resources.

Table 4.14 Critical zones in the Black Sea by pollution factors and anthropogenic impact (Egorov et al. 2013)

Aquatic area	Factor
Dnieper-Bug estuary	^{90}Sr, ^{137}Cs, ^{239,240}Pu, ^{210}Po, Hg, Aroclor 1254, ΣPCB$_7$[a], ΣDDT[b]
Danube coastal area	^{137}Cs, ^{90}Sr, ^{239,240}Pu, ^{210}Po, Hg, ΣPCB$_7$, ΣDDT
Karkinitsky Bay	Hg, ^{239,240}Pu, ^{210}Po, ΣDDT
Cape Tarkhankut aquatic area	^{90}Sr
Prikerchensky district	^{137}Cs, ^{90}Sr, ^{210}Po, ΣDDT
Feodosia port aquatic area	Hg, ΣPCB$_7$, ΣDDT
Karadag aquatic area	^{210}Po
Balaklava Bay	^{239,240}Pu, ΣPCB$_7$, ΣDDT, hypereutrophication—according to estimates of primary production in the spring-summer period
Sevastopol Bay aquatic areas	
Inkerman	^{239,240}Pu, hypereutrophication—according to estimates of primary production in the spring-summer period
South	^{239,240}Pu, ^{10}Po, ΣPCB$_7$, ΣDDT
North	Hg, ΣPCB$_7$, ΣDDT, hypereutrophication—according to estimates of primary production in the spring-summer period
Gollandiya	^{239,240}Pu, Hg, ΣPCB$_7$, ΣDDT
Konstantinovsky Ravelin	^{239,240}Pu, Hg
Streletskaya	^{239,240}Pu, Hg, ΣPCB$_7$, ΣDDT
Kruhla	Hypereutrophication—according to estimates of primary production in the spring-summer period
Yalta (deep water discharge)	Hg, ^{210}Po

[a]Sum of seven congeners of polychlorinated biphenyls (PCBs): 28, 52, 101, 138, 153, 180, 209
[b]Sum of p,p'-DDT and its metabolites p,p'-DDE and p,p'-DDD

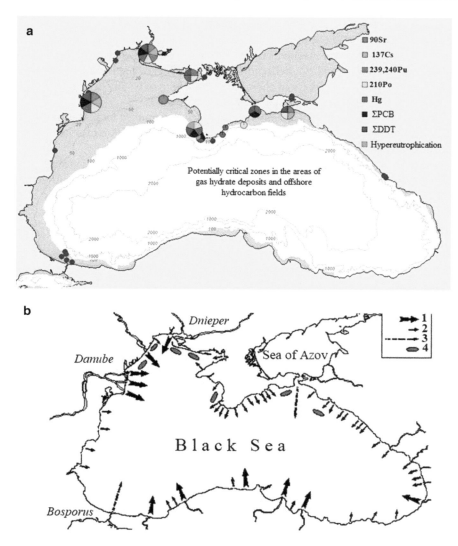

Fig. 4.50 Critical zones in the Black Sea for radioactive and chemical contamination (**a**) (Egorov et al. 2013) and the main routes (Zaitsev 1992) of pollutants (**b**) with river waters (1), slopes (2) and sewer (3) runoff, dumping (4)

As is known, from a biogeochemical point of view, pollutants and eutrophying substances enter the shelf ecosystems from the sea catchment basin, from the pelagic zone, from the underlying deep layers, as well as from the atmosphere. The living matter produced within the photic layer, along with allochthonous suspensions, absorbs radioactive and chemical contaminants from the aquatic environment, exhibits a negative effect of toxicants and, as a result, is deposited in the layers of bottom sediments as a result of sedimentation processes. At certain time scales, bottom sediments for the ecosystem of the oxic layer comprise a geological depot, while the pelagial zone consists of a water depot. The distribution of critical zones presented in Fig. 4.50 shows that the main contemporary anthropogenic pollution in the Black Sea is sustained by the coastal ecosystems of the oxygen zone. The confinement of critical zones directly to sources of pollution (see Fig. 4.50) testified that the intensity of hydrological processes did not ensure the leveling of gradients in the fields of distribution of radioactive and chemical contamination throughout the Black Sea. For this reason, the main factor in the formation of critical zones should evidently include the ratio between the fluxes of pollutants entering them as well as those of their elimination in geological depots.

The assessment the fluxes of sedimentation self-purification of waters was made possible by the use of data on the rate of sedimentation alongside materials on the concentration of pollutants by bottom sediments. The results of calculating fluxes in critical zones and in relatively clean areas are summarised in Table 4.15.

Aquatic areas comprised of critical zones presented in Table 4.15 are marked with asterisks. The profiles of sedimentation rate changes constructed from these data depending on the sea depth are shown in Fig. 4.51. They showed that there was a significant variability of sedimentation processes occurring in the Black Sea regions, against the background of which a general tendency to decrease in their intensity with increasing sea depth was manifested.

Data on the deposition of pollutants in the bottom sediments are graphically reflected in Fig. 4.52. These indicate that the fluxes of deposition of both radioactive and chemical substances into the strata of bottom sediments in critical zones were in all cases higher than in relatively clean areas. This phenomenon is explained by the fact that when the intensity of sedimentation fluxes is equal, the content of contaminants in sediments is proportional to their concentration in the aquatic environment.

4.6.4 Aquatic Areas with Liquid External Boundaries

In many cases, the problem of regulation of anthropogenic pollution of the marine environment refers to aquatic areas having liquid boundaries. The formation characteristics of the ecological state in such areas are determined by the external influx of contaminants, as well as the elimination of such contaminants from the aquatic environment as a result of the influence of biogeochemical processes and water exchange with adjacent areas. It is clear that, against the background of the influence of the processes of biotic self-purification of the marine environment, the formation of fields of pollutants in these waters is largely influenced by hydrodynamic processes. For this reason, the criteria for the maximum allowable anthropogenic impact should be developed for those scales of space and time at which the concentration of pollutants in the water of these areas may exceed the MAC. In this section, the solution to the problems of regulation of anthropogenic impact is considered on the examples of the

Table 4.15 Fluxes of sedimentation self-purification of the Black Sea waters from radioactive and chemical contamination (Egorov et al. 2013)

Area	Coordinates	Depth (m)	Sedi-men-tation (mm per year/g/(m² year))	^{90}Sr, Bq/(m² year)	^{137}Cs, Bq/(m² year)	^{239,240}Pu Bq/(m² year)	^{210}Po, Bq/(m² year)	Hg, μg/(m² year)	ΣPCB Aro-clor 1254, μg/(m² year)	ΣPCB$_7$, μg/(m² year)	ΣDDT, μg/(m² year)
Western halistat. zone	43.25° N, 32.08° E 43° 22.50′ N, 32° 11.50′ E	1983 2041	0.4/70	0.02	3.9–8.4	0.018	6.2	1.2 3.6	0.6		
Continental slope	44° 39.4′ N, 31° 46.2′ E	607	2.2/138	8.1	2.8–11.0	0.050	18.4		12.0		
Danube coastal area	45° 12.4′ N, 29° 51.0′ E 45° 12.4′ N, 29° 46.0′ E 45° 04.0′ N, 29° 46.0′ E	23 15 21	11.5/3994 11.5/3994	47.5	800–1000[a]	0.811[a]	480.0[a]	511.0[a] 1310.0[a] 2184.0[a]		195.0[a]	94.0[a]
Dnieper-Bug estuary	46° 33.0′ N, 31° 25.0′ E	13	9.2/3670	297.6[a]	370–550[a]	4.107[a]	220.0[a]		126.0[a]		
Çoruh estuary	41° 39.7′ N, 41° 33.2′ E	70	5.3/3071	1.54	90–150		146.3[a]				
Karadag	44° 54.68′ N, 35° 24.69′ E 44° 41.55′ N, 35° 33.088′ E	45 545	2245/4.3 1133/6.6	4.3 6.6			86.9[a]			2.4	1.7
Cape Khersones	44° 25.15′ N, 33° 06.196′ E	725	2512/4.8	4.8			53.5				

(continued)

Table 4.15 (continued)

Area	Coordinates	Depth (m)	Sedi-men-tation (mm per year/g/(m² year))	^{90}Sr, Bq/(m² year)	^{137}Cs, Bq/(m² year)	$^{239,240}Pu$ Bq/(m² year)	^{210}Po, Bq/(m² year)	Hg, µg/(m² year)	ΣPCB Aro-clor 1254, µg/(m² year)	ΣPCB7, µg/(m² year)	ΣDDT, µg/(m² year)
Balaklava Bay	44° 30.08′ N, 33° 35.848′ E	7.7	2219/6.3	6.3		0.482	59.2	610[a]		169.0[a]	158.0[a]
Balaklava Bay	44° 29.92′ N, 33° 35.964′ E	12	3519/6.7	6.7		1.211[a]	99.6[a]	6774[a]			
Balaklava Bay	44° 29.74′ N, 33° 35.711′ E	27	2131/4.3	4.3		0.718[a]	68.2	1342[a]			
Cossack Bay	44° 35.05′ N, 33° 24.48′ E	16	653/1.2	1.2			84.4				
Streletskaya Bay	44° 36.54′ N, 33° 28.086′ E	4	888/8.7	8.7	57.7	0.910[a]	64.4	610.0[a]		108.0[a]	45.0[a]
Outer roadstead		25	664/0.4	0.4	43		14.4				
Konst. ravelin	44° 37.5′ N, 33° 31.3′ E	13	3253/6.0	6.0	292	1.496[a]	38.1			94[a]	32[a]
Pavlovsky Cape	44° 37.1′ N, 33° 32.1′ E	15	607/1.5 607/7.1	1.5 7.1	59.7	0.322	29.3		455[a]	101[a]	28[a]
Gollandiya Bay	44° 37.3′ N, 33° 33.7′ E	15	1727		245.9	0.850[a]	20.7			706[a]	40[a]
Inkerman	44° 36.4′ N, 33° 36.0′ E	4	7094/9.5	9.5	989[a]	2.149[a]				546[a]	92[a]

[a] Critical aquatic zones

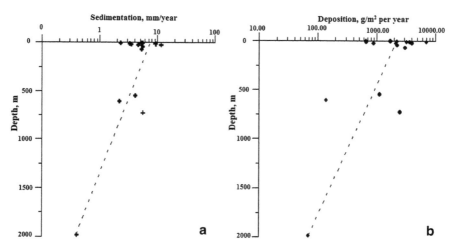

Fig. 4.51 Intensity of sedimentation (**a**) and deposition (**b**) at different depths of the Black Sea (Egorov et al. 2013)

Cape Martyan marine protected area in the vicinity of the Nikitsky Botanical Garden (Egorov et al. 2018b) and the estuary area of the Vodopadnaya River in Yalta.

Cape Martyan Protected Area. The purpose of research in the Cape Martyan conservation area was to determine the dynamic characteristics of the aquatic area (to estimate its area with reference to satellite coordinates, to digitise the contours of depths and boundaries of the computational grid, to calculate the volume of water and the intensity of their exchange), to measure the concentration and content (pool) ^{90}Sr, labile Cu, Zn, Cd, Pb, total Hg, lindane, DDT and $\sum$PCB in water, as well as to estimate the maximum permissible fluxes of pollutants entering the protected area with water stagnation, based on the condition of limiting the concentration of pollutants according to sanitary and hygienic criteria (MAC).

A bathymetric map of the Cape Martyan strict nature reserve is shown in Fig. 4.53.

In order to determine the dynamic characteristics of the aquatic area, we used studies of currents in the Yalta region obtained during the period 1960–1980 from free-floating drifters and autonomous buoy stations fitted with current meters. The results showed that the predominant transport was oriented parallel to the coastline (Fig. 4.54).

In general, it was established that the protected area is characterised by a gently undulating seabed topography, having an average depth of 10 m and a maximum of 20 m. Stretching along the coast for 1.6 km, the area extends to a water border 500 m away from the land. The aquatic area comprises 560,000 m^2, while the total volume of water is 5,300,000 m^3. An assessment of the hydrodynamic characteristics of the region showed that the coastal strip of the protected area is characterised by weak currents. In 80–90% of cases, their measured velocities in this region did not exceed

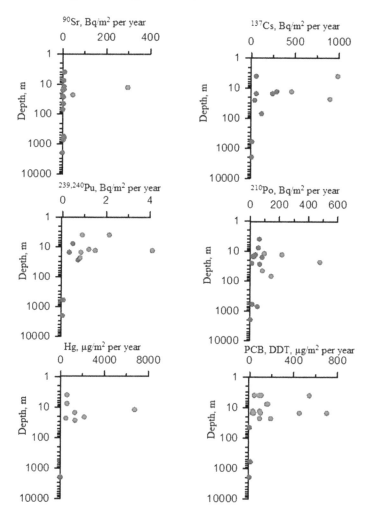

Fig. 4.52 Intensity of deposition of pollutants in the layers of bottom sediments in conditionally clean (blue circles) and critical (red circles) zones of the Black Sea (Egorov et al. 2013)

20 cm/s. With distance from the coast, the velocities of currents 5 miles farther from the sea increased to 100–150 cm/s.

It is obvious that the exchange of waters with those from the open sea has a great influence on the processes of water reclamation and the ecological state of individual aquatic areas and coastal areas. The lower the intensity of water renewal, the lower the fluxes of pollutants entering the aquatic area whose concentration will reach the maximum allowable (MAC) level. This implies that the estimates of the maximum allowable fluxes of pollutants should be normalised based on the implementation of scenarios of the most stagnant state of the waters of the aquatic area. According

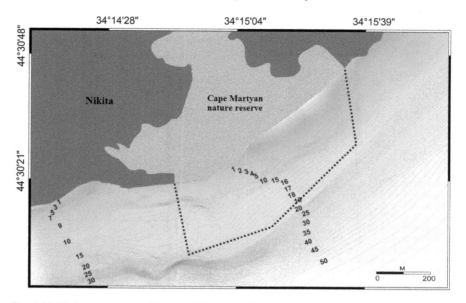

Fig. 4.53 Bathymetric map of the Cape Martyan strict nature reserve (Egorov et al. 2018b)

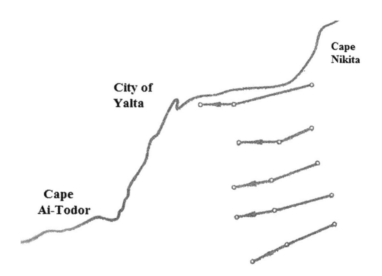

Fig. 4.54 Drift pattern of untethered floats in the Yalta region (Egorov et al. 2018b)

to estimates, such a critical hydrodynamic mode corresponds to a minimum current velocity of 5 cm/s, at which a complete renewal of the waters of the protected zone will occur in 5–6 h.

The characteristics of radioactive and chemical contamination of the aquatic area of the Cape Martyan strict nature reserve are presented in Table 4.16.

Table 4.16 Concentrations, MACs and maximum allowable fluxes of pollutant discharge into the aquatic area of the Cape Martyan strict nature reserve (Egorov et al. 2018b)

Conta-minant	Concen-tration in water	MAC In water	Concen-tration in water (% of MAC)	Pool in the aquatic area Cape Martyan	Maximally acceptable pool in water	Maximally acceptable discharge to the aquatic area in 5 h
1	2	3	4	5	6	7
^{90}Sr	(3.6 ± 1.00) Bq/m^3	5000 Bq/m^3	0.07	19.10 MBq	26,500.0 MBq	132,400 MBq
Cu (labile)	(1.45 ± 0.04) μg/L	5[a] μg/L	29.00	7.70 kg	26.5 kg	94 kg
Zn (labile)	(1.35 ± 0.38) μg/L	50[a] μg/L	2.70	7.20 kg	265.0 kg	1289 kg
Cd (labile)	(0.41 ± 0.01) μg/L	10[a] μg/L	4.10	2.20 kg	53.0 kg	254 kg
Pb (labile)	(2.76 ± 0.05) μg/L	10[a] μg/L	27.60	14.60 kg	53.0 kg	192 kg
Hg (total)	(158.5 ± 7.90) ng/L	100 ng/L	158.50	840.05 g	530.0 g	1550 g
Lindane	(0.34 ± 0.09) ng/L	<10[b] ng/L	3.50	1.85 g	53.0 g	255 g
$\sum$DDT	(0.22 ± 0.23) ng/L	<10[b] ng/L	2.25	1.19 g	53.0 g	259 g
$\sum$PCB	(1.33 ± 1.60) ng/L	<10[b] ng/L	0.15	7.76 g	53.0 g	226 g

[a] "On approval of quality standards for water bodies of fishery significance, including standards for maximum permissible concentrations of substances in water bodies harmful to fish", Decree of 13th December 2016 No. 552 of the Ministry of Agriculture of the Russian Federation
[b] According to the Rospotrebnadzor schedule of fisheries regulation

According to the materials given in Table 4.16, with the exception of mercury, the concentration of all pollutants in the aquatic area is significantly lower than the MAC. The fifth column of the table presents data on the content of pollutants in the aquatic area at present comprising 5,300,000 m^3. The sixth column gives the results of calculations of the limiting assimilation capacity of the aquatic area if the concentration of contaminants in the water reaches the MAC. The seventh column gives estimates of the limiting fluxes of pollutants entering the aquatic area, which is in a "stagnant" hydrodynamic regime characterised by a water exchange period of 5 h. These estimates are clearly suitable for use as criteria for the implementation of the sustainable development of the aquatic area according to the factor of radioactive and chemical pollution of its waters. In general, the studies performed indicate a satisfactory ecological state of the aquatic area of the Cape Martyan strict nature reserve by most of the factors considered. The adoption of measures to reduce the discharge of mercury and nitrogen compounds into the aquatic area will make it possible to implement the concept of sustainable development of this recreational zone based on the factor of the assimilation capacity of its aquatic area.

Estuary zone of the Vodopadnaya river. The aim of the study was to study the seasonal content (in winter, spring and summer) of mineral forms of nitrogen (nitrites, nitrates and ammonium), mineral phosphorus (PO_4), as well as the primary production of phytoplankton, in order to predict and normalise water eutrophication in the area of Bolshoi Yalta near the mouth of the Vodopadnaya River (44° 29.17′ N, 34° 09.53′ E). A map of the mouth and estuary zone of the Vodopadnaya river is shown in Fig. 4.55.

To the south of the mouth of the Vodopadnaya River, a city beach is located. Since the general transport of water masses in this area is directed from the northeast to the southwest (Fig. 4.54), the runoff of the Vodopadnaya River is a source of eutrophication of this recreational zone.

The definitions of phytoplankton production and hydrochemical characteristics of the estuary zone of the Vodopadnaya (Uchan-Su) river are presented in Table 4.17. Here, the results of observations in 2012–2019 are ranked in ascending order of dates by seasons. The data demonstrate that the primary production of phytoplankton in winter, spring and summer seasons in the estuary zone of the river Vodopadnaya ranged from 18.5 to 201.0 mgC/m^3 per days; with, the sum of nitrogen compounds $\sum N_i$ was within 33.0–175.4 $\mu g/L$: NO_2—0.6–6.6 $\mu g/L$; NO_3—9.2–101.3 $\mu g/L$; NH_4—10.0–71.3 $\mu g/L$; the concentration of mineral phosphorus (PO_4) was 2.7–76.1 $\mu g/L$. The values of the Redfield parameter (R_{atom}) (Redfield 1958), characterising the limitation of the primary production of phytoplankton by biogenic elements was rated at 8.3–107.2 for the estuarine zone of the river Vodopadnaya. These data indicated that, although the primary production of phytoplankton in the region could be limited by nitrogen ($R_{atom} = 8.34$), in most cases it was limited by PO_4 (R_{atom} ranged from 22.18 to 107.4).

It should be noted that the concentration of biogenic elements at the mouth of the Vodopadnaya River (44° 29.40′ N, 34° 09.85 ′E) was 14–16 times higher than their concentration in the mouth zone in terms of the amount of nitrogen compounds, while the concentration of phosphates could vary between the same order of values

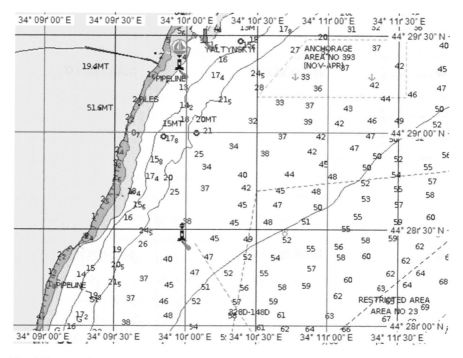

Fig. 4.55 Map of the estuary and mouth zone of the Vodopadnaya river

and up to 25 times higher. This clearly indicates the eutrophying effect of the river runoff on the coastal waters of the Yalta aquatic area of the Black Sea. It follows from publicly available reference manuals that the flow of the Vodopadnaya River varies from high flooding to complete drying up. On average, it flows at 0.384 m^3/s. At the maximum recorded concentrations of nutrients at the river mouth ($\sum N_i = 2804$ µg/L and $PO_4 = 66.6$ µg/L) the influx of nitrogen compounds into the coastal aquatic area of Yalta amounts to 93.03 kg/day, while the influx of PO_4 is 2.204 kg/day. Considering that the synthesis of 1000 g of organic matter by weight requires 80 g of carbon, 2 g of phosphorus, and 14 g of nitrogen (Zilov 2009), we can conclude that with nitrogen limitation of production processes, new products can amount to $(93.03/14) \times 80 = 531$ kg C_{org}/day, while with phosphorus limitation, the amount will be $(2.204/2) \times 80 = 88.16$ kg C_{org} /day.

Waters are considered to be eutrophic when the primary production of phytoplankton amounts to 100 mgC$_{org}$/m^3 per day. With an average depth of the photosynthesis layer of 10 m, the eutrophic primary production in it will be 1 gC$_{org}$/m^3 per day. From the above estimates, it can be seen that, with nitrogen limitation of production processes, the eutrophication of waters to a level of 100 mgC$_{org}$/m^3 per day due to the intake of 531 kg/day being distributed to the aquatic area of 531,000 m^2 (about 0.5 km^2). With phosphorus limitation, the area of eutrophication due to the runoff of the Vodopadnaya River will be approximately 88,000 m^2 Hence it follows

Table 4.17 Production and hydrochemical parameters of the estuarine zone of the Vodopadnaya river

No. in order	Date	Primary products, mgC/m³ per day	NH$_4$		NO$_2$		NO$_3$		∑Ni	PO$_4$		R$_{atom}$
			x (µg/L)	δ	x (µg/L)	δ	x (µg/L)	δ		x (µg/L)	δ	
River Vodopadnaya (44° 29.17′ N, 34° 09.53′ E)												
1	09.01.2018	31.5	10.0	0.50	1.7	0.100	21.3	0.600	33.0	4.0	0.100	22.18
2	19.04.2017	18.5	25.0	1.20	0.6	0.100	36.8	1.100	62.4	5.0	0.100	37.82
3	29.04.2019		71.3	3.41	2.8	0.042	101.3	0.039	175.4	5.0	0.075	107.20
4	30.06.2017	201.0	52.2	2.50	1.6	0.100	9.2	0.300	63.0	76.1	1.100	8.34
5	05.07.2019		13.5	0.65	0.8	0.020	13.3	0.460	27.6	2.7	0.400	34.46
6	04.09.2019		70.0	3.00	6.6	0.100	73.3	2.2	149.9	8.8	0.100	56.10
Fresh water from the mouth of the Vodopadnaya River (44° 29.40′ N, 34° 09.85′ E)												
1	05.07.2019		45.0	2.16	44.0	0.66	2370.0	71.1	2459	66.6	1.0	
2	04.09.2019		80.0	4.80	6.6	0.10	2718.0	3.0	2804	60.2	1.5	

that the runoff of the Vodopadnaya River can significantly reduce the recreational water quality of the aquatic area of Yalta due to the eutrophication of its waters.

The above approach can be used as a method for assessing and forecasting the areas of eutrophication of recreational zones due to the input of nutrients from slope or sewage runoff.

Conclusion

It has been established that, as applied to the problems of control theory, homeostasis of natural marine ecosystems by biotic and abiotic factors is achieved due to the existence of negative feedbacks according to the Le Chatelier–Braun principle. It has been demonstrated that contemporary analytical approaches to assessing the balance of matter and energy in biogeocenoses in combination with semi-empirical methods for describing biogeochemical interactions of living and inert matter in the marine environment can be used as a theoretical basis for describing the radioisotopic and chemical homeostasis of marine ecosystems. It has been shown that the limiting fluxes of self-purification of waters as a result of biogeochemical interaction mechanisms are determined by the "ecological carrying capacity", "assimilation capacity", as well as the radiocapacity of the marine environment. It is noted that optimal solutions to problems arising in connection with applying the ecocentric principle in marine nature management to ensure the sustainable development of aquatic areas are obtained by maintaining a balance between the consumption and reproduction of marine resources. It is proposed that the measure of the restoration of water quality resources be assessed according to biogeochemical criteria that refer to their ability to self-purify, while the measure of their consumption should be measured by the ratio of the concentration of pollutants to water in comparison with the maximum allowable concentration (MAC).

In general terms, the performed studies show that the development of the theory of radioisotopic and chemical homeostasis of marine ecosystems, based on the semi-empirical theory of mineral and radioisotopic exchange, confirm the hypothesis of Academician Vladimir Vernadsky that the necessary conditions of the habitat for living matter are perpetuated through the processes of its reproduction. In providing for the assessment and prediction of the anthropogenic evolution of marine ecosystems, the proposed theoretical framework supports the development of principles for optimal marine nature management, taking into account the intensity of natural biogeochemical processes of radioactive and chemical conditioning of the marine environment. The interpretation of the results of experimental and natural observations alongside theoretical analysis from the point of view of anthropogenic evolution allows us to predict long-term trends associated with the increased intensity of biogeochemical cycles involving the material, energetic and chemical composition of the components of marine ecosystems.

References

Alekin OA, Lyakhin YI (1984) Khimiya okeana. Gidrometeoizdat, Leningrad, 343p (in Russian)

Aleksakhin RM, Fesenko SV (2004) Radiatsionnaya zashchita okruzhayushchei sredy: antropotsentricheskii i ekotsentricheskii printsipy, Radiatsionnaya biologiya. Radioekologiya 44(1):93–103 (in Russian)

Barinov GV (1965) Obmen ^{45}Ca, ^{137}Cs, ^{144}Ce mezhdu vodoroslyami i morskoi vodoi. Okeanologiya 5(1):111–116 (in Russian)

Belyaev VI (1978) Teoriya slozhnykh geosistem. Naukova dumka, Kiev, 155p (in Russian)

Bodeanu N (1992) Algal blooms and development of the main phytoplanktonic species at the Romanian Black Sea littoral in conditions of intensification of the eutrophication process. In: Marine coastal eutrophication: proceedings of an international conference, Bologna, Italy, 21–24 Mar 1990. Elsevier Science, Amsterdam, pp 891–906. https://doi.org/10.1016/B978-0-444-89990-3.50077-8

Bogdanov YA, Kopelevich OV (1974) Granulometricheskie issledovaniya tonkodispersnogo veshchestva okeanskoi sredy. In: Formy elementov i radionuklidov v morskoi vode. Nauka, Moscow, pp 119–123 (in Russian)

Bogdanov YA, Gurvich EG, Lisitsyn AP (1983) Mekhanizm okeanskoi sedimentatsii i differentsiatsii khimicheskikh elementov v okeane. In: Biokhimiya okeana. Nauka, Moscow, pp 165–200 (in Russian)

Bogdanova AK (1959) Vodoobmen cherez Bosfor i ego rol' v peremeshivanii vod Chernogo morya. Tr Sevastopol Biol Stantsii 12:401–420 (in Russian)

Boltachev AR, Eremeev VN (2011) Rybnyi promysel v azovo-chernomorskom basseine: proshloe, nastoyashchee, budushchee. In: Eremeev VN, Gaevskaya AV, Shulman GE, Zagorodnyaya YA (eds) Promyslovye bioresursy Chernogo i Azovskogo morei. EKOSI-Gidrofizika, Sevastopol, pp 7–26 (in Russian)

Britton G (1983) The biochemistry of natural pigments. Cambridge University Press, Cambridge, 366p. Russian edition: Britton G (1986) Biokhimiya prirodnykh pigmentov. Mir, Moscow, 422p

Brzezińska A, Trzosińska A, Zmijewska W, Wódkiewicz L (1984) Trace metals in suspended matter and surficial bottom sediments from the Southern Baltic. Oceanologia 18:59–77

Bufetova MV (2015) Zagryaznenie vod Azovskogo morya tyazhelymi metallami. Yug Rossii Ekol Razvit 10(3):112–120 (in Russian)

Clerck R, Guns M, Vyncke W, Van Hoeyweghen P (1984) La teneur en métaux lourds dans le cabillaud, le flet et la crevette des eaux cotières belges. Rev Agric 37(4):1079–1086

Droop MR (1974) The nutrient status of algal cells in continuous culture. J Mar Biol Assoc UK 54(4):825–855. https://doi.org/10.1017/S002531540005760X

Egorov VN (2001) Normirovanie potokov antropogennogo zagryazneniya chernomorskikh regionov po biogeokhimicheskim kriteriyam. Ekol Morya 57:75–84 (in Russian)

Egorov VN (2012) Biogekhimicheskie mekhanizmy realizatsii kompensatsionnogo gomeostaza v chernomorskikh ekosistemakh. Mor Ehkol Zh 11(4):5–17 (in Russian)

Egorov VN, Erokhin VE (1998) Empiricheskaya model' kinetiki adaptivnoi ustoichivosti pigmentnoi sistemy makrofitov pri intoksikatsii fenolom. Ekol Morya 47:90–95 (in Russian)

Egorov VN, Ivanov VN (1981) Matematicheskoe opisanie kinetiki obmena tsinka-65 i margantsa-54 u morskikh rakoobraznykh pri nepishchevom puti postupleniya radionuklidov. Ekol Morya 6:37–43 (in Russian)

Egorov VN, Popovichev VN, Burlakova ZP, Krupatkina DK, Kovalenko TP, Aleksandrov BG (1992) Matematicheskaya model' biosedimentatsionnoi funktsii ekosistemy foticheskogo sloya zapadnoi khalistatiki Chernogo morya. In: Polikarpov GG (ed) Molismologiya Chernogo morya. Naukova dumka, Kiev, pp 50–62 (in Russian)

Egorov VN, Polikarpov GG, Kulebakina LG, Stokozov NA, Evtushenko DB (1993) Model' krupnomasshtabnogo zagryazneniya Chernogo morya dolgozhivushchimi radionuklidami ^{137}Cs i ^{90}Sr v rezul'tate avarii na Chernobyl'skoi AES. Vod Resur 20(3):326–330 (in Russian)

Egorov VN, Povinec PP, Polikarpov GG, Stokozov NA, Gulin SB, Kulebakina LG, Osvath I (1999) ^{90}Sr and ^{137}Cs effects in the Black Sea after the Chernobyl NPP accident inventories, balance and tracer applications. In: International scientific co-operation radiological impact assessment in the South-Eastern Mediterranean area. Thessaloniki, vol 2, pp 104–129

Egorov VN, Stokozov NA, Mirzoeva NY (2001) Long-term post-Chernobyl ^{90}Sr and ^{137}Cs profiles as the indicators of the large-scale vertical water mixing in the Black Sea. In: International conference on the study of environmental change using isotope techniques: book of extended synopses, Vienna, 23–27 Apr 2001, pp 182–184

Egorov VN, Polikarpov GG, Stokozov NA, Mirzoeva NY (2008) Balans i dinamika polei kontsentratsii [137]Cs i [90]Sr v vodakh Chernogo morya. In: Polikarpov GG, Egorov VN (eds) Radioekologicheskii otklik Chernogo morya na chernobyl'skuyu avariyu. EKOSI-Gidrofizika, Sevastopol, pp 217–250 (in Russian)

Egorov VN, Artemov YG, Gulin SB (2011) Metanovye sipy v Chernom more: sredoobrazuyushchaya i ekologicheskaya rol'. EKOSI-Gidrofizika, Sevastopol, 405p (in Russian)

Egorov VN, Malakhova LV, Malakhova TV, Todorenko DA (2012) Adaptatsionnye kharakteristiki chernomorskoi zelenoi vodorosli *Ulva rigida* Ag. pri khronicheskom i impaktnom vozdeistvii polikhlorbifenilov Naukovii Visnik Uzhgorods'kogo Universitetu. Ser Biol 32:12–18 (in Russian)

Egorov VN, Gulin SB, Popovichev VN, Mirzoeva NY, Tereshchenko NN, Lazorenko GE, Malakhova LV, Plotitsyna OV, Malakhova TV, Proskurnin VY, Sidorov IG, Gulina LV, Stetsyuk AP, Marchenko YG (2013) Biogeokhimicheskie mekhanizmy formirovaniya kriticheskikh zon v Chernom more v otnoshenii zagryaznyayushchikh veshchestv. Mor Ehkol Zh 12(4):5–26 (in Russian)

Egorov VN, Gulin SB, Malakhova LV, Mirzoeva NY, Popovichev VN, Tereshchenko NN, Lazorenko GE, Plotitsyna OV, Malakhova TV, Proskurnin VY, Sidorov IG, Stetsyuk AP, Gulina LV (2018a) Normirovanie kachestva vod Sevastopol'skoi bukhty po potokam deponirovaniya ^{90}Sr, ^{137}Cs, [239,240]Pu, [210]Po, Hg, PKhB i DDT v donnye otlozheniya. Vodnye Resursy 45(2):188–195 (in Russian)

Egorov VN, Plugatar' YV, Malakhova LV, Mirzoeva NY, Gulin SB, Popovichev VN, Sadogurskii SE, Malakhova TV, Shchurov SV, Proskurnin VY, Bobko NI, Marchenko YG, Stetsyuk AP (2018b) Ekologicheskoe sostoyanie akvatorii osobo okhranyaemoi prirodnoi territorii "Mys Mart'yan" i problema realizatsii ee ustoichivogo razvitiya po faktoram evtrofikatsii, radioaktivnogo i khimicheskogo zagryazneniya vod. Nauchnye zapiski prirodnogo zapovednika "Mys Mart'yan" 9:36–40 (in Russian)

Egorov VN, Popovichev VN, Gulin SB, Bobko NI, Rodionova NY, Tsarina TV, Marchenko YG (2018c) Vliyanie pervichnoi produktsii fitoplanktona na oborot biogennykh elementov v pribrezhnoi akvatorii Sevastopolya (Chernoe more). Biol Morya 44(3):207–214 (in Russian)

Evans DW, Kathman RD, Walker WW (2000) Trophic accumulation and depuration of mercury by blue crabs (*Callinectes sapidus*) and pink shrimp (*Penaeus duorarum*). Mar Environ Res 49(5):419–434. https://doi.org/10.1016/S0141-1136(99)00083-5

Finenko ZZ, Krupatkina-Akinina DK (1974) Vliyanie neorganicheskogo fosfora na skorost' rosta diatomovykh vodoroslei. In: Biologicheskaya produktivnost' yuzhnykh morei. Naukova dumka, Kiev, pp 120–135 (in Russian)

Finenko ZZ, Lanskaya LA (1971) Rost i skorost' deleniya vodoroslei v limitirovannykh ob"emakh vody. In: Ekologicheskaya fiziologiya morskikh planktonnykh vodoroslei. Naukova dumka, Kiev, pp 22–49 (in Russian)

Finenko ZZ, Churilova TY, Suslin VV (2011) Otsenka biomassy fitoplanktona i pervichnoi produktsii v Chernom more po sputnikovym dannym. In: Eremeev VN, Gaevskaya AV, Shulman GE, Zagorodnyaya YA (eds) Promyslovye bioresursy Chernogo i Azovskogo morei. EKOSI-Gidrofizika, Sevastopol, pp 237–248 (in Russian)

Finkel'shtein MS, Pronenko SM (1991) Tendentsiya mnogoletnikh izmenenii kontsentratsii fosfatov v zapadnoi chasti Chernogo morya. Ekol Morya 39:1–5 (in Russian)

Fowler SW, Benayen J (1979) The influence of factors of surroundings on the flux of selenium through the marine organisms. In: Interaction between water and living matter: proceedings of the international symposium, vol 1, Odessa, 6–10 Oct 1975

Fowler SW, Small LF (1972) Sinking rates of euphausiid fecal pellets. Limnol Oceanogr 17(2):293–296. https://doi.org/10.4319/lo.1972.17.2.0293

Garkavaya GP, Bogatova YI, Bulanaya ZT (1997) Mnogoletnyaya dinamika biogennykh veshchestv Kiliiskogo girla del'ty Dunaya. In: Materials of the II congress of the hydroecological society of Ukraine, vol 1. Naukova dumka, Kiev, pp 23–24 (in Russian)

Goldberg ED (1981) The oceans as waste space: the argument. Oceanus 24(1):2–9

Goncharov VP, Emel'yanova LP, Mikhailov OV, Tsyplev YI (1965) Ploshchadi i ob"emy Sredizemnogo i Chernogo morei. Okeanologiya 5(6):918–925 (in Russian)

GOST 17.1.5.01-80 (2002) Okhrana prirody. Gidrosfera. Obshchie trebovaniya k otboru prob donnykh otlozhenii vodnykh ob"ektov dlya analiza na zagryaznennost'. IPK Izd-vo standartov, Moscow, 7p (in Russian)

GOST R 51592-2000 (2008) Voda. Obshchie trebovaniya k otboru prob. Standartinform, Moscow, 48p (in Russian)

Granstrom ML, Ahlert KC, Wlesenfeld I (1984) The relationships between the pollutants in the sediments and in water of the Delaware and Raritan Canal. Water Sci Technol 16(5–7):375–380. https://doi.org/10.2166/wst.1984.0144

Greze II (1977) Amfipody Chernogo morya i ikh biologiya. Naukova dumka, Kiev, 155p (in Russian)

Greze VN (ed) (1979) Osnovy biologicheskoi produktivnosti Chernogo morya. Naukova dumka, Kiev, 387p (in Russian)

Greze VN (1982) Ekosistema Yuzhnoi Atlantiki i problema energeticheskogo balansa pelagicheskogo soobshchestva okeana. Okeanologiya 22(6):996–1001 (in Russian)

Gudiksen PH, Harvey TF, Lange R (1989) Chernobyl source term, atmospheric dispersion and dose estimation. Health Phys 57(5):697–706. https://doi.org/10.1097/00004032-198911000-00001

Gudiksen PH, Harvey TF, Lange R (1991) Chernobyl source term estimation. In: Proceedings of the seminar on comparative assessment of the environmental impact of radionuclides released during three major nuclear accidents: Kyshtym, Windscale, Chernobyl, vol 1. Luxembourg, 1–5 Oct 1990

Gulin SB (2000) Recent changes of biogenic carbonate deposition in anoxic sediments of the Black Sea: sedimentary record and climatic implication. Mar Environ Res 49(4):319–328. https://doi.org/10.1016/S0141-1136(99)00074-4

Gulin SB, Artemov YG (2007) Issledovanie struinykh vykhodov metana iz dna Chernogo morya v mezhdunarodnoi ekspeditsii nauchno-issledovatel'skogo sudna "Meteor" (Germaniya) v fevrale 2007 g. Mor Ehkol Zh 6(2):98–100 (in Russian)

Gulin SB, Egorov VN, Stokozov NA, Mirzoeva NY (2008) Opredelenie vozrasta donnykh otlozhenii i otsenka skorosti osadkonakopleniya v pribrezhnykh i glubokovodnykh akvatoriyakh Chernogo morya s ispol'zovaniem prirodnykh i antropogennykh trasserov. In: Polikarpov GG, Egorov VN (eds) Radioekologicheskii otklik Chernogo morya na chernobyl'skuyu avariyu. EKOSI-Gidrofizika, Sevastopol, pp 499–502 (in Russian)

Gulin SB, Egorov VN, Polikarpov GG, Stokozov NA, Mirzoyeva NY, Tereschenko NN, Osvath I (2012) General trends in radioactive contamination of the marine environment from the Black Sea to Antarctic Ocean. In: Burlakova EB, Naydich VI (eds) The lessons of Chernobyl: 25 years later. Nova Science Publishers, USA, New York, pp 281–299

Gulin SB, Egorov VN, Mirzoeva NY, Proskurnin VY, Bei ON, Sidorov IG (2017) Radioemkost' kislorodnoi i serovodorodnoi zon Chernogo morya v otnoshenii ^{90}Sr i ^{137}Cs. Radiatsionnaya biologiya. Radioekologiya 57(2):191–200 (in Russian)

Hutchinson GE (1969) Limnologiya. Mir, Moscow, 591p (in Russian)

Hubert M (1984) Dynamik und Stabilität Biologischer Populationen. Allg Forst Z 39(6):116–121

Ivanov MV, Polikarpov GG, Lein AY, Gal'chenko VF, Egorov VN, Gulin SB, Gulin MB, Rusanov II, Miller YM, Kuptsov VI (1991) Biogeokhimiya tsikla ugleroda v raione metanovykh gazovydelenii Chernogo morya. Dokl AN SSSR 3(5):1235–1240 (in Russian)

Ivanov VA, Ovsyany EI, Repetin LN, Romanov AS, Ignat'eva OG (2006) Gidrologo-gidrokhimicheskii rezhim Sevastopol'skoi bukhty i ego izmeneniya pod vozdeistviem klimatich-eskikh i antropogennykh faktorov. EKOSI-Gidrofizika, Sevastopol, 90p (in Russian)

Ivanov VN, Egorov VN, Popovichev VN, Shevchenko MM (1986) Matematicheskoe modelirovanie kinetiki obmena mikroelementov u morskikh rakoobraznykh pri pishchevom i parenteral'nom putyakh ikh postupleniya. Ekol Morya 23:68–77 (in Russian)

Ivlev VS (1955) Eksperimental'naya ekologiya pitaniya ryb. Pishchepromizdat, Moscow, 252p (in Russian)

Izrael YA, Tsyban AV (1983) Ob assimilyatsionnoi emkosti Mirovogo okeana. Dokl AN SSSR 272(3):702–705 (in Russian)

Izrael YA, Tsyban AV (2009) Antropogennaya ekologiya okeana. Flinta Nauka, Moscow, 532p (in Russian)

Jernelov A (1983) "Receiving capacity"—some argument against the concept as an ocean management tool. Siren News UNEP Reg Seas Program 21:231

Keck G, Raffenot I (1979) Etude éco-toxicologique de la contamination chimique par les PCB dans la rivière du Furans (Ain). Rev Méd Vét 130(3):339–358

Kinne O (1997) Ethics and eco-ethics. Mar Ecol Prog Ser 153:1–3. https://doi.org/10.3354/meps15 3001

Kitaev SP (1984) Ekologicheskie osnovy bioproduktivnosti ozer raznykh prirodnykh zon. Nauka, Moscow, 208p (in Russian)

Klenkin AA, Korpakova IG, Pavlenko LF, Temerdashev ZA (2007) Ekosistema Azovskogo morya: antropogennoe zagryaznenie. Prosveshchenie-Yug, Krasnodar, 324p (in Russian)

Kostova SK, Egorov VN, Popovichev VN (2001) Mnogoletnie issledovaniya zagryazneniya rtut'yu sevastopol'skikh bukht (Chernoe more). Ekol Morya 56:99–103 (in Russian)

Kowal NE (1971) Model of elemental assimilation by invertebrates. J Theor Biol 31(3):469–474. https://doi.org/10.1016/0022-5193(71)90022-1

Kuenzler EJ (1965) Zooplankton distribution and isotope turnover during operation swordfish. US AEC Document NYO-3145-1. New York Operations Office. New York, 12p

Lein AY, Ivanov MV (2009) Biogeokhimicheskii tsikl metana v okeane. Nauka, Moscow, 575p (in Russian)

Lisitsyn AP (1982) Lavinnaya sedimentatsiya. In: Lavinnaya sedimentatsiya v okeane. Izdatel'stvo Rostovskogo universiteta, Rostov-on-Don, pp 3–58 (in Russian)

Lowman FG, Rice TR, Richards FA (1971) Accumulation and redistribution of radionuclides by marine organisms. In: Radioactivity in the marine environment. The National Academies Press, Washington, DC, pp 162–199. https://doi.org/10.17226/18745

Magnusson K, Ekelund R, Grabic R, Bergqvist PA (2006) Bioaccumulation of PCB congeners in marine benthic infauna. Mar Environ Res 61(4):379–395. https://doi.org/10.1016/j.marenvres.2005.11.004

Maher WA (1985) Characteristics of selenium in marine animals. Mar Pollut Bull 16(1):33–34. https://doi.org/10.1016/0025-326X(85)90257-7

Malakhova LV (2006) Soderzhanie i raspredelenie khlororganicheskikh ksenobiotikov v kompo-nentakh ekosistem Chernogo morya. Dissertation abstract, Sevastopol, 24p (in Russian)

Marei AN, Zykova AS (eds) (1980) Metodicheskie rekomendatsii po sanitarnomu kontrolyu za soderzhaniem radioaktivnykh veshchestv v ob"ektakh vneshnei sredy. MZ SSSR, Moscow, 356p (in Russian)

Matishov GG, Pol'shin VV, Il'in GV, Novenko EY, Karageorgis A (2006) Zakonomernosti litokhimii i palinologii sovremennykh donnykh otlozhenii Azovskogo morya. Vestn Yuzhnogo Nauchnogo Tsentra 2(4):38–51 (in Russian)

Matishov GG, Bufetova MV, Egorov VN (2017) Normirovanie potokov postupleniya tyazhelykh metallov v Azovskoe more. Nauka Yuga Rossii 13(1):44–58 (in Russian)

Mee LD (ed) (1997) Black Sea transboundary diagnostic analysis. UN Publ, New York, p 142

Menshutkin VV, Finenko ZZ (1975) Matematicheskoe modelirovanie protsessa razvitiya fitoplank-
tona v usloviyakh okeanicheskogo apvellinga. Tr Inst Okeanol AN SSSR 102:175–183 (in
Russian)
Milchakova NA, Mironova NV, Ryabogina VG (2011) Morskie rastitel'nye resursy. In: Eremeev
VN, Gaevskaya AV, Shulman GE, Zagorodnyaya YA (eds) Promyslovye bioresursy Chernogo i
Azovskogo morei. EKOSI-Gidrofizika, Sevastopol, pp 237–248 (in Russian)
Minicheva GG (1996) Reaktsiya mnogokletochnykh vodoroslei na evtrofirovanie ekosistem.
Al'gologiya 6(3):250–257 (in Russian)
Mirzoeva NY, Polikarpov GG, Egorov VN (2008) Dozovye nagruzki ot izluchenii avariinykh ^{137}Cs
i ^{90}Sr na gidrobiontov pruda-okhladitelya ChAES, Kievskogo i Kakhovskogo vodokhranilishch,
Severo-Krymskogo kanala i Chernogo morya. In: Polikarpov GG, Egorov VN (eds) Radioeko-
logicheskii otklik Chernogo morya na chernobyl'skuyu avariyu. EKOSI-Gidrofizika, Sevastopol,
pp 362–369 (in Russian)
Mirzoeva NY, Gulin SB, Arkhipova SI, Korkishko NF, Migal' LV, Moseichenko IN, Sidorov IG
(2013) Potoki migratsii i deponirovaniya posleavariinykh radionuklidov ^{90}Sr i ^{137}Cs v razlichnykh raionakh Chernogo morya (elementy biogeokhimicheskikh tsiklov). Naukovo-metodichnii
zhurnal "Naukovi pratsi". Seriya "Tekhnogenna Bezpeka" 198(210):45–51 (in Russian)
Morozov NP, Petukhov SA (1979) Trace elements in hydrobionts and biotope of the surface layer
of seawater in the North Atlantic and the Mediterranean. In: ICES council meeting, 7 p
M-MVI-80-2008 (2008) Metodika vypolneniya izmerenii massovoi doli elementov v probakh
pochv, grunta i donnykh otlozheniyakh metodami atomno-emissionnoi spektrometrii i atomno-
absorbtsionnoi spektrometrii. Monitoring, Saint Petersburgh, 17p (in Russian)
M-MVI-539-03 (2003) Metodika vypolneniya izmerenii massovoi kontsentratsii Al, Fe, Cd, Co,
Mn, Cu, Ni, Pb, Ti, Cr, Zn v prirodnoi, pit'evoi i stochnykh vodakh atomno-absorbtsionnym
metodom. Monitoring, Moscow, 16p (in Russian)
Nifon E, Coussins I (2007) Modelling PCB bioaccumulation in a Baltic food web. Environ Pollut
148(1):73–82. https://doi.org/10.1016/j.envpol.2006.11.033
Odum EP (1971) Fundamentals of ecology. Saunders Company, Philadelphia, 574p. Russian edition:
Odum E (1986) Ekologiya (trans: Sokolov VE). Mir, Moscow, vol 1, 328p; vol 2, 376p
Osterberg C, Pirsi W (1971) Radioaktivnost' i pishchevye tsepi v okeane. In: Voprosy radioekologii.
Atomizdat, Moscow, pp 240–252 (in Russian)
Ovsyany EI, Romanov AS, Min'kovskaya RY, Krasnovid II, Ozyumenko BA, Tsymbal IM (2001)
Osnovnye istochniki zagryazneniya morskoi sredy Sevastopol'skogo regiona. In: Ekologich-
eskaya bezopasnost' pribrezhnoi i shel'fovoi zon i kompleksnoe ispol'zovanie resursov shel'fa:
collection of scientific papers, vol 2. EKOSI-Gidrofizika, Sevastopol, pp 138–152 (in Russian)
Ovsyany EI, Artemenko VM, Romanov AS, Orekhova NA (2007) Stok reki Chernoi kak faktor
formirovaniya vodno-solevogo rezhima i ekologicheskogo sostoyaniya Sevastopol'skoi bukhty.
In: Ekologicheskaya bezopasnost' pribrezhnoi i shel'fovoi zon i kompleksnoe ispol'zovanie
resursov shel'fa: collection of scientific papers, vol 15. EKOSI-Gidrofizika, Sevastopol, pp 57–65
(in Russian)
Parkhomenko AV, Krivenko OV (2011) Otsenka biomassy fitoplanktona v Chernom more za
period 1948–2001 gg. In: Eremeev VN, Gaevskaya AV, Shulman GE, Zagorodnyaya YA (eds)
Promyslovye bioresursy Chernogo i Azovskogo morei. EKOSI-Gidrofizika, Sevastopol, pp
237–248 (in Russian)
Patton A (1968) Energetika i kinetika biokhimicheskikh protsessov. Mir, Moscow, 159p (in Russian)
Pentreath RJ (1973) The roles of food and water in the accumulation of radionuclides by
marine teleost and elasmobranch fish. In: Radioactive contamination of the marine environment:
proceeding of a symposium, Seattle, 10–14 July 1972
Pererva VM, Lyal'ko VI, Shpak VF (1997) Flyuidoprovidni struktury i naftogazonosnist' Azovo-
Chornomors'kogo regionu. Dopov NAN Ukr 4:136–139 (in Ukrainian)
Petipa TS (1981) Trofodinamika kopepod v morskikh planktonnykh soobshchestvakh: zakonomer-
nosti potrebleniya pishchi i prevrashcheniya energii u osobi. Naukova dumka, Kiev, 242p (in
Russian)

Phillips GR, Lenhart TE, Gregory RW (1980) Relation between trophic position and mercury accumulation among fishes from the Tongue River Reservoir, Montana. Environ Res 22(1):73–80. https://doi.org/10.1016/0013-9351(80)90120-6

Pimenov NV (2006) Mikrobnye protsessy tsikla ugleroda na gidrotermal'nykh polyakh i kholodnykh metanovykh sipakh. Dissertation abstract, Moscow, 69p (in Russian)

Piontkovsky SA, Egorov VN, Ivanov VN (1983) Opyt ispol'zovaniya tsinka-65 dlya izucheniya ritma pitaniya planktonnykh rakoobraznykh. Ekol Morya 15:84–88 (in Russian)

PND F 14.1:2:4.140-98 (2013) Kolichestvennyi khimicheskii analiz vod. Metodika vypolneniya izmerenii massovykh kontsentratsii Be, V, Bi, Cd, Co, Cu, Mo, As, Ni, Sn, Pb, Se, Ag, Sb, Cr v pit'evykh, prirodnykh i stochnykh vodakh metodom atomno-absorbtsionnoi spektrometrii s elektrotermicheskoi atomizatsiei. Rosprirodnadzor, Moscow, 22p (in Russian)

PND F 16.1:2.23-2000 (2005) Opredelenie soderzhaniya rtuti v pochve, donnykh otlozheniyakh i gornykh porodakh. Rosprirodnadzor, Moscow, 8p (in Russian)

Polikarpov GG (1961) Materialy po koeffitsientam nakopleniya P^{32}, S^{35}, Sr^{90}, Y^{91}, Cs^{137} i Ce^{144} v morskikh organizmakh. Tr Sevastopol Biol Stantsii 14:314–328 (in Russian)

Polikarpov GG (1964) Radioekologiya morskikh organizmov. Atomizdat, Moscow, 295p (in Russian)

Polikarpov GG (1966) Radioecology of aquatic organisms. Reinhold Publ. Co., New York, p 314

Polikarpov GG (2006) Radiatsionnaya zashchita biosfery, vklyuchaya *Homo sapiens*: vybor printsipov i poiski resheniya. Mor Ehkol Zh 5(1):16–34 (in Russian)

Polikarpov GG (2012) Ekstremal'naya zhizn' i sozdavaemaya eyu samoi sebe oblast' zhizni v batiali Chernogo morya. Mor Ehkol Zh 11(3):5–14 (in Russian)

Polikarpov GG, Egorov VN (1986) Morskaya dinamicheskaya radiokhemoekologiya. Energoatomizdat, Moscow, 176p (in Russian)

Polikarpov GG, Egorov VN (eds) (2008) Radioekologicheskii otklik Chernogo morya na chernobyl'skuyu avariyu. EKOSI-Gidrofizika, Sevastopol, 667p (in Russian)

Polikarpov GG, Egorov VN, Levchenko IA (1987) Dinamicheskie aspekty nadezhnosti morskikh ekologicheskikh system. In: Nadezhnost' i gomeostaz biologicheskikh sistem: collection of scientific papers. Kiev, pp 175–180 (in Russian)

Polikarpov GG, Lazorenko GE, Tereshchenko NN, Mirzoeva NY, Egorov VN (2007) Dozovye nagruzki izluchenii ^{210}Po, ^{90}Sr, ^{137}Cs i $^{238,239,240}Pu$ na morskuyu i presnovodnuyu biotu v Ukraine. In: Migunov VI, Trapeznikov AV (eds) Problemy radioekologii i pogranichnykh distsiplin, vol 11. Izd-vo Ural'skogo un-ta, Ekaterinburg, pp 3–16 (in Russian)

Polikarpov GG, Egorov VN, Gulin SB (2012) Vernadskologiya: rol' v formirovanii morskoi radiokhemoekologii i v realizatsii programmy ustoichivogo razvitiya akvatorii. In: Korogodina VL, Mothersill C, Seymour C (eds) Sovremennye problemy genetiki, radiobiologii, radioekologii i evolyutsii: Crimean symposium, Alushta, vol 2. OIYaI, Dubna, 9–14 Oct 2010, pp 83–113 (in Russian)

Popov EP (1978) Teoriya lineinykh sistem avtomaticheskogo regulirovaniya i upravleniya. Nauka, Moscow, 256p (in Russian)

Popovichev VN, Egorov VN (1992) Pogloshchenie mineral'nogo fosfora vzveshennym veshchestvom foticheskogo sloya. In: Molismologiya Chernogo morya. Naukova dumka, Kiev, pp 62–69 (in Russian)

Popovichev VN, Egorov VN (2000) Bioticheskii obmen mineral'nogo fosfora v evfoticheskoi zone zapadnoi chasti Chernogo morya. In: Chteniya pamyati N. V. Timofeeva-Resovskogo: 100-letiyu so dnya rozhdeniya N. V. Timofeeva-Resovskogo posvyashchaetsya. EKOSI-Gidrofizika, Sevastopol, pp 140–158 (in Russian)

Popovichev VN, Egorov VN (2003) Fosfornyi obmen prirodnoi vzvesi v zone smesheniya Dunai— Chernoe more. In: Ekologicheskaya bezopasnost' pribrezhnoi i shel'fovoi zon i kompleksnoe ispol'zovanie resursov shel'fa: collection of scientific papers, vol 8. EKOSI-Gidrofizika, Sevastopol, pp 98–104 (in Russian)

Popovichev VN, Egorov VN (2008) Obmen mineral'nogo fosfora vzveshennym veshchestvom v foticheskoi zone Chernogo morya. In: Polikarpov GG, Egorov VN (eds) Radioekologicheskii otklik Chernogo morya na Chernobyl'skuyu avariyu. EKOSI-Gidrofizika, Sevastopol, pp 548–574 (in Russian)

Pospelova NV, Nekhoroshev MV (2003) Soderzhanie karotinoidov v sisteme "vzveshennoe veshchestvo—midiya—biootlozheniya midii." Ekol Morya 64:62–66 (in Russian)

Provasoli L, Shiraishi K (1959) Axenic cultivation of the brine shrimp *Artemia salina*. Biol Bull 117(2):347–355

Raymont JEG (1980) Plankton and productivity in the oceans. Pergamon Press. Russian edition: Raymont JEG (1983) Plankton i produktivnost' okeana. Legkaya i pishch. Prom-st', Moscow, 568p

Redfield AC (1958) The biological control of chemical factors in the environment. Am Sci 46(3):205–221

Reimers NF (1994) Ekologiya (teorii, zakony, pravila, printsipy i gipotezy). Izd-vo zhurnala "Rossiya molodaya", Moscow, 367p (in Russian)

Renfro WC, Fowler SW, La Rosa J (1975) Relative importance of food and water in long-term zinc-65 accumulation by marine biota. J Fish Res Board Can 32(8):1339–1345. https://doi.org/10.1139/f75-154

Revkov NK (2011) Makrobentos ukrainskogo shel'fa Chernogo morya. In: Eremeev VN, Gaevskaya AV, Shulman GE, Zagorodnyaya YA (eds) Promyslovye bioresursy Chernogo i Azovskogo morei. EKOSI-Gidrofizika, Sevastopol, pp 140–162 (in Russian)

Rice TR (1965) The role of plants and animals in the cycling of radionuclides in the marine environment. Health Phys 11(9):953–964

Romankevich EA (1977) Geokhimiya organicheskogo veshchestva v okeane. Nauka, Moscow, 256p (in Russian)

Romero-Romero S, Herrero L, Fernández M, Gómara B, Acuña JL (2017) Biomagnification of persistent organic pollutants in a deep-sea, temperate food web. Sci Total Environ 605–606:589–597. https://doi.org/10.1016/j.scitotenv.2017.06.148

Roots O, Kukk H. (1988) Polychlorinated biphenyls and chlororganic pesticides in algae from the Baltic Sea. Proc Est Acad Sci Chem 37(3):224–226

Rozhanskaya LI (1983) Mikroelementy v planktone i rybakh zapadnoi chasti Chernogo morya. Ekol Morya 12:22–30 (in Russian)

Rudyakov YA, Tseitlin VB (1980) Skorost' passivnogo pogruzheniya planktonnykh organizmov. Okeanologiya 20(5):732–738 (in Russian)

Rusanov II (2007) Mikrobnaya biogeokhimiya tsikla metana glubokovodnoi zony Chernogo morya. Dissertation abstract, Moscow, 24p (in Russian)

Ryabukhina EV, Kukleva OF, Stoikova OA (compliers) (2005) Predel'no dopustimye sbrosy (IDS) zagryaznyayushchikh veshchestv v vodnye ob"ekty: metod, ukazaniya po ekologo-toksikologicheskomu normirovaniyu. YarGU, Yaroslavl, 40p (in Russian)

Shnyukov EF, Ziborov AP (2004) Mineral'nye bogatstva Chernogo morya. In: Otdelenie morskoi geologii i osadochnogo rudoobrazovaniya NAN Ukrainy. Karbon LTD, Kiev, p 277 (in Russian)

Shnyukov EF, Starostenko VI, Gozhik PF, Kleshchenko SA et al (2001) O gazootdache dna Chernogo morya. Geol Zh 4:7–14 (in Russian)

Shulgina EF, Kurakova LV, Kuftarkova EA (1978) Khimizm vod shel'fovoi zony Chernogo morya pri antropogennom vozdeistvii. Naukova dumka, Kiev, 124p (in Russian)

Skopintsev BA (1975) Formirovanie sovremennogo khimicheskogo sostava vod Chernogo morya. Gidrometeoizdat, Leningrad, 336p (in Russian)

Small LF, Fowler SW (1973) Turnover and vertical transport of zinc by the euphausiid *Meganyctiphanes norvegica* in the Ligurian Sea. Mar Biol 18(4):284–290

Solomonova ES, Akimov AI (2014) Sootnoshenie mertvoi i zhivoi komponenty vzvesei v kul'turakh mikrovodoroslei v zavisimosti ot stadii rosta i osveshchennosti. Mor Ehkol Zh 13(1):39–44 (in Russian)

Sorokin YI (1982) Chernoe more: priroda, resursy. Nauka, Moscow, 215p (in Russian)

Sorokin YI, Avdeev VA (1991) Potreblenie i vremya oborota fosfata v vodakh Chernogo morya. In: Izmenchivost' ekosistemy Chernogo morya: estestvennye i antropogennye faktory. Nauka, Moscow, pp 153–157 (in Russian)

Sorokina NA (1979) Vliyanie soderzhaniya biomassy planktona na korrelyatsiyu mezhdu prozrachnost'yu i plotnost'yu vod okeana. In: Vzaimodeistvie mezhdu vodoi i zhivym veshchestvom: proceedings of the international symposium, Odessa, vol 1. Nauka, Moscow, 6–10 Oct 1975, pp 34–38 (in Russian)

Stelmakh LV (2013) Microzooplankton grazing impact on phytoplankton blooms in the coastal seawater of the Southern Crimea (Black Sea). Int J Mar Sci 3(15):121–127. https://doi.org/10.5376/ijms.2013.03.0015

Stelmakh LV (2017) Zakonomernosti rosta fitoplanktona i ego potrebleniya mikrozooplanktonom v Chernom more. Dissertation abstract, Sevastopol, 42p (in Russian)

Stelmakh LV, Babich II (2006) Sezonnaya izmenchivost' otnosheniya organicheskogo ugleroda k khlorofillu "a" i faktory, ee opredelyayushchie v fitoplanktone pribrezhnykh vod Chernogo morya. Mor Ehkol Zh 5(2):74–87 (in Russian)

Stetsyuk AP, Egorov VN (2018) Sposobnost' morskikh vzvesei kontsentrirovat' rtut' v zavisimosti ot ee soderzhaniya v akvatoriyakh shel'fa. Sist Kontrolya Okruzhayushchei Sredy 13(33):123–132 (in Russian)

Stokozov NA (2003) Dolgozhivushchie radionuklidy ^{137}Cs i ^{90}Sr v Chernom more posle avarii na Chernobyl'skoi AES i ikh ispol'zovanie v kachestve trasserov protsessov vodoobmena. Dissertation abstract, Sevastopol, 21p (in Russian)

Stokozov NA (2010) Morfometricheskie kharakteristiki Sevastopol'skoi i Balaklavskoi bukht. In: Ekologicheskaya bezopasnost' pribrezhnoi i shel'fovoi zon i kompleksnoe ispol'zovanie resursov shel'fa: collection of scientific papers, vol 23. EKOSI-Gidrofizika, Sevastopol, pp 198–208 (in Russian)

Stokozov NA, Gulin SB, Mirzoeva NY (2008) Soderzhanie ^{90}Sr i ^{137}Cs na vzveshennom veshchestve i v donnykh otlozheniyakh Chernogo morya posle avarii na Chernobyl'skoi AES. In: Polikarpov GG, Egorov VN (eds) Radioekologicheskii otklik Chernogo morya na chernobyl'skuyu avariyu. EKOSI-Gidrofizika, Sevastopol, pp 266–274 (in Russian)

Sushchenya LM (1972) Intensivnost' dykhaniya rakoobraznykh. Naukova dumka, Kiev, 193p (in Russian)

Sun YX, Hu YX, Zhang ZW, Xu XR, Li HX, Zuo LZ, Zhong Y, Sun H, Mai BX (2017) Halogenated organic pollutants in marine biota from the Xuande Atoll, South China Sea: levels, biomagnification and dietary exposure. Mar Pollut Bull 118(1–2):413–419. https://doi.org/10.1016/j.marpolbul.2017.03.009

Tsunogai S, Henmi T (1971) Iodine in surface water of the ocean. J Oceanogr Soc Jpn 27(2):67–72. https://doi.org/10.1007/BF02109332

Unifitsirovannye metody monitoringa fonovogo zagryazneniya prirodnoi sredy (1986) Gidrometeoizdat, Moscow, pp 82–95 (in Russian)

Van der Velden S, Dempson JB, Evans MS, Muir DCG, Power M (2013) Basal mercury concentrations and biomagnification rates in freshwater and marine food webs: effects on Arctic charr (*Salvelinus alpinus*) from Eastern Canada. Sci Total Environ 444:531–542. https://doi.org/10.1016/j.scitotenv.2012.11.099

Vandamme K, Baeteman M (1982) Teneur des organismes niarins des eaux côtières beiges en PCB et enpesticides organochlorés. Rev Agric (bruxelles) 35(2):1951–1958

Vandamme K, Maertens D (1983) Les teneurs en composés organochlorés dans les organismes marins de différents niveaux trophiques. Rev Agric (bruxelles) 36(6):1677–1682

Vanhaek P, Siddall SE, Sorgellos P (1984) International study on *Artemia*. XXXII. Combined effects of temperature and salinity on the survival of *Artemia* of various geographical origin. J Exp Mar Biol Ecol 80(3):259–275. https://doi.org/10.1016/0022-0981(84)90154-0

Vedernikov VI (1991) Osobennosti raspredeleniya pervichnoi produktsii i khlorofilla v Chernom more v vesennii i letnii periody. In: Izmenchivost' ekosistemy Chernogo morya: estestvennye i antropogennye faktory. Nauka, Moscow, pp 128–147 (in Russian)

Vernadsky VI (1965) Khimicheskoe stroenie biosfery Zemli i ee okruzheniya. Nauka, Moscow, 374p (in Russian)

Vernadsky VI (1978) Zhivoe veshchestvo. Nauka, Moscow, 358p (in Russian)

Vinogradov AP (1967) Vvedenie v geokhimiyu okeana. Nauka, Moscow, 213p (in Russian)

Vinogradov ME, Simonov AI (1989) Izmeneniya ekosistemy Chernogo morya. In: III congress of soviet oceanographers: plenary reports. Osnovnye problemy issledovaniya Mirovogo okeana. Gidrometeoizdat, Leningrad, pp 61–76 (in Russian)

Vinogradov ME, Sapozhnikov VV, Sushkina EA (1992) Ekosistema Chernogo morya. Nauka, Moscow, 110p (in Russian)

Vodyanitsky VA (1948) Osnovnoi vodoobmen i istoriya formirovaniya solenosti Chernogo morya. Tr Sevastopol Biol Stantsii 6:386–432 (in Russian)

Vodyanitsky VA (1958) Dopustim li sbros otkhodov atomnykh proizvodstv v Chernoe more? Priroda 12:46–52 (in Russian)

Voitsekhovitch OV, Kanivets VV, Kristhuk BF (1998) Project RER/2/003 status report of the Ukrainian Research Hydrometeorological Institute for 1993–1995. In: Marine environmental assessment of the Black Sea: working material of regional co-operation project RER/2/003, 1–3 Mar 1998, Vienna. IAEA, Vienna, 60 p

Water quality criteria summary, concentration NS (in μg/l) (1991) U.S. Environmental Protection Agency Office of Science and Technology, Health and Ecological Criteria Division, Ecological Risk Assessment Branch (WH-S85), Human Risk Assessment Branch (WH-550 D), Washington

Weers AW (1975a) The effects of temperature on the uptake and retantion of ^{60}Co and ^{65}Zn by the common shrimp *Crandon crandon* (L.). In: Combined effects of radioactive, chemical and thermal releases to the environment. Stockholm, 2–5 June 1975

Weers AW (1975b) Uptake of cobalt-60 from sea water and from labeled food by the common shrimp *Crangon crangon* (L). International symposium on impacts of nuclear releases into the aquatic environment. IAEA, Vienna, pp 359–361

Young ML (1975) The transfer of ^{65}Zn and ^{59}Fe along a *Fucus serratus* (L.) → *Littorina obtusata* (L.) food chain. J Mar Biol Assoc UK 55(3):583–610. https://doi.org/10.1017/S00253154000 17276

Yunev OA (2011) Otsenka mnogoletnikh izmenenii godovoi pervichnoi produktsii fitoplanktona razlichnykh raionov chernomorskogo shel'fa. In: Ekologicheskaya bezopasnost' pribrezhnoi i shel'fovoi zon i kompleksnoe ispol'zovanie resursov shel'fa: collection of scientific papers, vol 25. EKOSI-Gidrofizika, Sevastopol, pp 311–326 (in Russian)

Yunev OA, Moncheva S, Carstensen J (2005) Long-term variability of vertical chlorophyll *a* and nitrate profiles in the open Black Sea: Eutrophication and climate change. Mar Ecol Prog Ser 294:95–107

Zagorodnyaya YA, Moryakova VK (2011) Zooplankton kak kormovaya baza promyslovykh pelagicheskikh ryb. In: Eremeev VN, Gaevskaya AV, Shulman GE, Zagorodnyaya YA (eds) Promyslovye bioresursy Chernogo i Azovskogo morei. EKOSI-Gidrofizika, Sevastopol, pp 257–269 (in Russian)

Zaika VE (1972) Udel'naya produktsiya bespozvonochnykh. Naukova dumka, Kiev, 145p (in Russian)

Zaika VE (1981) Emkost' sredy—soderzhanie ponyatiya i ego primenenie v ekologii. Ekologiya Morya 7:3–10 (in Russian)

Zaitsev YP (1992) Ekologicheskoe sostoyanie shel'fovoi zony Chernogo morya u poberezh'ya Ukrainy (obzor). Gidrobiologicheskii Zhurnal 28(4):3–18 (in Russian)

Zaitsev YP (1998) Samoe sinee v mire. United Nations Publications, New York City, 142p (in Russian)

Zaitsev YP (2006) Vvedenie v ekologiyu Chernogo morya. Evan, Odessa, 224p (in Russian)

Zaitsev YP, Polikarpov GG (2002) Ekologicheskie protsessy v kriticheskikh zonakh Chernogo morya (sintez rezul'tatov dvukh napravlenii issledovanii s serediny XX do nachala XXI vekov). Mor Ehkol Zh 1(1):35–55 (in Russian)

Zherko NV, Egorov VN, Gulin SB, Malakhova LV (2001) Polikhlorbifenily v komponentakh ekosistemy Sevastopol'skoi bukhty. In: Ekologicheskaya bezopasnost' pribrezhnoi i shel'fovoi zon i kompleksnoe ispol'zovanie resursov shel'fa: collection of scientific papers, vol 2. EKOSI-Gidrofizika, Sevastopol, pp 153–158 (in Russian)

Zilov EA (2009) Gidrobiologiya i vodnaya ekologiya (organizatsiya, funktsionirovanie i zagryaznenie vodnykh ekosistem). Izd-vo Irkutskogo gos. Un-ta, Irkutsk, 147p (in Russian)

Conclusion

Contemporary changes in climatic conditions under significant anthropogenic impact have influenced the natural cycles of circulation in the biosphere of chemicals having varying levels of biological significance. Such changes have increased the relevance of work related to the conceptual development of optimal marine ecosystem management approaches in the context of the effect of pollutants both on the human population and on aquatic organisms—or hydrobionts. The pressing need to solve outstanding problems associated with this conceptual development required an in-depth study of the biogeochemical laws governing the interaction of living and inert matter with the radioactive and chemical components of the marine environment.

The consideration presented in Chap. 2 of problems associated with chemical and radioactive interactions taking place at different scales in the marine environment revealed the following basic patterns.

The total annual transport of chemical elements as a result of natural and anthropogenic processes can range from $n \times 10^{-5}$ to $n \times 10^{1}\%$ of their content in the waters of the world's oceans. Meteorological processes comprise a significant factor in the formation of the distribution fields of chemical elements and their isotopic carriers on a global spatial scale. When testing nuclear weapons in open environments, the maximum radioactive fallout was shown to be confined to the middle latitudes of the northern and southern hemispheres of the earth. The migration of chemical and radioactive contaminants dissolved in water across both spatial and depth dimensions under the simultaneous action of wave phenomena, currents, advection and turbulence always conduces towards decreasing gradients in the fields of their distribution in the marine environment.

The chemical and radioisotopic composition of the components of marine ecosystems depends on the influence of both abiotic and biotic factors. In inert matter, this composition is formed mainly as a result of geochemical processes, sorption interactions and the waste products of living matter. For living matter, the functional basis of marine ecosystems is comprised of chemical elements and their compounds. Various chemical compounds participate in the composition of organs and tissues of marine organisms, which determine the flux of complex of biochemical reactions

V. Egorov, *Theory of Radioisotopic and Chemical Homeostasis of Marine Ecosystems*, Springer Oceanography, https://doi.org/10.1007/978-3-030-80579-1

responsible for the energetic nutrition, somatic and generative growth, as well as the mineral metabolism of marine organisms.

The most representative indicator of the concentration of chemical substances by hydrobionts is the accumulation coefficient or concentration factor (CF), first proposed by the Academician Vladimir Ivanovich Vernadsky, which is equal to the ratio of the concentrations of elements and their isotopic and non-isotopic carriers in marine organisms and the aquatic environment. The CF characterises the accumulation of a radionuclide or its stable carrier by a hydrobiont, established as a result of dynamic equilibrium in the simultaneously occurring processes of the absorption of a chemical element or its radioisotope into the bodies of organisms along with its subsequent excretion. For this reason, $CF = const$ only when the chemical and isotopic composition of the aquatic environment is stable. Under all other conditions, the dependence of a change in the concentration factor over time reflects the kinetic patterns inherent in processes that conduce to achieving a stationary or limiting level of CF. Despite changes in the dynamics of accumulation of radionuclides by hydrobionts at different temperatures, their limiting concentration factors were not observed to differ significantly. It was noted (Rice 1965) that an increase in the concentration factor may be due to the formation process of new protoplasm, synthesised as a result of the absorption of photosynthetically active radiation by algae rather than an increase in illumination. With a decrease in pH, the chemical form of elements or their isotopic and non-isotopic carriers changes in such a way that bioassimilation processes tend to increase, while sorption processes tend to weaken. Within the limits of a particular species, a change in the size of marine organisms does not affect the metabolism of their individual cells; rather, the absorption of elements depends on the surface-mass ratio. The inhibition of organotrophy with an increase in population density is a result of the accumulation of metabolic products in the environment or the lack of a limiting chemical substrate. At the same time, it is noted that, since an increase in population density only inhibits the course of physiological processes when a certain limit is reached, which rarely occurs under natural conditions, there is no need to take this factor into account when using existing methods for assessing the production characteristics of populations.

The role of the biotic factor in the formation of fields of contamination is determined by the processes of interaction of living and inert matter with the radioactive and chemical components of the marine environment. In the open waters of seas and oceans, radioactive and chemical pollutants are distributed into aquatic areas and underlying layers under the influence of both abiotic and biotic factors. Within the photic layer, they are absorbed by primary producers through parenteral sorption and ingestion of nutriment. Simultaneously with absorption, the mineralisation and consequent transformation of their physicochemical forms take place as a result of desorption and the metabolic processes of hydrobionts. By being repeatedly passing through food webs, pollution is removed from the photic layer. Many authors point to a lack of knowledge of the mechanisms that control the accumulation, retention and excretion of inorganic substances by hydrobionts and the migration of radionuclides and their isotopic carriers along the food chain, as well as processes of sedimentation and spatial biotic transport. The solution to this problem is associated with the need to model the interactions of living and inert matter in marine ecosystems.

In Chap. 3, a description of a semi-empirical theory of radioisotope and mineral exchange of living and inert matter in the marine environment on the time scale of sorption, metabolic processes and trophic interactions is presented. This semi-empirical theory, which is parametrically compatible with contemporary methods for assessing the balance of matter and energy in aquatic ecosystems, accounts for the observed kinetic regularities of the concentration and exchange of radionuclides, biogenic elements, heavy metals and organochlorine compounds by inert matter and hydrobionts along the parenteral and alimentary pathways of mineral nutrition.

According to the data of natural observations interpreted along with the results of experiments with a radioactive label in the context of the developed methods for parametrising mathematical models, it is shown that the absorption of chemicals of various biological significance and their radionuclides by living and inert matter of the marine environment is described by linear functions, as well as by the Langmuir or Freundlich equations, while the rates of mineral and sorption exchange are described by the laws discovered by Michaelis–Menten and Droop, which correspond to the rates of metabolic reactions of the first- or zero order. Parameters of the dependences of changes in the concentrating function of hydrobionts on their specific surface, dimensional and trophic characteristics, as well as on the concentration of chemical elements and their radioactive carriers and chemical analogues in the aquatic environment, were determined. It is shown that, regardless of the biological significance of mineral elements, the kinetics of the exchange of their radionuclides by hydrobionts occupying different trophic levels can be interpreted in terms of a process of concentration of elements in one or two exchange pools (or compartments) of hydrobionts, exchanged at rates of first-order or zero-order metabolic reactions. Within the variation interval of concentrations of radionuclides in the aquatic environment, the relative volumes of pools and the rates of their exchange remain unchanged. When modelling the kinetic regularities of the exchange of radionuclides by marine organisms having different individual masses, the magnitude of the exchange resource pool compartments occurs in a power relationship on the individual mass of hydrobionts. However, the rate of exchange of an element in the compartments does not depend on the size characteristics of the specimens of this taxonomic group.

The development of compartment theory based on the assumption of new exchange pools arising during the growth of the hydrobiont helped to account for the decrease in the concentration and accumulation coefficients of radionuclides by hydrobionts during the production of organic matter observed in radiolabel experiments. It is shown that the kinetics of the concentration and exchange of radioactive and chemical elements by hydrobionts in furtherance of the production of organic matter can be described by a model, two compartments of which correspond to the accumulations of an element in hydrobionts during growth, exchanging indicators of the rates of first-order metabolic reactions with the environment. The third compartment only reflects the entry of the element into the main, non-exchangeable or weakly exchanging structures during the somatic growth of the marine organism. Thus, using the example of an empirical study and a mathematical description of the concentration kinetics of mono- and pentavalent ^{131}I by *Ulva*, it was shown that the kinetics of exchange of different physicochemical forms of pollutants by hydrobionts can be considered as a result of their entry into the exchange pools of hydrobionts corresponding to each physicochemical form. The interaction of chemical resource exchange pools with the environment is carried out at rates of first-order metabolic reactions. The numerical values of the rates of intake into the resource pools are determined by the physical and chemical forms of pollutants. The rates of elimination of different physicochemical forms of inorganic substances from the corresponding exchange pools in hydrobionts do not vary.

The final part of Chap. 3 analyses the field of application of the proposed semi-empirical theory of the mineral and sorption interaction of living and inert matter with radioactive and chemical components of the marine environment, as well as with the biological interpretation of its parameters. It is shown that the parameters of the generalised empirical model are independent, while the structure of the model allows its modification and development due to the detailed consideration of interaction characteristics and need to take new factors into account. For this reason, the equations that implement the theoretical basis of the model can be said to satisfy the requirements for semi-empirical models used in a systematic approach to solving problems of the interaction of living and inert matter with radioactive and chemical components of the marine environment.

In Chap. 4, the theory of radioisotope and chemical homeostasis of marine ecosystems is substantiated along with a description of its practical applications for the sustainable development of aquatic areas, taking into account the relevant factors of radioactive and chemical pollution of the marine environment. It is shown that these problems belong to the class of complex geosystems, whose effective study and management must be carried out on the basis of corresponding mathematical models. Only on the basis of theoretical models that describe not one object, but a certain class of phenomena, among which there can be both observable and unobservable phenomena, can the behaviour of geosystems that are not fully accessible for observation be predicted. The adequacy of the description of complex geosystems by theoretical models is determined by the degree to which they reflect phenomena observed in nature, as well as by the objectivity of theoretical premises in relation to unobservable phenomena.

The first section of the chapter substantiates the use of mathematical models for studying the characteristics of the functioning of natural marine ecosystems. The adequacy of such models consists in their incorporating both semi-empirical blocks that describe the observed regularities of mineral and radioisotope exchange, as well as theoretical blocks that reflect contemporary concepts about the material, trophic and energetic mechanisms of their functionality.

Sections 4.1.1 and 4.1.2 present a study of the dynamic characteristics of the interaction of ecosystems of the photic layer with radioactive and chemical components of the marine environment. It is shown that, when primary production links are limited in terms of biogenic elements and physiologically active radiation, the states of the system are determined by damped oscillatory processes whose periods of oscillations correspond to timescales of ecological succession processes. The stability of the system with respect to the concentration characteristics of the distribution of pollutants in the environment is determined by the flux of the pollutant as a result of anthropogenic impact, the biosedimentation productivity of the ecosystem of the photic layer, as well as the accumulation of the pollutant by hydrobionts linked to metabolic rate. The biosedimentation capacity of an ecosystem and degree of concentration of a pollutant by hydrobionts corresponds to the maximum rate of anthropogenic impact at which the system is stationary. Presented observational materials testify to the existence of natural biogeochemical mechanisms for the realisation of radioisotope and chemical homeostasis of marine ecosystems.

From the point of view of the theory of their management, it is shown that homeostasis of marine ecosystems is realised by negative feedbacks between their components according to the Le Chatelier–Braun principle. Section 4.4 demonstrates the natural biogeochemical mechanisms of homeostasis due to the concentrating function of living and inert matter, as well as the adaptive, trophic and population characteristics of hydrobionts. Observed regularities of natural homeostasis occurring in the photic layer of seawater in terms of biogenic elements, as well as those applying to conservative radioactive and chemical substances in the estuarine and critical zones of the Black Sea, are presented.

It is shown that pollution of aquatic areas can be normalised according to the criteria of the ecological carrying capacity, assimilation capacity and radiocapacity of water masses. The substantiated theory of radioisotope and chemical homeostasis of marine biogeocenoses is aimed at the development of methods for determining the ecological carrying capacity of the marine environment as a result of biotic interaction mechanisms, assessing the assimilation capacity of geological depots due to biogeochemical processes and calculating the characteristics of the radiocapacity of water depots, which is equal to their absorptive capacity for pollutants taking into account the maximum allowable concentration (MAC).

Sections 4.5 and 4.6 consider problems inherent in taking an ecocentric approach to marine ecosystem management. These problems are associated with the implementation of the principle of sustainable development of aquatic areas according to the factor of radioactive and chemical pollution of waters. Biogeochemical criteria for assessing the self-cleaning ability of the marine environment as a result of the impact of natural biogeochemical processes are substantiated. Methods of environmental regulation according to sanitary-hygienic and biogeochemical criteria are demonstrated on the examples of Sevastopol Bay, the Azov Sea and the Black Sea, as well as aquatic areas of recreational zones having liquid boundaries.

In general terms, the performed studies demonstrate that the development of the theory of radioisotopic and chemical homeostasis of marine ecosystems, based on the semi-empirical theory of mineral and radioisotopic exchange, confirm the hypothesis of Academician Vladimir Vernadsky that the necessary conditions of the habitat for living matter are perpetuated through the processes of its reproduction. In providing for the assessment and prediction of the anthropogenic evolution of marine ecosystems, the proposed theoretical framework supports the development of principles for optimal marine ecosystem management, taking into account the intensity of natural biogeochemical processes of radioactive and chemical conditioning of the marine environment.

The interpretation of the results of experimental and natural observations alongside theoretical analysis from the point of view of anthropogenic evolution allows us to predict long-term trends associated with the increased intensity of biogeochemical cycles involving the material, energetic and chemical composition of the components of marine ecosystems.

Lightning Source UK Ltd.
Milton Keynes UK
UKHW020805120822
407214UK00002B/3